The Role of GIS in COVID-19 Management and Control

Geographic information system (GIS) is one of the most important tools to help us understand public health and many aspects of our lives. Because of COVID-19, GIS has been brought into the spotlight more than ever before. People and civic leaders worldwide are turning to maps and real-time surveillance data to make sense of what has been happening in the world and to get answers to important questions on every aspect of this pandemic. This book examines the role of GIS in managing and controlling the spread of COVID-19 through 12 global projects and a multidisciplinary approach. It explains the innovative uses of GIS not only limited to data organization and data access, but also how improved GIS tools are used to make decisions, plan, and communicate various measures of control in both local and full-scale outbreaks during the COVID-19 pandemic.

Features:

- Provides cutting-edge GIS visualization, spatial temporal pattern, and hot spot tracking applications used for predictive modeling of COVID-19.
- Includes real-world case studies with broad geographical scope that reflect COVID-19 trends in cases, deaths, and vaccinations.
- Provides lifestyle segmentation analysis on the risk of transmission of COVID-19 and spatial patterns of vaccination hesitancy.
- Highlights real-world issues brought to light with the help of GIS, such as social discrimination and inequalities in women's access to mental health care, and analyzes the risk of transmission due to vaccination hesitancy.
- Shows the use of GIS and spatial analysis in pandemic mapping, management, and control from masking and social distancing to testing site locations accounting for at-risk and vulnerable populations.
- Discusses facilitating policy making with GIS.

Edited by a very talented medical geographer and GIS professor, Dr. Esra Ozdenerol, this book highlights key GIS capabilities and lessons learned during the COVID-19 response that can help communities prepare for the next crisis. It is a great resource for industry professionals and experts in health care, public health and safety, disaster management, and for students, academics, and researchers interested in applying GIS and spatial analysis to the study of COVID-19 and other pandemics.

The Role of GIS in COVID-19 Management and Control

Edited by
Esra Ozdenerol

CRC Press is an imprint of the
Taylor & Francis Group, an informa business

Designed cover image: © COVID-19 Dashboard by the Center for Systems and Science Engineering (CSSE) at John Hopkins University (JHU)

First edition published 2023
by CRC Press
6000 Broken Sound Parkway NW, Suite 300, Boca Raton, FL 33487-2742

and by CRC Press
4 Park Square, Milton Park, Abingdon, Oxon, OX14 4RN

CRC Press is an imprint of Taylor & Francis Group, LLC

Library of Congress Cataloging-in-Publication Data
Names: Ozdenerol, Esra, editor.
Title: The role of GIS in Covid-19 management and control / edited by Esra Ozdenerol.
Description: First edition. | Boca Raton : CRC Press, 2023. |
Includes bibliographical references.
Identifiers: LCCN 2022055126 (print) | LCCN 2022055127 (ebook) |
ISBN 9781032129754 (hardback) | ISBN 9781032129761 (paperback) |
ISBN 9781003227106 (ebook)
Subjects: LCSH: COVID-19 Pandemic, 2020–Geographic information systems. |
COVID-19 (Disease)–Epidemiology–Data processing. |
Medical mapping. | Disease management.
Classification: LCC RA644.C67 R65 2023 (print) |
LCC RA644.C67 (ebook) | DDC 362.1962/4144–dc23/eng/20230202
LC record available at https://lccn.loc.gov/2022055126
LC ebook record available at https://lccn.loc.gov/2022055127

ISBN: 978-1-032-12975-4 (hbk)
ISBN 978-1-032-12976-1 (pbk)
ISBN 978-1-003-22710-6 (ebk)

DOI: 10.1201/9781003227106

Typeset in Times New Roman
by Newgen Publishing UK

Dedication

To Derin and Deniz

This book is dedicated to my sons, Derin and Deniz, who kept me company during COVID-19. Their love gave me strength and helped me lead my family from a sense of vulnerability to a place of resilience. With this book, I also honor the lives of loved ones and all those who were lost during the pandemic.

Contents

Foreword

COVID-19 overwhelmed my practice much the way an advancing hurricane's surge suddenly swamps a shoreline. Apart from news from media outlets, the first week of March 2020 was utterly unremarkable; I supervised radiation treatment for my cancer patients just as I always had. By close of the second week, I had drifted deep into (for me) uncharted waters of Infectious Disease and Disaster Medicine. Across the blur of 72 hours I helped oversee the creation of the first community-level pandemic testing site Memphis, Tennessee, had ever seen. I was lobbing countless emails to colleagues across the city and the country, trying to best choose who needed first access to COVID-19 testing, cancer treatment, or both. The only consolation I held onto was the fact that I knew just as little as everyone else around me. There was no map, no data to sharpen navigation through decision points. There was simply "dead reckoning," shaped by blunt instinct and reasonable guesses.

Before the pandemic, I had been deeply interested in the role geography plays in the treatment of my patients. Radiotherapy requires expensive machines, highly trained specialty staff, and lots of upkeep. Cancer patients have to travel to clinical facilities that can support this. If patients can't make it in for their daily sessions, they simply don't get treated, so location matters. With the help of Dr. Ozdenerol, my team had begun formally mapping the Memphis metropolitan area, identifying communities most at risk for their residents to experience treatment interruptions due to social or transportation hardship. We have consistently found our most economically deprived neighborhoods—typically majority African American—to be most impacted by such lapses. This can directly increase cancer deaths.

Interestingly, we found a similar dynamic at play at our large centralized COVID-19 testing site. Although it had been created with best of intentions to provide testing to underserved Memphians, its high overhead of course required everyone to travel to us. Within weeks, less frequent testing and greater COVID-19 infection rates became evident in poorer zip codes. Crude early data collected from health department resources pointed us toward a vulnerable area where we reproduced a second testing site, which was soon followed by two more testing centers. Yet the original geographical disparities seen in COVID-19 testing and treatment stubbornly outlasted the lifespan of all four of our sites. We had expected different results after repeating the same actions, the exact definition of insanity.

What had been needed all along was a steadfast field map, a detailed topographic description of the healthcare and social risk landscape of Memphis to guide rational deployment of specific resources matched to neighborhoods in need. This requires disciplined data collection and analytic infrastructure. Although too late to impact COVID-19, this is exactly what Dr. Ozdenerol continues to spearhead alongside me and our colleagues at the University of Tennessee Health Science Center. Happily, the book you now hold contains an introduction to our ongoing collaboration to prepare for future public health challenges to Memphis. But it offers so much more.

Reading through the many examples of geographically informed, data-driven population health responses to COVID-19 inspires me with hope that we can collectively, as a global community, be better to do better when the next storm hits. This text shines and plots our course toward this seaworthy future.

David Schwartz, MD, FACR
Memphis, Tennessee

Preface

With the COVID-19 pandemic, data visualization has taken on new importance in our daily lives. We watched circles multiply and swell on a map as the virus spread around the globe. We saw lines on time-series charts turn nearly vertical during surges in cases. The maps and visualizations created with geographic information systems (GIS) clarified the extent and impact of the pandemic and aided decision-making, planning, and community action. GIS have been used for data preparation, platform construction, model construction, and map production. Because no story can be captured by data alone, I decided to edit this volume to showcase global projects with innovative use of GIS, not only focusing on data organization and access but also how GIS tools are used during the pandemic. Integration of GIS in the COVID-19 pandemic in many countries is largely unknown. In this edited volume, I tried to provide stories with context and caveats along with data from Mexico to India to Indonesia. This book consists of 13 chapters of selected papers presenting in-depth case studies that adopt GIS and spatial analysis for mapping and analyzing the pandemic in various regions and countries in the world. The global case studies provided in each chapter explore the spread of the pandemic and get directly involved with some GIS and mapping applications coupled with data science techniques.

Chapter 1 provides an overview of the current role of GIS in pandemic control and management.

Chapter 2 explains why quality information and suitable Internet access are considered important to understand the COVID-19 pandemic. Unfortunately, a large segment of the population in the world lacks opportunities to develop their information skills and health literacy and the inequalities are more evident during the COVID-19 crisis. This chapter includes a description of information access and health literacy as a prerequisite for health, and defines information skills, disinformation, and health literacy.

Chapter 3 discusses how understanding the spatial pattern of COVID-19 in metropolitan cities is essential for the implementation of appropriate policy responses. This chapter utilizes GIS to understand the dynamic patterns of COVID-19 cases in Jakarta and Surabaya, two metropolitan cities in Indonesia, in the context of their local restriction policies. The Hot Spot and Anselin Local Moran's I analyses were cleverly applied to assess the risk of spread and reveal the different factors that drove the transmission of COVID-19 in both cities.

Chapter 4 examines the relationships between the coronavirus disease, age demographics, and comorbidities across counties in the United States. The analysis is conducted using geographically weighted regression to analyze the impacts of age, heart disease, obesity, and diabetes on coronavirus incidence and death-case ratios during the early stages of the pandemic. The findings reveal regional and temporal differences in coronavirus incidence and mortality among high-risk populations. Model results, standardized residual maps, and cluster-outlier maps are presented to highlight key findings related to this nationwide study.

Chapter 5 analyzes and maps the spatial and temporal spread of COVID-19 across the 14 districts of Kerala, India, during January 2020 to December 2021. GIS mapping and descriptive statistical analysis of total infections, those quarantined, those hospitalized, positive cases, those under treatment, samples tested, those turning negative, and daily and cumulative deaths suggest that Kerala had peaked to a dangerously high level of infections and deaths by May 2021, whereby December 2021 netted a cumulative death of 10,944. Insights gathered from academic and archival sources along with available travel data at domestic and international scales suggest that phases of travel relaxation during August 2020–March 2021 provided opportunities for social and public gatherings—which eventually contributed to a steep rise in COVID-19.

Chapter 6 analyzes the impacts of the three main hurricanes in 2020 in Latin American and Caribbean (LAC) countries in the context of the global health crisis caused by the COVID-19 pandemic. During the pandemic, the tropical cyclones Eta, Laura, and Iota caused more than 2.7 million new internal displacements in 13 countries of the region. The temporal coincidence of these three hurricanes with the first pandemic wave meant increased exposure of internally displaced people (IDPs) to COVID-19 infections. This way, in 2020, a singularly adverse scenario of great relevance emerged for the LAC region. This paper highlights the differential impacts of COVID-19 that internally displaced persons faced in Latin America and the Caribbean region during much of the year 2020; and the lessons learned for Disaster Risk Reduction including displaced people.

Chapter 7 presents the Hot Spot Tracker: detecting and visualizing the types of spatiotemporal hot spots of COVID-19 in the United States. The authors developed models using space-time cluster analysis to identify the hot spots and created a dashboard that provides interactive visualizations of the results. The models use the Getis-Ord Gi statistics to identify clusters based on county-level daily confirmed cases, and Mann-Kendall tests combined with an adapted rule-based process to classify the type of the clusters. The output clusters are classified as emerging, intensifying, persistent, diminishing, historical, and sporadic hot spots. They then developed an interactive web dashboard that displays the statistically tested spatiotemporal trends of COVID-19 breakouts in concise and straightforward graphics.

Chapter 8 provides unique insights into where and when COVID-19 impacted different households through geodemographic segmentation, typically used for marketing purposes, to identify consumers' lifestyles and preferences. The aim of the study is to associate lifestyle characteristics with COVID-19 infection and mortality rates at the US county level and sequentially map the impact of COVID-19 on different lifestyle segments. The results suggest that prevention and control policies can be implemented to those specific households exhibiting spatial and temporal pattern of high risk.

Chapter 9 presents a science mapping analysis of the role of location-based social media conversations in COVID-19 research. By reviewing the current scholarship, the author provides an overview and data visualization of the main themes and trends covered by the relevant literature. The author presents a summary of the intellectual, conceptual, and social structures of research involving geosocial COVID-19 conversations, as well as their evolution and dynamical aspects.

Chapter 10 highlights the use of GIS to shape policies regarding masking and distancing in a partially vaccinated population, specifically focusing on school openings and mask wearing in schools. This was based on maps from the Shelby County Health Department in Tennessee showing vaccination rates and active COVID-19 case rates in Memphis, Tennessee. The authors specifically delineate what information is obtained in the GIS analyses and what additional assumptions/knowledge must be added to make their maps useful tools for guiding policy decisions. They highlight the strengths and pitfalls of GIS tools in the context of public health decision-making and specifically based on their local experience with the presentations in the Joint Task Force briefings and to the City of Memphis and county mayors' offices.

Chapter 11 demonstrates the use of GIS to map women's health, well-being, and economic opportunities in the context of COVID-19. Emerging data show that the ongoing COVID-19 pandemic has exacerbated existing gender inequalities in all spheres of life including employment, unpaid care work, as well as in access to essential health services. In this chapter, the authors make use of data collected in the first year of the pandemic to explore the impact of the crisis on subnational inequalities, including the provision of critical health services for different subgroups of women. In Kenya, they show how lockdowns impacted women's access to maternal care services, both in urban and rural areas. In the USA, they explore the association between existing spatial inequalities in women's access to mental health care and spatial patterns with regard to reported increases in depression and anxiety outcomes during COVID-19. Finally, in Rwanda, they highlight the disproportionate impact of the economic fallout on women's employment and access to resources. The findings from these three case studies point to major disruptions in access to services and highlight the importance of spatial analysis to appropriately target national and local COVID-19 recovery and mitigation responses.

Chapter 12 presents a study that correlates lifestyle characteristics to COVID-19 vaccination rates at the US county level and provides where and when COVID-19 vaccination impacted different households. The authors grouped counties by their dominant LifeMode, and the mean vaccination rates per LifeMode were calculated. High-risk Lifestyle segments and their locations are the areas in the United States where clearly the public might benefit from a COVID-19 vaccine. They then used logistic regression analysis to predict vaccination rates using ESRI's tapestry segmentation and other demographic variables. Their findings demonstrate that vaccine uptake appears to be highest in the urban corridors of the Northeast and the West Coast and in the retirement communities of Arizona and Florida and lowest in the rural areas of the Great Plains and the Southeast. Looking closely at other parts of the West, such as the Dakotas and Montana, counties that contain Native American reservations have higher vaccination rates. Racial/ethnic minorities also adopt the vaccine at higher rates. The most effective predictor of vaccination hesitancy was Republican voting habits, with Republican counties less likely to take the vaccine. The other predictors in order of importance were college education, minority race/ethnicity, median income, and median age. The results suggest that prevention and control policies can be implemented to those specific households.

Chapter 13 examines the impact of COVID-19 on domestic workers in Titwala, India. Delving into the nexus of the pandemic, gender, and livelihoods, this study uses semistructured interviews and other secondary data from journals, news reports, and books to explore the impact of the COVID-19 pandemic-induced lockdown on 38 domestic workers in Titwala, an extended suburb of Mumbai (India), which has recently emerged as a preferred residential area of the Mumbai Metropolitan Region. With the fast growth of residential real estate within the last decade, this area has witnessed a tremendous influx of households where many female domestic workers have found employment—a typical lifestyle maintained by most middle-class families in the Indian society. However, following the pandemic and the lockdowns across India, many of these domestic workers not only lost their livelihoods, but their returns after the end of the lockdown were largely marked by underemployment, reduced wages, loss of bargaining power, accumulated debts, COVID-19-induced social discrimination, and loss of access to education by their school-going children due to unaffordable Internet data. This has exacerbated the already prevalent insecurities among the domestic workers, further threatening their livelihoods. This chapter proposes legal recognition of domestic work and informal workers such that their basic human and income rights can be protected.

These global case studies provided in each chapter explored how GIS played an important role in controlling and managing the COVID-19 virus in their respected countries. The development of GIS technology has provided a platform to produce spatial data and perform complex spatial analysis. The World Health Organization (WHO) has extensively applied GIS and a real-time mapping dashboard to illustrate the distribution of COVID-19 cases and deaths by country in the world. Locally, every country from coarser to finer scales, at the county and district levels, mapped the location of people infected with the virus, those suspected of having the virus, hospitals to which patients with COVID-19 were referred, the route of travel that patients took daily and weekly, and the place of the distribution of masks and disinfectants to combat the COVID-19 crisis effectively. They were able to prioritize service delivery based on the characteristics of each area according to the prevalence of the disease or the need for medical services.

A notable feature of "GIS's Role in COVID-19 Management and Control" is its attempt to tell a fuller story of how COVID-19 changed the world; the rising numbers and maps have been critical for informing our behaviors during the COVID-19 crisis. It is impossible to capture the full significance of the crisis and its many snowball effects in one book. Because of the fallout—from personal and collective traumas to profound economic disruption—we can transform this crisis into a historic opportunity for once-in-a-lifetime reforms of global health systems based on science, technology, equity, and solidarity.

I hope that I have conveyed the amazing breadth of GIS use in COVID-19's management and control with this book. Our eagerness to learn and adopt these technologies, and our optimism for the world and our work together are among the greatest sustaining forces toward combating future pandemics.

Esra Ozdenerol
Memphis, Tennessee

Acknowledgments

Globally, the COVID-19 pandemic has caused an enormous loss of life. With vast and growing global interdependency, intercontinental travel, and mass migration, the realities of globalization and climate change have fueled rapid spread of this disease. The COVID-19 crisis urged global cooperation and solidarity. I am most thankful to my contributors from all over the world working cooperatively in responding to their national health emergencies and building resilience during the pandemic response and applying GIS technology to strengthen their domestic public health capacities. The GIS dashboards they created with the emergence of potential trends, sudden pivots, and troubling setbacks maps with multiplying circles of new cases; they weaved together vulnerable, brave storytelling and compelling data from their countries during the COVID-19 crisis. GIS technology continues to be the glue that holds this whole effort together to limit the virus's spread. When governments issued lockdowns and other restrictions in response to COVID-19, GIS researchers and professionals applied GIS in responding swiftly and decisively to the COVID-19 pandemic and demonstrated just how valuable geospatial thinking is. As we enter year three of the pandemic, GIS will continue to play a key role in quantifying the waves of change that ripple through the global society. Some of this data will help us make personal risk assessments in our daily lives, whereas others might inform policy decisions. Maps and charts will also highlight emerging trends that might otherwise get lost as we navigate the daily noise of an ongoing crisis.

I owe a great deal to the many people who have provided inspiration, help, and support for what would eventually become this book: Irma Britton, Senior Editor of CRC Press, for patience and guidance throughout this entire project. I owe my perpetual gratitude to Dr. David Schwartz, chair of the Department of Radiation Oncology at the University of Tennessee Health Science Center, who wrote the Foreword to this book. I thank him and my other colleagues from the University of Tennessee Health Sciences Center (UTHSC) College of Medicine, Drs. Fridtjof Thomas, Arash Shaban-Nejad, Altha Stewart, Bob Davis, and Karen C. Johnson, for their collaboration on a multidisciplinary initiative to create a centralized COVID-19 data registry at UTHSC—the Memphis Pandemic Health Informatics System (MEMPHI-SYS)—linking COVID-19 testing data across community health providers serving the full Memphis metro region. We navigated remote work with relentless Zoom meetings during the COVID-19 crisis. These medical experts, public health scholars, epidemiologists, and statisticians have quickly adapted GIS and included my GIS expertise in this multidisciplinary initiative. The increasing utilization of GIS with real-time web-based dashboards and space-time cubes of COVID-19 testing results has given *new eyes* to public health professionals, researchers, and hospital and health system employees—as well as the public they serve.

I thank my student Jacob Seboly for coauthoring two chapters with me and Joanna Gobel for coming up with a fantastic cover design. I offer very special thanks to my program coordinator, Mekensie Ivy, for her contagious smile. I would also like to thank all my friends who have lent their support and listened to my complaints for the past years. Special thanks go to Yasemin Tuzun, Maribel Medina, Ashley Fowler and Funda Cam. *I would like to thank* University Middle and Christian Brothers High School teachers for their commitment and dedication to my sons' education, especially when the pandemic's shift from *in*-person to *remote* learning posed massive challenges. I owe a debt of gratitude to my family in Turkey, particularly my parents, Erdogan and Umran Ozdenerol, calling me over the ocean and offering constant encouragement every single day during COVID-19.

This book is dedicated to my sons, Derin and Deniz. Their love gave me strength and helped me lead my family from a sense of vulnerability to a place of resilience. A sudden change in our routine was overwhelming! Our family life, health, work, education, and wider support networks were affected by life in lockdown. Focusing on keeping myself and my family safe, healthy, and happy was enough.

Our ancestors went through catastrophes, including wars, famines, and plagues. They were not as lucky with vaccines, telehealth, online education, and technological advances. In the twenty-first century, we have the means to deal with crisis and we have GIS … Go to GI-Yes!

Esra Ozdenerol
Memphis, Tennessee

Editor

Esra Ozdenerol (PhD, 2000, Louisiana State University) is a Dunavant University Professor of Geographic Information Science in the Department of Earth Sciences and the director of Graduate Certificate in Geographic Information Systems (GIS) at the University of Memphis. She also serves as adjunct professor of Preventive Medicine and Health Outcomes Policy at the University of Tennessee Health Science Center and adjunct professor of Biology at Arkansas State University.

Dr. Ozdenerol has been with the University of Memphis since 2003, where she teaches GIS and its applications to health disparities. Her research interests entail the use of geospatial technologies in a diverse range of environmental health issues: social determinants of health, infectious disease epidemiology, birth health outcomes, opium epidemic, and children's lead poisoning. She served as associate director of Benjamin L. Hooks Institute for social change between 2010 and 2013. In 2010, she served as president of the Memphis Area Geographic Information Council, a nonprofit organization of GIS professionals. She was invited by the National Academies of Science on the Contribution of Remote Sensing for Decisions about Human Welfare and was also recognized as a "Health Research Fellow" by the University of Memphis. She launched the first interactive mapping website featuring the pivotal events of the civil rights movement. Her textbook *Spatial Health Inequalities: Adapting GIS Tools and Data Analysis* has been recognized by leading institutions such as American Association of Geographers and adopted at various geography and public health departments. Her book *Gender Inequalities: GIS Approaches to Gender Analysis* fosters engagement with the newest mapping and GIS application in contemporary issues regarding gender inequalities and nurtures recognition of how institutional, global, everyday, and intimate spaces are inherently gendered, classed, raced, and sexualized. It demonstrates the spatiality of the politics of gender difference, and the contributions of GIS and spatial analysis to the struggles for equality and social justice. Dr. Ozdenerol has lectured nationally and internationally on GIS and remote sensing and issues of health disparities. With her keynote speech, she opened the Refugee Week at the Refugee Therapy Center in London in 2013. She has developed GIS workshop series for professionals and campus communities that address and support different aspects of the data life cycle and data used with different GIS software and tools. Dr. Ozdenerol has visited 50 countries, lived in three continents, and brings her travels into her World Geography and GIS courses. She has published a bilingual children's book for teaching languages and changing attitudes about other cultures and geographies. Dr. Ozdenerol is the lead singer of the Memphis City Sound Chorus, an award-winning women's chorus of Sweet Adelines International. She has an unstoppable passion for singing harmony as well as teaching geography.

Contributors

Roberto Ariel Abeldaño Zuñiga
University of Sierra Sur
Oaxaca, Mexico

Gregory T. Armstrong
St. Jude Children's Research Hospital
Memphis, Tennessee, USA

Ginette Azcona
UN Women

Danang Azhari
Resilience Development Initiative (RDI)
Bandung, Indonesia

Antra Bhatt
UN Women

Sujayita Bhattacharjee
International Institute of Population Sciences (IIPS)
Mumbai, India

Julia Brauchle
UN Women

Emily A. Fogarty
Farmingdale State College
Farmingdale, New York, USA

Alisa Haushalter
Department of Community & Population Health
Shelby County Health Department and The University of Tennessee Health Science Center
Memphis, Tennessee, USA

Manoj Jain
Emory University
Rollins School of Public Health
Atlanta, Georgia, USA

Karen C. Johnson
Department of Preventive Medicine
The University of Tennessee Health Science Center
Memphis, Tennessee, USA

Jennifer M. Kmet
Shelby County Health Department
Memphis, Tennessee, USA

Rajesh Kumar Abhay
Department of Geography
Dyal Singh College
University of Delhi
India

Shimod Kunduparambil
Kannur University
Kerala, India

Naomi W. Lazarus
Department of Geography and Planning
California State University
Chico, California, USA

Dongying Li
Department of Landscape Architecture and Urban Planning
Texas A&M University
College Station, Texas, USA

Liang Li
Shelby County Health Department
Memphis, Tennessee, USA

Guize Luan
School of Earth Sciences
Yunnan University, China

Fathia Lutfiananda
Resilience Development Initiative (RDI)
Bandung, Indonesia

Motomi Mori
St. Jude Children's Research Hospital
Memphis, Tennessee, USA

Gabriela Narcizo de Lima
Department of Geography
Faculty of Letters
University of Porto
Portugal

Lilian A. Nyindodo
Baptist Health Sciences University
Memphis, Tennessee, USA

Esra Ozdenerol
Earth Science
University of Memphis
Memphis, Tennessee, USA

Allison P. Plaxco
Shelby County Health Department and
The University of Memphis
School of Public Health

Rachel Rice
Shelby County Health Department
Memphis, Tennessee, USA

Saut Sagala
Bandung Institute of Technology (ITB)
Bandung, Indonesia

Indah Salsabiela
Resilience Development Initiative (RDI)
Bandung, Indonesia

Jacob Daniel Seboly
Mississippi State University
Starkville, Mississippi, USA

Madhuri Sharma
University of Tennessee
Knoxville, Tennessee, USA

Irma Singarella
University of Memphis
Memphis, Tennessee, USA

Jesse Smith
St. Jude Children's Research Hospital
Memphis, Tennessee, USA

Chaitra Subramanya
Shelby County Health Department
Memphis, Tennessee, USA

David Sweat
Shelby County Health Department
Memphis, Tennessee, USA

Li Tang
St. Jude Children's Research Hospital
Memphis, Tennessee, USA

Michelle Taylor
Shelby County Health Department
Memphis, Tennessee, USA

Fridtjof Thomas
Department of Preventive Medicine
The University of Tennessee Health
Science Center
Memphis, Tennessee, USA

Fei Zhao
School of Earth Sciences
Yunnan University, China

Xiaolu Zhou
Department of Geography
Texas Christian University
Fort Worth, Texas, USA

1 The Role of GIS in COVID-19 Management and Control

Esra Ozdenerol

CONTENTS

The field of geographic information systems (GIS), which is already involved in so many aspects of our daily lives, is in the spotlight now more than ever. During this coronavirus pandemic, people are turning to maps to make sense of what is happening in the world. The maps and visualizations created with GIS clarified the extent and impact of the pandemic and aided decision-making, planning, and community action. We watched circles multiply and swell on a map as the virus spread around the globe. We saw lines on time-series charts turn nearly vertical during surges in cases. For information about case numbers, demographics, causes, or any other aspect of the virus to really make sense, you need to know about the geographical component. Without that geographical anchor point for coronavirus information, the information does not have the context it needs to truly be meaningful and helpful. In responding to this public health crisis and its knock-on economic and social effects, GIS has demonstrated the pivotal role geographic technology plays in understanding, responding to, recovering from, and mitigating these threats. After the Johns Hopkins University's (JHU's) Center for Systems Science and Engineering (CSSE, 2020) dashboard went viral on January 22, 2020, accumulating over 140 million views (Johns Hopkins, 2020) (Figure 1.1), GIS-based dashboards became crucial resources to guide the understanding of the SARS-CoV-2 virus, prepare policy makers to direct a response, and inform the public (ESRI China, 2020; Dong et al., 2020). The JHU CSSE dashboard's interactive map locates and tallies confirmed infections, fatalities, and recoveries. Graphs detail virus progress over time. Viewers can see the day and time of the most recent data update and data sources. The dashboard's five authoritative data sources include the World Health Organization (WHO, 2020), the US Centers

DOI: 10.1201/9781003227106-1

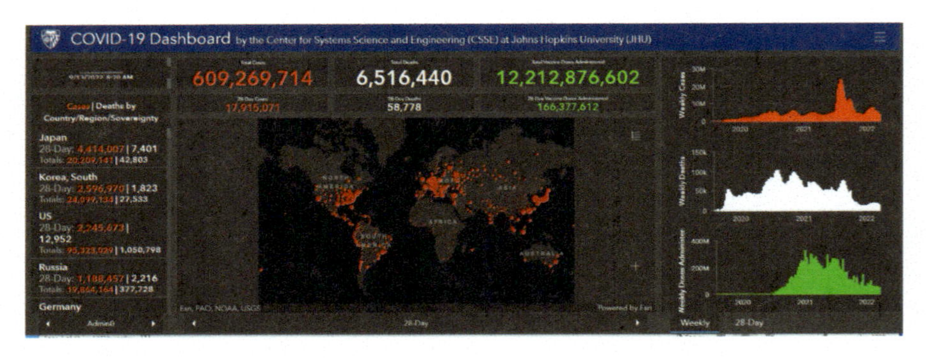

FIGURE 1.1 Johns Hopkins University. CSSE Coronavirus COVID-19 Global Cases (dashboard). 2020. (Source: https://gisanddata.maps.arcgis.com/app. Screenshot date: September 13, 2022.)

for Disease Control and Prevention (CDC, 2020), the National Health Commission of the People's Republic of China (NHC, 2020), the European Centre for Disease Prevention and Control (ECDC, 2020), and the Chinese online medical resource (DXY, 2021). The dashboard provides links to these sources and others. Dr. Lauren Gardner, director of CSSE, explains this work in detail on her blog (Gardner, 2020). The corresponding data repository is accessible in GitHub (Johns Hopkins CSSE GitHub, 2020). ESRI's ArcGIS Living Atlas (Johns Hopkins CSSE, 2019) continues to assist data distributors in adopting a semi-automated living data stream strategy and makes data feature layers freely accessible to update their dashboards. The intense response to the Johns Hopkins University CSSE GitHub and other dashboards shows how eager people are to track health threats and the mapping of coronavirus disease. In a few short clicks, a tremendous amount of information about the COVID-19 virus from these Internet resources can be accessed. Geospatially derived knowledge is more readily available than ever through the mobile apps and cloud-based analysis tools. Geospatial infrastructure plays an important role as driver and beneficiary of digital transformation to improve our understanding of the virus across the world to meet unprecedented challenges at a scale and a pace that can let us respond effectively to rapidly changing conditions (Pratt, 2020).

The COVID-19 pandemic and the events of 2020 were a stress test for our world, bringing widespread social and racial inequities, such as housing, food insecurity, and unequal access to technology, into stark relief and exacerbating the already trying situations of many. For example, COVID-19 single-handedly reversed rising life expectancy trends in 27 countries. The pandemic's toll was greater among males (Aburto et al., 2021). The pandemic's disproportionate impact on communities of color underscores the series health effects of racial and ethnic inequality (Andrasfay et al., 2021). The pandemic caused an immediate spike in undernourishment both globally and regionally, primarily because of job losses and reduction in work hours amid lockdowns (FAO, IFAD, UNICEF, WFP and WHO, 2021). A sudden steep drop in labor force participation happened worldwide from 2019 to 2020; it fell from 61% to less than 59% (International Labour Organization, 2021). The burden of job loss is not shared evenly. Although men's global employment had returned to prepandemic levels in mid-2021, there were still 13 million fewer women in the workforce than in 2019. Vaccination rates against diseases other than COVID-19 have also fallen and doses of childhood vaccines (e.g., measles, diphtheria-tetanus-pertussis) were missed, dipping lowest in April 2020 (Causey et al., 2021). One in four COVID-19 deaths in the United States deprived a child of a primary or secondary caregiver, a disproportionate number of whom were children of color. High infection rates and decreased mobility had an impact on increasing global prevalence of depressive disorders. (Santomauro et al., 2021). In 2020, cigarette sales in the United States increased for the first time in nearly 20 years (Federal Trade Commission, 2020). An unprecedented number of drug overdoses followed the onset of the pandemic. One hundred thousand overdose deaths were reported in a single year from April 2020 to April 2021 (McKnight-Eily et al., 2020; Brulhart, 2021). Collective behavioral changes such as governments imposing lockdowns and other restrictions in response to COVID-19 produced swift results on reducing emissions of carbon dioxide, nitrogen dioxide, and fine particulates and a striking decrease in air pollution in urban areas. (Berman & Ebisu, 2020). Remote care such as telehealth became more accessible to more people. COVID-related disruptions caused the worst education crisis on record due to school closures, and low- and middle-income countries will likely experience longer-lasting effects than those in high-income nations (World Bank, UNICEF and UNESCO, 2020).

As COVID-19's myriad impacts are far from over, GIS will continue to play a key role in quantifying the waves of change that ripple through society. Some WebMaps, GIS-based data dashboards, and applications that are receiving data updates in near-real-time (at the time of writing), help us make personal risk assessment in our daily lives, whereas other dashboards might inform policy decisions and highlight emerging trends that might otherwise get lost as we navigate the daily noise of an ongoing crisis. The most compelling result arising from published studies on exploring the geographical tracking and mapping of COVID-19 is it is still on top of the public health research agenda and is being intensively studied. Exciting things took place and are still taking place in terms of online real- or near-real-time mapping of COVID-19 cases and of social media reactions to its spread. Predictive risk mapping using population travel data, and tracing and mapping of superspreader trajectories and contacts across space and time, are proving indispensable for timely and effective pandemic monitoring and response. ESRI supports its users and community at large with the ESRI Global COVID-19 website (ESRI Global COVID-19 Website, 2020) and the ESRI COVID-19 GIS Hub that provide relevant and authoritative community-driven resources from around the world (ESRI COVID-19 GIS HUB, 2020).

Dashboards and web maps that bring together location and time-sensitive public events in relationship to a spreading disease give travelers and officials the potential to reduce exposure. More web services have allowed GIS users to consume and display disparate data inputs without central hosting or processing, to ease data sharing and speed information aggregation. Some revealed global trends (Figure 1.2), and some were local GIS dashboards revealing COVID-19 case and death status daily.

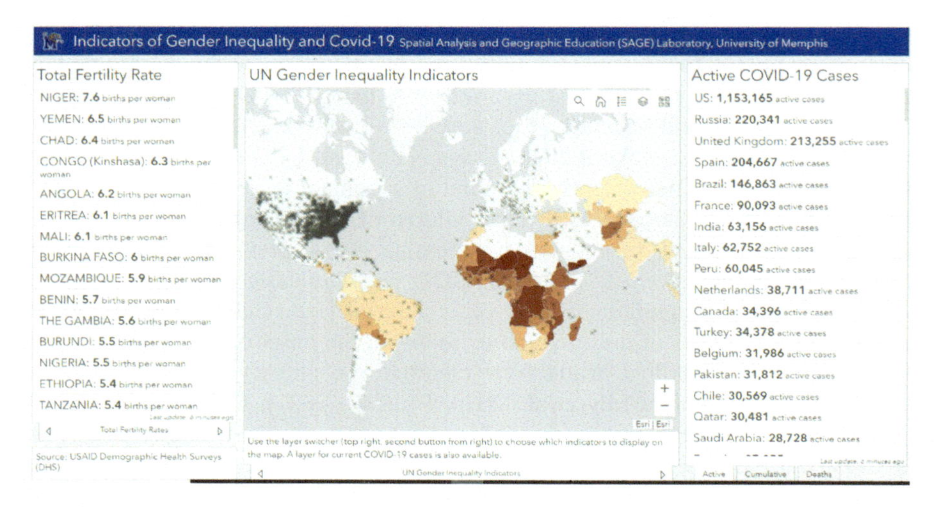

FIGURE 1.2 A screenshot of an online GIS dashboard of COVID-19 and Gender Inequalities. (Source: Ozdenerol, E. COVID-19 and Gender Inequalities, Spatial Analysis and Geographic Education Laboratory, https://cpgis.maps.arcgis.com/apps/opsdashboard/index.html#/f4000bb07cfc4ba4989fa72e1d056b6c)

The GIS-based dashboards are just the beginning of how GIS and location technologies can support the fight against COVID-19. For example, digital supply chain maps have proven foundational to planning and ensuring geographical diversity in suppliers as well as aligning needs with equitable and speedy vaccine distribution (Boulos & Geraghty, 2020). Site selection, whether for emergency treatment units, COVID-19 testing centers, or permanent infrastructure is a common and high-value application of GIS technology. (Boulos & Geraghty, 2020). Apps and maps displayed information and navigation to testing sites, hospitals with available beds, clinics offering medical aid along with current wait times, grocery stores and pharmacies that are open, and places to purchase personal protective equipment. (Boulos & Geraghty, 2020). Drones are also being used for broad disinfectant operations in China (Brickwood, 2020). Integrated drone and GIS technologies can help target and speed up efforts in places they are needed most. In highly impacted areas, drones reduce human contact with lab samples and free up ground transport assets and personnel (Huber, 2020). ESRI's story maps, an innovative use of storytelling during the COVID-19 pandemic, served to understand and analyze the COVID-19 pandemic critically. In ESRI's story maps, community stories relevant to people's COVID-19 pandemic experience are shared, such as stories of loss, stories of recovery, and vaccination stories. They also provided crowdsourced stories to connect communities, identified vulnerable populations and their needs, and communicated wellness and safety messages (ESRI's StoryMaps team, 2020)

The Coronavirus Science-of-Science Map (SOS) and Guided Tour, shown in Figure 1.3, originates from the Center for Information Convergence and Strategy and communicates scientific information and research by using network analysis to find

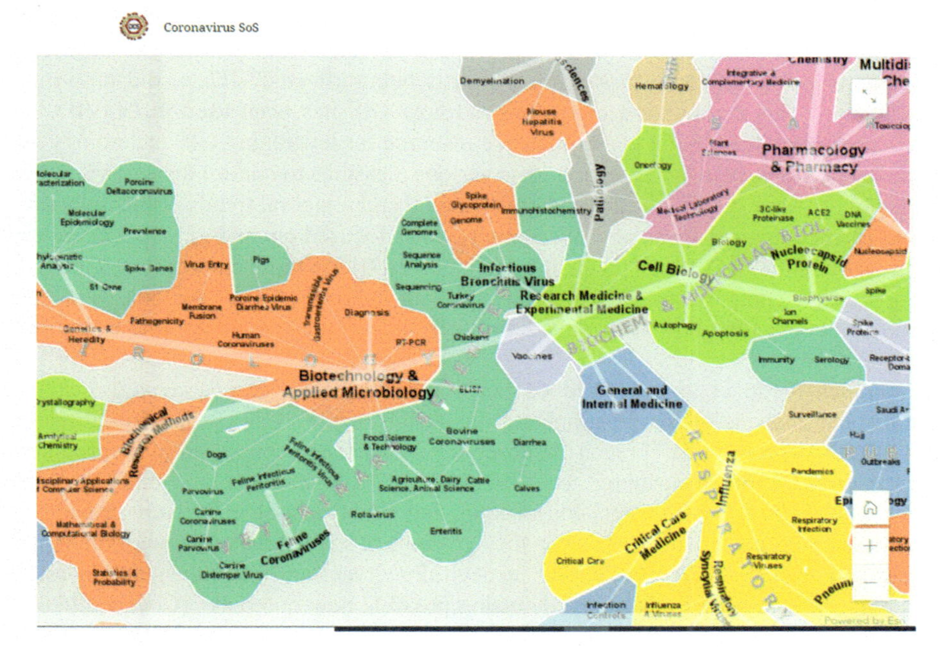

FIGURE 1.3 A screenshot of the Coronavirus Science-of-Science Map (SOS). (Source: Skupin A. Coronavirus SoS: Science-of-Science Map and Guided Tour. https://bigdata.sdsu.edu/coronavirus/)

patterns and relationships in the data set of 15,500 research papers, and about fifty years of coronavirus research (Skupin, 2020).

1.1 METHODS

A literature search was conducted to identify review articles discussing COVID-19 and the use of GIS and spatial modeling. Several online databases were queried, including Google Scholar, Journal Storage (JSTOR), Web of Science, ScienceDirect, and Public/Publisher Medline (Pubmed). The following key words were used individually and in combination as inclusion criteria for articles to be considered for this chapter: GIS, Coronavirus, COVID-19, Spatial analysis, Systematic review, contact tracing, COVID-19 diagnoses, emerging technology, pandemic, screening, surveillance, tracking. The abstracts of these publications were reviewed to confirm applicability.

1.2 RESULTS

The interdisciplinary nature of how geographic and spatial analysis was used in COVID-19 research was notable among the published reviews. Geospatial techniques, especially WebGIS, have even been widely used to visualize the data on a map and were critical to informing the public regarding the spread of the virus, the overall trends of COVID-19 data such as the following: What are the case numbers for a *certain area*? Who in a *defined area* is being affected? *Where* did the virus originate? This compilation of reviews on how GIS has been used in COVID-19 research provided a good comprehensive resource for leveraging GIS technology to make decisions, plan, and communicate various measures to control the spread of the pandemic. Most reviews attempt to cover spatial and temporal trends of the outbreak and predictive analysis using GIS; Some reviews focused on emerging technologies for tackling the COVID-19 pandemic and included GIS in these current technologies. Some made comparisons with the body of work published in various phases of the pandemic revealing how GIS played a role presenting and visualizing the evolution of the pandemic.

Franch-Pardo et al. reviewed 63 scientific articles on geospatial and spatial-statistical analysis of the geographical dimension of the 2019 coronavirus disease (COVID-19) pandemic (Franch-Pardo et al., 2020). They found the most pressing issue these studies contributed to COVID-19 and GIS literature was that GIS is a better social mobilization, decision-making, and planning tool, effectively providing feedback to public health officials on the dynamics of spatial transmission of the virus (Desjardins et al., 2020); facilitating successful interventions in public health by estimating the number of infections, detecting possible new outbreaks (Coccia, 2020); improving patient outcomes by locational analysis of health services (Kuupiel et al., 2019); formulating appropriate scientific, policy, and social measures by mapping human mobility (Mollalo et al., 2020); and predicting the spatial and temporal trends of the outbreak (Giuliani et al., 2020). Franch-Pardo et al. published two reviews: the first one in the first half of 2020 (Franch-Pardo et al., 2020) and the second one from June 2020 to December 2020 (Franch-Pardo et al., 2021). Their contribution was to make comparisons with the body of work published during the two phases, revealing the evolution of the pandemic. They found that in the first half of 2020, the research focused mostly on climatic and weather factors; later, dominant topics studied were the relationship between pollution and COVID-19 dynamics, which enhance the impact of human activities on the pandemic's evolution. The spatial statistical models were among the most widely used and thus could be considered among the most popular tools for studying the COVID-19 pandemic during the second half of 2020. Spatial regression and autocorrelation as well as multicriteria analysis were found in most GIS-based studies, although they found the study of socioeconomic variables more common compared to the first half of the year (Franch-Pardo et al., 2021). They found an increase in remote sensing approaches, which are now widely applied in studies around the world. Lockdowns and associated changes in human mobility have been extensively examined using spatiotemporal techniques in the second half of 2020. Overall, their results showed that there was an increase in the use of both spatial statistical tools (e.g., geographically weighted regression, Bayesian models, spatial regression) applied to socioeconomic variables and analysis at finer spatial and temporal scales. Another common method utilized to study COVID-19 is hot spot analysis, which facilitated targeted interventions by local, state, and federal agencies. The interpolation methods were also used to address the spatial and spatiotemporal patterns of the pandemic not only associated with atmospheric themes such as pollution and climate but also with socioeconomic ones.

Numerous spatial and aspatial models and methods, in conjunction with GIS, were also mentioned due to its geographical contribution to the study of the pandemic. With GIS tools, multicriteria decision analysis (MCDA) has also been used to create a population vulnerability model where the susceptibility to the risk of contagion of COVID-19 was evaluated and mapped in various countries. Sánchez-Sánchez et al. (2020) created a population vulnerability model where they evaluate and map the susceptibility to the risk of contagion of COVID-19 in Chetumal, Mexico. Other similar examples are found in Kampala, Uganda (Bamweyana et al., 2020), and in Nepal (Maharjan et al., 2020).

Spatiotemporal analyses of pollution levels during the pandemic tend to focus on the consequence of reductions in mobility and economic production (Filonchyk et al., 2020; Sandifer et al., 2020; Wu et al., 2020). Remote sensing (RS) and UAVs have been used for studies of land-use change, with significant examples such as the case of Yao et al. (2020), who analyzed the potential distribution of soya crops and how COVID-19 has affected the soya market. Studies using data mining to study human mobility tend to focus on areas where lockdowns have been established. WebGIS platforms for the dissemination of information to the public and data accessibility have already been addressed in other previous studies (Boulos & Geraghty, 2020; Franch-Pardo et al., 2020; Koller et al., 2020; Zúñiga et al., 2020). The data are combined with spatial data for the geographic regions (states and municipalities) and with 2020 population counts for the given regions. Then, the incidence and mortality statistics are calculated at the national, state, and municipal levels. Web mapping is used to disseminate public information about COVID-19. Most spatial analyses were conducted within the territories of China and the United States. In comparison with the previous six months, global-scale studies have mostly declined. India accounted for the greatest increase in studies using spatial analysis along with a consistent amount of research coming from Brazil and Italy.

Zhou et al. analyzed the spatial representation of the disease, material, population, and social psychology at three scales: individual, group, and regional (Zhou et al., 2020). At the individual scale, they compared spatial epidemic tracking and the spatiotemporal trajectories of patients. At the group scale, they estimated to population flow with the spatial distribution. At the regional scale, they carried out the segmentation of spatial risk, the analysis of balance between the supply and demand of medical resources, and the spatial differentiation analysis of material transportation capacity and social sentiment. In assessing the contribution of GIS and spatial big data technology to the containment of the COVID-19 epidemic, they found many challenges remain to be studied. For example, the status of big data source restrictions in commercial enterprises restricted the data supply needed for social management, which resulted in the lack of a mature scheme for big data aggregation, and caused web application performance issues such as slow servers and loading time.

Bolulos et al. (2020) in their review describes a range of practical online/mobile GIS and mapping dashboards such as the WHO dashboard (2020) and HealthMap (2019) and John Hopkins University's dashboard ((Johns Hopkins CSSE, 2019) for tracking the 2019/2020 coronavirus epidemic. A near-real-time (at the time of writing) application, the "close contact detector," is meant for individual users in China to check whether the app user has had any close contact with a person confirmed or suspected to have been infected with SARS-CoV-2 in the recent past (BBC News, 2020). Northeastern University developed EpiRisk, a predictive tool estimating the probability that infected individuals would spread the disease to other parts of the world via air travel and tracking the effectiveness of travel bans as part of the Global Epidemic and Mobility Model (GLEAM) project (EpiRisk, 2020).

Mastaneh and Mouseli reviewed technologies and their solutions in the era of the COVID-19 crisis (Mastaneh & Mouseli, 2020). They divided technologies into two categories: those that support the diagnostic process and case findings, and those with therapeutic and logistic applications. The former included noncontact thermometers, artificial intelligence (AI), drones, self-assessment applications, and virus genome sequencing. GIS was in the latter category with logistic applications providing support services along with telemedicine big data and blockchain. They concluded that the threat of COVID-19 has become an opportunity for a variety of technologies to test themselves in real conditions. Technologies with the ability to reduce human contact through teleservices, as well as those that quickly enable decision-making via in-depth analysis received more attention among health authorities and organizations (Mastaneh & Mouseli, 2020).

Mbunge et al. (2020) also reviewed emerging technologies for tackling the COVID-19 pandemic. He included geospatial technology, AI, big data, telemedicine, blockchain, 5G technology, smart applications, Internet of Medical Things (IoMT), robotics, and additive manufacturing as emerging technologies that are substantially important for COVID-19 detecting, monitoring, diagnosing, screening, surveillance, mapping, tracking, and creating awareness (Mbunge et al., 2020). They aimed at providing a comprehensive review of these technologies for tackling COVID-19 with emphasis on the features, challenges, and country of domiciliation. Their results show that performance of the emerging technologies is not yet stable due to nonavailability of enough COVID-19 data sets, inconsistency in some of the data sets available, and non-aggregation of data sets due to contrasting data format, missing data, and noise. Further research is required to strengthen these technologies and there is a strong need for robust computationally intelligent models for early differential diagnosis of COVID-19.

Saran et al. (2020), in their comprehensive review of geospatial technology and services for infectious diseases surveillance, illuminate the path for decision makers and public health officials for implementation of a web-based spatiotemporal health information system and epidemiological modelling tools. Their review concluded that geospatial applications and dynamic modelling algorithms can offer a well-timed solution to the all-time historic challenge of humankind understanding the disease outbreaks, vulnerabilities to population health, and adaption of upcoming generations (Saran et al., 2020).

Ozdenerol et al. (2021, 2022) with her recent publications correlated lifestyle characteristics to COVID-19 infection and mortality (Ozdenerol & Seboly, 2021) as well as vaccination hesitancy (Ozdenerol & Seboly, 2022). These two publications give insight into behavioral social science research into how COVID-19 has arisen and why it became so completely widespread. They look at this situation nationwide in the United States. They provided unique insights into where and when COVID-19 impacted different households with certain lifestyle traits. The results suggest that prevention and control policies can be implemented for those specific households exhibiting spatial and temporal patterns of high risk. They sequentially mapped and graphically illustrated when and where each LifeMode had above/below average risk for COVID-19 infection/death on specific dates. A strong northwest-to-south and northeast-to-south gradient of COVID-19 incidence was identified, facilitating an empirical classification of the United States into several epidemic subregions based on household lifestyle characteristics. High risk lifestyle segments and their locations were the areas in the United States where clearly the public might benefit from a COVID-19 vaccine. Their findings demonstrate that vaccine uptake appears to be highest in the urban corridors of the Northeast and the West Coast and in the retirement communities of Arizona and Florida, and lowest in the rural areas of the Great Plains and the Southeast. Looking closely at other parts of the West, such as the Dakotas and Montana, counties that contain Native American reservations have higher vaccination rates. Racial/ethnic minorities also adopt the vaccine at higher rates. The most effective predictor of vaccination hesitancy was Republican voting habits, with Republican counties less likely to take the vaccine. The other predictors, in order of importance were college education, minority race/ethnicity, median income, and median age. Ozdenerol et al.'s (2022) approach correlating lifestyle characteristics to COVID-19 vaccination rate at the US county level provided unique insights into where and when COVID-19 vaccination impacted different households.

1.3 CONCLUSION

This chapter is a compendium that identifies the themes and analyses that are being carried out with GIS and spatial-statistical tools in the story of COVID-19 and its control and management. It is a good addition to recent reviews on COVID-19 that offers a good perspective on GIS and how it will continue to play a key role in streamlining efforts for future and more effective pandemic control. It can also serve as support for the evolution in the use of these tools in recurrent topics such as spatiotemporal analysis and disease mapping, health and social geography, environmental variables, data mining, and web-based mapping. The online GIS platform facilitates access to information, knowledge, and experience, as well as best practices for science, technology and innovation facilitation, and initiatives and policies. The online platform also facilitates the dissemination of relevant open access data generated worldwide. As key health indicators are collected and mapped by GIS at subnational levels, and finer spatial resolution of COVID-19 data become available, big data will lead to new avenues for detailed spatial investigations. Successful emerging technology solutions, such as the cloud, social media, cell phones, and Internet usage, will require creative and transformative approaches to pandemic control and management. The world's milestones and performance metrics to manage and control the COVID-19 pandemic are a catalyst for the control of future pandemics and have great potential to accelerate human progress, to bridge the digital divide, and to develop knowledge societies to reach health equality. In our globally interconnected world, connectedness meant conversation—a great deal of conversation. We became each other's databases, servers, and maps; leaning on each other, multiplying, amplifying, and anchoring the things we could imagine by sharing our suffering, our successes, and failures during the COVID-19 era. With this book, I share these successes and failures and how GIS users and developers transformed this crisis into a historic opportunity for once-in-a-lifetime reforms of the national and global health systems based on science, equity, and solidarity and, most importantly, integrated with GIS.

REFERENCES

Aburto, J. M., Schöley, J., Kashnitsky, I., Zhang, L., Rahal, C., Missov, T. I., Mills, C. M., Dowd, J. B., & Kashyap, R. (2022). Quantifying impacts of the COVID-19 pandemic through life-expectancy losses: A population-level study of 29 countries, *International Journal of Epidemiology, 51*(1), 63–74. https://doi.org/10.1093/ije/dyab207

Andrasfay, T., & Goldman N. (2021). Reductions in 2020 US life expectancy due to COVID 19 and the disproportionate impact on the black and Latino populations. *Proceedings of the National Academy of Sciences (PNAS), 118*(5), 1–6.

Bamweyana, I., Okello, D. A., Ssengendo, R., Mazimwe, A., Ojirot, P., Mubiru, F., & Zabali, F. (2020). Socio-economic vulnerability to COVID-19: The spatial case of Greater Kampala Metropolitan Area (GKMA). *Journal of Geographic Information System, 12*(4), 302–318. https://doi.org/10.4236/jgis.2020.124019

BBC News (2020). China launches coronavirus "close contact detector" app. www.bbc.co.uk/news/techn ology -51439 401

Berman J. D., & Ebisu, K. (2020). Changes in U.S. air pollution during the COVID-19 pandemic. *Science of the Total Environment, 739*, 139864.

Boulos, M. N. K., & Geraghty, E. M. (2020). Geographical tracking and mapping of coronavirus disease COVID-19/severeacute respiratory syndrome coronavirus 2 (SARS-CoV-2) epidemic and associated events around the world: How 21st-century GIS technologies are supporting the global fight against outbreaks and epidemics. *International Journal of Health Geographics, 19* (1), 8. https://doi.org/10.1186/s1294 2-020-00202-8

Brickwood, B. (2020). XAG introduces drone disinfection operation to fight the coronavirus outbreak. Health Europa. www.healtheuropa.com/xag-introduces-drone-disinfection-operation-to-fight-the-coronavirus-outbreak/97265/

Brulhart, M. (2021). Mental health concerns during the COVID 19 pandemic as revealed by helpline calls. *Nature, 600*, 121–126.

Causey, K. (2021). Estimating global and regional disruptions to routine childhood vaccine coverage during the COVID-19 pandemic in 2020: A modeling study. *Lancet, 398*, 522–534.

Coccia, M. (2020). Factors determining the diffusion of COVID-19 and suggested strategy to prevent future accelerated viral infectivity similar to COVID. *Science of the Total Environment, 729*, 138474. https://doi.org/10.1016/j.scitotenv.2020.138474

Desjardins, M. R., Hohl, A., & Delmelle, E. M. (2020). Rapid surveillance of COVID-19 in the United States using a prospective space-time scan statistic: Detecting and evaluating emerging clusters. *Applied Geography, 118*, 102202. https://doi.org/10.1016/j.apgeog.2020.102202

Dong, E., Du, H., & Gardner, L. (2020). An interactive web-based dashboard to track COVID-19 in real time. *The Lancet. Infectious Diseases, 20*(5), 533–534. https://doi.org/10.1016/S1473-3099(20)30120-1

DXY.cn. (2021). Chinese online medical resource, online platform for providers, healthcare organizations and health consumers. https://portal.dxy.cn/.

EpiRisk. (2020). EpiRisk Interactive Covid-19 Map. https://tinyurl.com/EpiRisk

ESRI China (Hong Kong). List of Novel Coronavirus Dashboards. https://storymaps.arcgis.com/stories/a1746ada9bf48c09ef76e5a788b5910 and https://go.esri.com/coronavirus Accessed March 20, 2020.

ESRI Covid-19 GIS Hub. (2020). https://coronavirus-resources.esri.com/

ESRI Global Covid-19. (2020). www.esri.com/en-us/covid-19/overview

ESRI's StoryMaps team. (2020). Mapping the Wuhan coronavirus outbreak. https://storymaps.arcgis.com/stories/4fdc0 d03d3 a34aa 485de 1fb0d 2650e e0.

European Centre for Disease Prevention and Control (ECDC). (2020). Weekly COVID-19 country overview. www.ecdc.europa.eu/en

FAO, IFAD, UNICEF, WFP and WHO. (2021). The state of food security and nutrition in the world 2021. Transforming food systems for food security improved nutrition and affordable healthy diets for all by FAO, IFAD, UNICEF, WFP and WHO. www.fao.org/publications/sofi/2021/en/

Federal Trade Commission. (2021). *Cigarette Report for 2020*. October. www.ftc.gov

Filonchyk, M., Hurynovich, V., Yan, H., Gusev, A., & Shpilevskaya, N. (2020). Impact assessment of COVID-19 on variations of SO2, NO2, CO and AOD over East China. *Aerosol and Air Quality Research, 20*(7), 1530–1540. https://doi.org/10.4209/aaqr.2020.05.0226

Franch-Pardo, I., Napoletano, B. M., Rosete-Verges, F., & Billa, L. (2020). Spatial analysis and GIS in the study of COVID-19. A review. *Science of the Total Environment, 79*, 140033. https://doi.org/10.1016/j.scito tenv.2020.140033

Franch-Pardo, I., Desjardins, M. R., Barea-Navarro, I., & Artemi Cerdà, A. (2021). A review of GIS methodologies to analyze the dynamics of COVID-19 in the second half of 2020. *Transactions in GIS, 25*(5), 2191–2239.

Gardner, L. (2020, January 23). Mapping 2019-nCoV (Updated February 11, 2020). https://systems.jhu.edu/research/public-healt h/ncov/

Giuliani, D., Dickson, M. M., Espa, G., & Santi, F. (2020). Modelling and predicting the spatio-temporal spread of coronavirus disease 2019 (COVID-19) in Italy. *BMC Infectious Diseases, 20*, 700. https://doi.org/10.1186/s12879-020-05415-7

HealthMap. (2019). Novel coronavirus 2019-nCoV (interactive map). https://healthmap.org/wuhan /

Huber M. (2020, February 7). Drones enlisted to fight corona virus in China. AIN Online. www.ainonline.com/aviation-news/general-aviation/2020-02-07/drones-enlisted-fight-corona-virus-china

International Labour Organization. (2021). ILOSTAT Database. Data retrieved on June 15, 2021, and presented by World Bank. https://ilostat.ilo.org/data/

Johns Hopkins CSSE. (2019). Corona Virus 2019-nCoV Cases (The Living Atlas). https://livin gatla s.arcgi s.com/en/brows e/#d=2&q=%22Cor ona%20Virus%20201 9%20nCoV%20Cas es%22

Johns Hopkins University. (2020). CSSE Coronavirus COVID-19 Global Cases (dashboard). 2020. https://gisanddata.maps.arcgis.com/app

Johns Hopkins University. (2019). CSSE GitHub–CSSEGISandData/COVID-19: Novel Coronavirus (COVID-19) Cases (data repository). https://githu b.com/CSSEG ISand Data/COVID -19

Koller, D., Wohlrab, D., Sedlmeir, G., & Augustin, J. (2020). Geografische Ansätze in der Gesundheitsberichterstattung. *Bundesgesundheitsblatt-Gesundheitsforschung-Gesundheitsschutz*, *63*, 1108–1117. https://doi.org/10.1007/s0010 3-020-03208-6

Kuupiel, D., Tlou, B., Bawontuo, V., Drain, P. K., & Mashamba-Thompson, T. P. (2019). Poor supply chain management and stock-outs of point-of-care diagnostic tests in Upper East Region's primary healthcare clinics, Ghana. *PLoS One*, 14, e0211498. https://doi.org/10.1371/journal.pone.0211498

Maharjan, B., Maharjan, A., Dhakal, S., Gadtaula, M., Shrestha, S. B., & Adhikari, R. (2020). Geospatial mapping of COVID-19 cases, risk and agriculture hotspots in decision-making of lockdown relaxation in Nepal. *Applied Science and Technology Annals*, *1*(1), 1–8. https://doi.org/10.3126/asta.v1i1.30263

Mastaneh, Z., & Mouseli, A. (2020). Technology and its solutions in the era of COVID-19 crisis: A review of literature. *Journal of Evidence-Based Health Policy, Management and Economics*, *4*(2), 138–149. https://doi.org/10.18502/jebhpme.v4i2.3438

Mbunge, E., Akinnuwesi, B., Fashoto, S. G., Metfula, A. S., & Mashwama, P. (2021). A critical review of emerging technologies for tackling COVID-19 pandemic. *Human Behavior and Emerging Technologies, 3*(1), 25–39. https://doi.org/10.1002/hbe2.237.

McKnight-Eily Lela, R. et al. (2020, February 5). Racial and ethnic disparities in the prevalence of stress and worry: Metal health conditions and increased substance use among adults during the COVID 19 pandemic—United States. *Morbidity and Mortality Weekly Report, 70*(5), 162–166.

Mollalo, A., Vahedi, B., & Rivera, K. M. (2020). GIS-based spatial modeling of COVID-19 incidence rate in the continental United States. *Science of the Total Environment*, *728*, 138884. https://doi.org/10.1016/j.scito tenv.2020.138884

Montanez, A., Christiansen, J., Devins, S., Surillo M., & Taylor, A. R. (2022, March). Data captured Covid's uneven toll. Visualizing ongoing stories of loss, adaptation and inequality. *Scientific America*.

National Health Commission of the People's Republic of China (NHC). (2020). Daily briefing on novel coronavirus cases. February 3. http://en.nhc.gov.cn/2022-09/20/c_86123.htm

Ozdenerol, E. (2020). Covid-19 and gender inequalities, spatial analysis & geographic education laboratory. https://cpgis.maps.arcgis.com/apps/opsdashboard/index.html#/f4000bb07cfc4ba4989fa72e1d056b6c)

Ozdenerol, E., & Seboly, J. (2021). Lifestyle effects on the risk of transmission of COVID-19 in the United States: Evaluation of market segmentation systems. *International Journal of Environmental Research and Public Health, 18*(9): 4826. https://doi.org/10.3390/ijerph18094826

Ozdenerol, E., & Seboly, J. (2022). The effects of lifestyle on COVID-19 vaccine hesitancy in the United States: An analysis of market segmentation. *International Journal of Environmental Research and Public Health,* 19(13), 7732. https://doi.org/10.3390/ijerph1913773

Pratt, M. (2020). Opportunity in adversity. Arc User. October. www.esri.com/about/newsroom/arcuser/opportunity-in-adversity/

Public Health England. (2020). Weekly national flu reports: 2019 to 2020 season. Octobe 1. www.gov.uk/government/statistics/weekly-national-flu-reports-2019-to-2020-season

Sánchez-Sánchez, J. A., Chuc, V. M. K., Canché, E. A. R., & Uscanga, F. J. L. (2020). Vulnerability assessing contagion risk of Covid-19 using geographic information systems and multi-criteria decision analysis: Case study Chetumal, México. In M. F. Mata-Rivera, R. Zagal-Flores, J. Arellano Verdejo, & H. E. Lazcano Hernandez (Eds.), *GIS LATAM: First Conference* (pp. 1–17). Cham, Switzerland: Springer. https://doi.org/10.1007/978-3-030-59872-3_1

Sandifer, P., Knapp, L., Lichtveld, M., Manley, R., Abramson, D., Caffey, R., & Singer, B. (2020). Framework for a community health observing system for the Gulf of Mexico region: Preparing for future disasters. *Frontiers in Public Health, 8*, 588. https://doi.org/10.3389/fpubh.2020.578463

Santomauro, D., Mantilla Herrera, A., Shadid, J., Zheng, P., Ashbaugh, C., Pigott, D., Abbafati, C., Adolph, C., Amlag, J., Aravkin, A., Bang-Jensen, B., Bertolacci, G., Bloom, S., Castellano, R., Castro, E., Chakrabarti, S., Chattopadhyay, J., Cogen, R., Collins, J., Dai, X. et al. (2021). Global prevalence and burden of depressive and anxiety disorders in 204 countries and territories in 2020 due to the COVID 19 pandemic. *Lancet, 398*(10312), 1700–1712.

Saran S., Singh, P., Kumar, V., & Chauhan, P. (2020). *Journal of the Indian Society of Remote Sensing, 48*(8), 1121–1138. https://doi.org/10.1007/s12524-020-01140-5(0123456789().,-volV)(0123456789,-().volV)

Skupin, A. (2020). Coronavirus SoS: Science-of-Science Map and Guided Tour. San Diego State University. https://bigdata.sdsu.edu/coronavirus/

US Centers for Disease Control and Prevention (CDC). (November 2020). 2019–2020 U.S. Flu Season: Preliminary Burden Estimates. www.cdc.gov/flu/about/burden/preliminary-in-season-estimates.htm.

US Centers for Disease Control and Prevention (CDC). 2020. COVID Data Tracker. Accessed November 20, 2021.

World Bank, UNICEF and UNESCO. (2021). The state of the global education crisis: A path to recovery. World Bank, UNICEF and UNESCO. 2021.

World Health Organization (WHO) (2020). WHO COVID 19 Dashboard. http://covid19.who.int/ data downloaded December 31, 2021.

Wu, X., Yin, J., Li, C., Xiang, H., Lv, M., & Guo, Z. (2020). Natural and human environment interactively drive spread pattern of COVID-19: A city-level modeling study in China. *Science of the Total Environment, 756*, 143343. https://doi.org/10.1016/j.scito tenv.2020.143343

Yao, H., Zuo, X., Zuo, D., Lin, H., Huang, X., & Zang, C. (2020). Study on soybean potential productivity and food security in China under the influence of COVID-19 outbreak. *Geography and Sustainability, 1*(2), 163–171. https://doi.org/10.1016/j.geo sus.2020.06.002

Zhou, C., Su, F., Pei, T., Zhang, A., Du, Y., Luo, B., Cao, Z., Wang, J., Yuan, W. Yuanjiang Zhu, Y., Song, C., Chen, J., Xu, J., Li, F., Ma, T., Jiang, L., Yan, F., Yi, J., Hu, Y., Liao, Y., Xiao, H. COVID-19: Challenges to GIS with big data. (2020). *Geography and Sustainability, 1*(1), 77–87. https://doi.org/10.1016/j.geosus.2020.03.005

Zúñiga, M., Pueyo, A., & Postigo, R. (2020). Herramientas espaciales para la mejora de la gestión de la información en alerta sanitaria por COVID-19. *Geographicalia, 72*, 141–145. https://doi.org/10.26754/ ojs_geoph/ geoph.20207 25005

RESOURCE GUIDE OF ORGANIZATIONS THAT PROVIDE COVID-19 MAPS AND DATA

This section lists organizations for which readers can visit their websites and view and download COVID-19 data. This list is a select group of websites.

COVID-19 Dashboard by the Center for Systems Science and Engineering (CSSE) at the Johns Hopkins University (JHU)

https://coronavirus.jhu.edu/map.html
An interactive map locates and tallies confirmed infections, fatalities and recoveries.

World Health Organization Coronavirus (COVID-19) dashboard

https://covid19.who.int/
An interactive dashboard locates and tallies confirmed cases and deaths by country in the world.

UN COVID Data Hub

https://covid-19-data.unstatshub.org/
United Nations Department of Economic and Social Affairs Statistics geospatial data web services, suitable to produce maps and other data visualizations and analyses, and easy to download in multiple formats.

ESRI Global COVID-19 website

https://www.esri.com/en-us/covid-19/overview
Maps and geographic information systems (GIS) provide valuable insights to help organizations respond to the crisis, maintain continuity of operations, and support the process of reopening. Maps and analyses provide a common frame of reference and integrate all types of relevant data of COVID-19.

ESRI COVID-19 GIS Hub

https://coronavirus-resources.esri.com/
This link provides information to get maps, data sets, applications, and more for coronavirus disease.
ESRI China (Hong Kong). List of Novel Coronavirus Dashboards. https://storymaps.arcgis.com/stories/a1746ada9bf48c09ef76e5a788b5910 and https://go.esri.com/coronavirus
ESRI's StoryMaps. Mapping the spread of COVID-19.
Https://storymaps.arcgis.com/stories/4fdc0d03d3a34aa485de1fb0d2650ee0

Google COVID-19 data

https://health.google.com/covid-19/open-data/explorer?loc=US
This website guides us to COVID-19 open data across the world with statistics and data visualizer.

University of Nebraska COVID-19 dashboard

https://covid19.unl.edu/covid-19-dashboard
This is an interesting website that guides us to university-developed COVID-19 dashboard with all the test results and positivity rates across the world with COVID-19 information.

WorldOmeter dashboard

www.worldometers.info/coronavirus/
This is a popular COVID-19 information dashboard with all the active and closed cases.

Our world in data

https://ourworldindata.org/coronavirus
This is a very fast updating website with global dashboard with COVID-19 data explorer and many active links on the dashboard.

IHME: Institute for Health Metrics and Evaluation

https://covid19.healthdata.org/global?view=cumulative-deaths&tab=trend
IHME provides the best COVID-19 projections globally with patterns and trends in COVID-19 data

New York Times

www.nytimes.com/interactive/2021/world/covid-cases.html (Paid)
The *New York Times* tracks the global outbreak with an interactive dashboard and hot spots.

Bloomberg

www.bloomberg.com/graphics/2020-coronavirus-cases-world-map/
Bloomberg provides an interactive COVID-19 dashboard with worldwide COVID-19 data, vaccine tracker and also holds three-year information cumulatively.

CNN health

www.cnn.com/interactive/2020/health/coronavirus-maps-and-cases/
CNN health provides a COVID-19 dashboard with global cases and deaths with a world map of spread per 100k cases.

ncov2019

https://ncov2019.live/
This is a live COVID-19 data provider which mainly focuses on data with a survival rate calculator.

NBC News

www.nbcnews.com/health/health-news/coronavirus-map-confirmed-cases-2020-n1120686
NBC News provides a COVID-19 live dashboard with new COVID-19 cases information in graphical manner.

Goverment of Canada countrywise COVID-19 data

https://health-infobase.canada.ca/covid-19/international/
This is Canada government's official website providing COVID-19 data across the globe from collaboration between public health agency of Canada, statistics Canada and natural resources Canada.

Statista

www.statista.com/topics/5994/the-coronavirus-disease-covid-19-outbreak/#topicHeader__wrapper
Statista provides COVID-19 data in a tree structure and with many active links of COVID-19 data across the world.

European Centre for Disease Prevention and Control

www.ecdc.europa.eu/en/data/dashboards
This website provides dashboards for COVID-19 cases for the EU/EEA, SARS-cov2-variants data, and COVID-19 vaccine tracker.

World Health Organization (Europe)

www.arcgis.com/apps/dashboards/ead3c6475654481ca51c248d52ab9c61
This dashboard illustrates the information by providing graphs, maps and total cases count of individual euro countries.

Euronews

www.euronews.com/my-europe/2020/11/13/europe-s-second-wave-of-coronavirus-here-s-what-s-happening-across-the-continent
Euronews website provides visualization charts on deaths and cases due to COVID-19 with respective countries.

Chinese online medical resource

Online platform for providers, healthcare organizations, and health consumers https://portal.dxy.cn/.

National Health Commission of the People's Republic of China

Daily briefing on novel coronavirus cases
http://en.nhc.gov.cn/2022-09/20/c_86123.htm.

2 Information Access as a Strategy to Overcome Health Inequities in the Context of COVID-19

Irma Singarella

CONTENTS

DOI: 10.1201/9781003227106-2

> We know that every outbreak will be accompanied by a kind of tsunami of information, but also within this information you always have misinformation, rumors [...] it goes faster and further, like the viruses that travel with people.
>
> *Zarocostas (2020), p. 676*

2.1 INTRODUCTION

Information is one of the essential commodities needed by human beings. Information access is the freedom or ability to identify, obtain, and make use of data and content effectively. It is crucial for the advancement of society, and an important empowerment tool for people to achieve their goals. *The Cengage Computer Encyclopedia* describes it as "vital to social, political, and economic advancement." Historically, information has been disseminated in diverse formats, including the media, libraries, social networks, and individual interactions. Technological developments transformed information access, making vast stores of business, education, health, government, and entertainment information accessible on the World Wide Web (Information Access, 2019).

Emergent equipment and devices were great allies during a pandemic. Other technologies like the Internet of Things (IoT), connect the physical world to the Internet. We are experiencing a growing trend of devices connected to the Internet and to each other, smartphones, wearable data collection devices, and other "connected" products that are becoming more common, useful, and popular (Mohn, 2020). Artificial intelligence proved to be essential during the pandemic for the functioning of society and economy. These technologies offer opportunities for better services and communication, but the COVID-19 pandemic functioned as an experiment that revealed different disparities including uneven access to digital infrastructures (Faraj et al., 2021).

Other connected technology devices enabled communication, education, and diverse forms of remote work. Another example of technology enhancement were the chatbots, computer programs designed to simulate conversations between computer (or other devices) and humans. Location data applications tracked and mapped the spread of the virus for health workers and researchers. These and other programs expanded the technological capabilities during the pandemic.

COVID-19 is a viral disorder characterized by high fever, cough, dyspnea, chills, persistent tremors, muscle pain, headache, sore throat, a new loss of taste and/or smell, and viral pneumonia and other severe symptoms (National Library of Medicine, 2020).The magnitude of the inequities during the pandemic made information access a hard goal to achieve. This book chapter comes from the urgent need to investigate the role of information literacies and information access in the quality of life of individuals and communities, which became more evident during the coronavirus (COVID-19) crisis. The purpose is to provide a broad definition of information and information literacies and an overview of the diverse efforts of different professionals who aim to construct a better world using information as their tool.

2.2 INFORMATION ACCESS IS A BASIC HUMAN RIGHT

In reaction to the disasters and horrors of World War II, the United Nations proclaimed that all humans have basic important rights in the document the Universal Declaration of Human Rights (Dawkins, 2017). One of these is the right to access information. Article 19 states that everyone has the right to freedom of opinion and expression; this right includes the opportunity to seek, receive, and impart information and ideas through any media, and regardless of frontiers (United Nations, 2021). The Declaration was the first international agreement on the basic principles of human rights, and the foundation for most of the protections we must have today.

Different organizations have defended the right to information access as a critical component of the democratic life. For example, the professional organization American Library Association (ALA) expresses the basic principles of librarianship in its Code of Ethics and in the Library Bill of Rights and its interpretations. Nobody should be restricted or denied access for expressing or receiving constitutionally protected speech (American Library Association-ALA, 1996). The ALA also states that

> Books and other library resources should be provided for the interest, information, and enlightenment of all people of the community the library serves. Educational materials should not be excluded because of the origin, background, or views of those contributing to their creation.

Attempts to restrict access to information resources violate the basic tenets of the *Library Bill of Rights* (American Library Association-ALA, 2006, 2007).

Individuals must be well informed on all the aspects and situations that affect their life, and they also need to find meaning in the information that they discover. An interesting case narrated by Bishop (2012) illustrates the importance of information access as a human right. In 1998, the Chilean government refused to release information about an extensive timber logging operation by a US company on the Island of Tierra del Fuego. Various human rights groups filed a request with the Inter-American Commission on Human Rights, in Washington, DC. The Commission urged Chile's government to guarantee citizens access to public information. This case, like numerous others, reflects the lack of access and information disparities that affect the quality of life of individuals, communities, and populations at large.

2.2.1 The Digital Divide

Technological advancement in the era of COVID-19 brought new means of interactions among governments, business, and citizens. Health organizations and introduced new policies intended to control the spread of the coronavirus. These new policies include lockdowns and social-distancing measures that resulted in technological progress and different relations among governments, businesses, and citizens. Such changes increased online activities and expanded interactions, including robotics delivery systems, distance learning, remote work payless and contactless payment services, online entertainment, and an increase in online shopping (Renu, 2021).

While technological advances have improved access to information and knowledge, they have also introduced new forms of exclusion and inequalities. Theorists have described these inequities pertaining to information access as the "digital divide." This situation refers to the breach between those who have technological access and those who do not. It addresses issues concerning equal opportunity and equity that influence the development of marginalized populations within and across systems nationally and internationally (United Nations Educational, Scientific and Cultural Organization-UNESCO, 2021). It is attributed to limitations imposed by educational backgrounds, socioeconomic level, gender, age, disabilities, and geography as well as restrictions experienced by some ethnic and racial groups. The concept of the "digital divide" became popular during the nineties, when the National Telecommunications and Information Administration (NTIA)-Department of Commerce published the report *Falling through the Net: A Survey of the "Have Nots" in Rural and Urban America* (1995), a research project on Internet use among the US population. The disparity between the digital information "haves" and "have-nots" is a persistent international problem, and is manifested in access, content, literacies, and training (Schweitzer, 2015).

During the pandemic, a large segment of the population did not have access to the Internet. For example, a considerable number of students did not own laptops or broadband access to Internet. In the medical context, patients did not have access to telehealth services and other technologies that they needed for their treatments. Even when the pandemic resulted in the rapid adoption of telehealth/telemedicine, Choi et al. (2022) discovered that only a fifth of older Medicare beneficiaries used telehealth services during "the most serious global pandemic in a century" when social distancing was an especially important safety measure. Telehealth use was negatively associated with elder age and lower income.

Beyond the digital divide, language barriers continue to hinder the right to information of minority language users. While only 5% of the languages of the world are present on the Internet, around 40% of the 7,000 languages spoken worldwide are at risk of disappearing, jeopardizing unique knowledge systems and cultures. Since 2003, the United Nations Educational Scientific and Cultural Organizations (UNESCO) has recommended the development and promotion of multilingual information and universal access to the cyberspace world. The organization has recognized the importance of promoting multilingualism and equitable access to information and knowledge. They encourage access to information for all, multilingualism, and cultural diversity on the global information networks. Other marginalized populations such as persons with disabilities, women, and girls also continue to face obstacles in accessing information, which compounds the inequalities they already face.

2.3 INFORMATION ACCESS AS A SOCIAL DETERMINANT OF HEALTH

Social determinants of health are

> the conditions in which people are born, grow, live, work, and age, including the health system. These circumstances are influenced by the distribution of money, power, and resources at global, national, and local levels, which are themselves influenced by policy choices. The social determinants of health are mostly responsible for health inequities—the unfair and avoidable differences in health status seen within and between countries. (World Health Organization-WHO, 2022)

Access is not enough to develop informed individuals and communities. Those able to access and use information may develop the necessary skills, knowledge, and economic ability to use resources, services, and to attain general wellness. Effectively addressing these challenges requires inclusive models of social interactions, as well as knowledge democratization supported by international adapted policy frameworks and innovative uses of digital technologies. The complexity of the information society requires that individuals can access, understand, and use a diversity of data transmitted in different formats. People need training and a lifelong learning mindset to analyze, understand, and interpret the world. Information literacy helps address social exclusion by providing the skills to analyze and interpret diverse types of information.

2.4 DIMENSIONS OF LITERACY

2.4.1 Information Literacy

The Association of College and Research Libraries (ACRL), a section of the American Library Association (ALA), defined information literacy as the ability to recognize when information is needed to locate, evaluate, and effectively use the needed information. It empowers us to reach and express informed views and to engage fully with society (2017). In the wider context, information skills are important because they help increase general literacy. It includes making balanced judgments about any content we find and use. It incorporates a set of skills and abilities that everyone needs, to undertake information-related tasks, such as how to discover, access, interpret, analyze, manage, create, communicate, store, and share information (Chartered Institute of Library and Information Professionals-CLIP, 2018).

2.4.2 Health Literacy

The World Health Organization stated that a low level of health literacy is a critical concern and maximizes inequities (Bin Naeem & Kamel Boulos, 2021). COVID-19's rapid development into a pandemic called for people to access health information, and more important, adapt their behavior. Most significant information was published in an easy-to-understand manner that offered simple and practical solutions, such as washing hands, maintaining physical distance, instructions on where to find information about the latest recommendations, and advice. A large amount of complex, contradictory, and false information was available during the pandemic. News, social media, websites, published literature, and other information outlets had intense repercussions for emotional and well-being. The COVID-19 pandemic highlighted that poor health literacy among a population is an underestimated public health problem globally (Paakkari & Okan, 2020).

2.4.3 Media Literacy

Mass media are instruments or technological means of communication that reach large numbers of people with a common message: press, radio, television, and others (National Library of Medicine, 1980). Media literacy is most described as a skill set that promotes critical engagement with messages produced by the media. It includes effective inquiry and critical thinking about the messages we receive and generate. These skills are especially important during a health crisis like the COVID-19 pandemic. Media literacy focuses on education to develop the skills to create (produce) with media technologies, as well as to critically analyze and evaluate media and media messages, and critical use of non-textual communication formats (Lewis, 2021). Media literacy can benefit society, especially vulnerable populations.

Covington (2021) identifies the diverse components of media literacy as follows:

- Medial and spatial analysis geography and information access
- Understanding and acknowledging of information needs
- Awareness of how we engage with the digital world

Media use increased and assumed a unique role during the pandemic, serving simultaneously as the gateway to work, education, social life, news, and public health information. A media literate population can help reduce the spread of misinformation and benefit people and public health (Guldin et al., 2021)

The health crisis provoked a precipitous expansion in information search and consumption (Rajasekhar et al., 2021). Access to general news websites and mobile apps also increased. Even when the availability of health-related websites increased, some users never verified sources. COVID-19 had a formidable impact on media consumption, but pandemic-related content may be emotionally unsettling and provoke anticipatory anxiety. This situation also presents serious public health challenges.

Researchers from the Australia National University (ANU) have found that the brain can be untrustworthy when it comes to deciphering fake news, and especially when headlines are repeated, presented with photos, or are easy to imagine. Their findings are outlined in an open-access e-book, *The Psychology of Fake News*, which analyses the psychological factors that lead us to believe and share misinformation and conspiracy theories, and interventions to correct false beliefs and reduce the spread of doubtful information (Rainer et al., 2021).

2.4.4 Digital Literacy and Digital Health Literacies

The ALA defines digital literacy as the ability to use information and communication technologies to find, evaluate, create, and communicate information requiring both cognitive and technical skills. When reading online information, an individual must have the skills to access content that has hyperlinks, audio clips, videos, and other formats. During the information era, we have seen a proliferation of user-generated content. These include any form of information resources (videos, texts, and audios among others) that have been created and posted by users of online platforms. Walker (2021) states the user-generated content published through personal webpages, blogs, and social media platforms has increased the amount of information available. These resources provide complementary and supplementary content that helps audiences understand and interpret information. It is challenging to filter reliable information, but digital literacy provides the competences to create, collaborate, understand, share, and evaluate electronic documents and other content.

Digital health literacy refers to the skills to search, choose, assess, and utilize online health information and health care-related digital applications to improve well-being. This literacy is positively associated with psychological well-being (Amoah et al., 2021). An important technological development during the pandemic is telehealth, which was rapidly integrated into care delivery. Lama et al. (2022) define telehealth as the exchange of medical information from one site to another through electronic communication to improve patients' health. During the coronavirus (COVID-19) pandemic, digital health literacy became an indispensable resource in promoting health and well-being. It empowers people to actively participate in managing their health by promoting access and appropriate utilization of accurate health information.

2.5 INFORMATION AND THE EMERGENCE OF COVID-19

The coronavirus pandemic brought an explosion of both accurate and inaccurate information about the disease, an infodemic. We may experience the proliferation of false or misleading content in digital and physical during an outbreak. It may cause misperceptions and risk-taking behaviors that can harm health. It may also lead to suspicion in health establishments and weaken the public health response. It may increase and extend epidemics when people are uncertain about what they need to do to protect their health and the health of the communities (World Health Organization-WHO, 2022a). The infodemic included rumors, frauds, schemes, and conspiracy theories from unfiltered sources often propagated through social media and other outlets.

Misinformation refers to unintentional mistakes such as inaccurate photo captions, dates, statistics, translations, or when satire is taken seriously. Misinformation can lengthen pandemics and increase the number of deaths. This scenario causes confusion and risk-taking behaviors that can intensify or lengthen outbreaks when people are unsure how to protect their health (Cochrane, 2021).

With the rapid growth of digital communication and social media platforms, information can spread faster than a virus. Disinformation refers to fabricated or deliberately manipulated data and audio/visual content, intentionally created conspiracy theories, or rumors. Naeem and Bhatti (2020) state that the excessive amount of misinformation made it more difficult to distinguish between evidence-based information and a broad range of unreliable content The crisis limited the capacity of health-care and health information professionals to promote public health awareness through trustworthy information.

Rajabifard et al. (2022), in their book *COVID-19 Pandemic, Geospatial Information, and Community Resilience: Global Applications and Lessons*, stated that the COVID-19 pandemic has left no country intact. All sectors of society were impacted. As a result, the world was forced to rapidly adapt to confronting social and economic changes and challenges. Nevertheless, the pandemic also provided the opportunity to rise to the challenges. It provides opportunities to gain experience and build resilient and sustainable communities.

2.6 METHODOLOGY

2.6.1 Literature Review

A literature review is an evidence-based, in-depth analysis of a subject. It is a critical appraisal of the publications and the current collective knowledge on a subject (Winchester & Salji, 2016). The purpose of this methodology is to report the current knowledge on a topic and base this summary on previously published research. This chapter provides an overview of relevant activities and projects related to diverse forms of information access during the pandemic.

The author of this chapter performed multiple searches in academic databases including PubMed, the Cumulative Index of Nursing and Allied Literature (CINAHL) and Scopus. Since COVID-19 is a recent topic, the literature review included documents from 2019 to 2022. The keywords searched included "information services," "COVID-19," and "information access," among others.

The literature review focused on two research questions:

- Which information strategies have been developed to facilitate information access during the COVID-19 pandemic?
- Which are the characteristics of successful information access projects?

2.6.2 Which Information Strategies Have Been Developed to Facilitate Information Access during the COVID-19 Pandemic?

For the purposes of this analysis, the word "strategy" is defined as plans and actions designed to provide information access. These strategies developed to provide information access are summarized and grouped in five different but overlapping categories: awareness services, collaborations, communication services and online teaching, research, and geographic information systems (GIS).

Awareness services. Awareness services constitute a way of letting people know about materials, information, or content that has been developed recently. These projects provide access to curated content quality information. Health information professionals (HIPs) are a diverse group of professionals who provide access to and deliver valuable information that improves patient care and supports education research and publication. HIPs are a group of specialists, often including informationists, bioinformationists, and librarians, or who have special knowledge in quality health information resources (Ma et al., 2018). Diverse organizations developed awareness information services about the coronavirus, while providing other services such as document delivery and research consultations. During the pandemic, they quickly increased awareness through public health education, which was considered the most effective tool to protect during the crisis. Information channels to facilitate information access include mobile apps, artificial intelligence, chat services, social media, and video-based lectures on YouTube, Vimeo, Daily motion, and other sources in which experts shared clips about the symptoms and preventive measures (Ali & Bhatti, 2020; Ali & Gatiti, 2020; Naeem & Bhatti, 2020).

Diverse information professionals compiled resources and research guides to help students, staff, faculty, researchers, and the greater communities. Its primary goal was to provide direct access to high-quality meta-analyses, literature syntheses, and clinical guidelines from a variety of trusted sources (Allee et al., 2021). This is also an excellent example of the next category found in the literature: Collaborations.

Collaborations. During the pandemic we experienced a broad commitment among researchers, health-care professionals, information professionals, information technology (IT) experts, policy makers, and social media experts to prevent the spread of misinformation (Naeem & Bhatti, 2020; Naeem & Boulos, 2021). The literature reveals expanded relationships between diverse professional groups to help stop the spread of misinformation. Fighting the infodemic became the new front in the COVID-19 battle. DeRosa et al. (2021) presented a project between public librarians and other HIPs. The staff members received training to respond to consumer health questions about COVID-19 and develop specialized consumer health reference programs, interactive presentations, and educational materials.

At the start of the COVID-19 pandemic, a group of HIPs built capacity collecting information and sharing resources quickly. They met challenges, including moving to telecommuting, and navigating a changing information landscape, and in doing so, improved relationships across the region. (Mazure et al., 2021). Other collaborations included proactively adapted programs to respond to the stay-at-home social-distancing order. Ongoing collaboration projects between diverse professional groups and disciplines proved to be essential during the pandemic. For example, Charney et al. (2021) described an ongoing partnership between HIPs and health-care professionals. A development team, which consisted of the hospital and medical school disaster preparedness medical director, the medical library director, librarians, and the IT and marketing departments in a multistate health-care system worked together to develop a shared website. They distributed and curated timely resources.

Public Services and Outreach. Lindsay et al. (2021) studied the websites of selected associations to identify their responses to the COVID-19 pandemic. These authors highlighted diverse roles and initiatives taken: the use of technologies to communicate, online research consultations, and expanded chat services. Other articles also described methods for communicating and promoting remote resources and services in response to the coronavirus, including telehealth (Rahaman, 2021).

Various organizations used innovative technologies and implemented virtual services models that resulted extremely effective that continued beyond the duration of this crisis. In response to the COVID-19 pandemic and the subsequent stay-at-home order, the Southern Illinois University Medical Library used modern technologies and implemented virtual service models to improve communication and provide services and information access remotely (Howes, 2021). Several of these platforms were Connectwise, Webex, and Zoom, and information providers adapted their websites to provide updated information. Charbonneau and Vardell (2022) researched and explained the scope and adaptive nature of reference services provided by over a one-year period (between March 2020 and March 2021). Librarians reported an increase in reference questions during the pandemic and responded in creative ways despite limited time and the unexpected reduction in resources.

Online learning or distance education. Online learning is defined as education via communication media (correspondence, radio, television, computer networks) with little or no in-person face-to-face contact between students and teachers. (*Medical Subject Headings* 1999—Originally *ERIC Thesaurus,* 1997). The COVID-19 pandemic disrupted diverse education missions. During the spring of 2020, university faculty members, teachers, librarians, and other educators were forced to move in-person services and resources to a virtual environment. This online mandate meant deciding between synchronous and asynchronous activities and learning new tools. (Ibacache et al., 2021; Orcutt et al., 2021; Lucey et al., 2022). Educators, health, and information providers transitioned to remote work online, and instruction became the norm for hundreds of higher education institutions. For some populations, this was the first time working remotely. To accomplish that goal and preserve the continuity of educational projects, different organizations all over the world compiled resources and created websites, interactive tutorials, and research guides for educational purposes. Diverse government offices and information centers expanded digital collections. Information providers and educators had to reach out to faculty, students, and other patrons using various technologies and tools (Mi et al., 2020). Different online learning efforts proliferated during the pandemic with the purpose of keeping communities engaged, and the education activities continued during the crisis. Various institutions provided laptops to their students and developed "hot spots" for Internet access.

Research. The acute crisis of COVID-19 defied the scientific community to produce timely evidence about the virus. pandemic yielded an unprecedented quantity of new publications. Ioannidis et al. (2021) state that we experienced a "Covidization" of research with a spread of publications from diverse disciplines. The growth of COVID-19 articles available was far more rapid and massive compared with cohorts of authors historically publishing on H1N1, Zika, Ebola, HIV/AIDS, and tuberculosis. There was a "rapid and massive growth" of information and this led to rapid dissemination (Gai, 202; Sahoo & Pandey, 2020). As a result of this increase in the amount of research activity, all the renowned databases and leading publishers provided free access to research and other literature related to COVID-19.

Geographic Information Systems (GIS). GIS have a history of providing important tools for research and information promotion and awareness, and proved to be effective during the pandemic. Information systems and research services are best approached from a geographical perspective (Hill, 2006). These include researchers on global climate modeling and ocean dynamics, and those who characterize and track the ecological and geological characteristics of an area or region; historians who need information on specific geographical areas at specific times, and information professionals, and students, among others. For example, geospatial analysis and web mapping applications and software such as ArcMap and ArcGIS were used to guide and support the search for probable and confirmed COVID-19 cases to improve detection and testing of suspected cases.

As practitioners use GIS to visualize, analyze, and model systems, they combine different literacies during their research, analysis, and interpretation process. Wells et al. (2021) state that given the global repercussions and consequences of COVID-19, there is an increasing reliance on technology to understand the impact of socioeconomic inequality on health outcomes.

As with other health crises, GIS systems and methods proved indispensable for our timely understanding of diseases sources and dynamics. Innovative methods to track disease, population needs, and current health and social service provision proved to be essential (Kamel Boulos & Geraghty, 2020). GIS was used to measure resource accessibility, unemployment, and food insecurity (Siegal et al., 2021). A range of practical online/mobile GIS and mapping dashboards include applications for tracking the 2019/2020 coronavirus epidemic and associated events as they unfolded around the world. Some of these received data updates in near-real-time. These and other articles discuss additional ways GIS can support the fight against infectious disease outbreaks and epidemics (Schmidt et al., 2021; Kamel Boulos & Geraghty, 2020).

2.7 WHAT ARE THE CHARACTERISTICS OF SUCCESSFUL HEALTH INFORMATION ACCESS PROJECTS?

The projects that were able to achieve their goals and were effective in the effort to provide information access services in the context of the COVID-19 pandemic had these characteristics:

- **Flexibility (adaptive behaviors)** - New roles emerged for educators, HIPs, and health-care providers. They became promoters, expanded cooperative agreements, and were required to be flexible while continuing to provide services and information access. They also include quick adaptation and combination of information technologies and virtual platforms to promote safety guidelines and social distancing but remain socially connected.
- **Promoted networking and cooperative behavior** - Interprofessional collaborations and resources sharing were essential to make science more accessible and understandable while promoting diverse literacies. The COVID-19 pandemic resulted in opposition and polarized perspectives and extraordinarily strong positions related to people's understanding and interpretation of the disease. Diverse professionals explored strategies to move forward in these divisions and work together.
- **Outreach efforts** - Besides networking and cooperative behavior, professionals found a way to develop outreach projects during the pandemic. The literature revealed intentional efforts to connect and establish strong relationship with the communities they serve.
- **Proactive (Initiative taking) actions to promote diverse types and levels of information** - The projects that proved to be successful promoted evidence-based and scientific content and created trusted sources of information that was "easy to read," as well as multicultural and multilingual educational materials. Information professionals have been highly active promoting information services and resources and providing diverse information literacy educational activities. These professionals developed guides, and information evaluation tools are useful and updated (See list at the end of the chapter).

2.8 CONCLUSION

The virus known as COVID-19 spread around the globe and brought unexpected challenges for information services providers. The lack of information access was a critical factor during the pandemic. Diverse teams collaborated on interprofessional projects by gathering information and using, organizing, and sharing resources easily accessed and distributed. As information services worldwide have gone through remarkable changes throughout history, different information literacies have become indispensable in promoting health and well-being, especially during a health crisis like the COVID-19 pandemic.

The pandemic disrupted political, economic, and social life around the world, disproportionally affecting underrepresented groups. Resolution of unequal access is important for developing nations because they cannot build and maintain economic success without adequate information. Access to reliable information and technology is crucial to contribute to the sustainable development of communities, and to hold their governments accountable. Underrepresented and vulnerable populations were excessively at risk of being affected by the virus and its consequences.

During the crisis information overload, misinformation, data manipulation, and rumors were detrimental to relief efforts. The pandemic was an unexpected test that challenged beliefs about technology use and information access. It revealed the severity of the inequities on information and technology access, but it also provided excellent opportunities to identify barriers and develop new strategies through collaboration, flexibility, and outreach.

Traditional disaster management strategies, as well as new ones, emerged, while health organizations played a pivotal role during pandemic preparedness, response, and recovery. These organizations adapted their services and assessed their users' perceptions during the process of providing information access. Numerous professionals collaborated to make science accessible and understandable and built resilience. They also embraced multicultural and multilingual information services.

During the COVID-19 pandemic, diverse professional groups transcended their traditional roles to offer literacy workshops and collaborate with other professionals. Some of the practices reflected in the literature represent an invitation to develop new projects and services that are especially relevant during health crises.

Martínez-Alcalá et al. (2021) state that the COVID-19 pandemic revealed that even in societies with high-level technological advances, there are still challenges that need to be addressed urgently. For example, psychological well-being has been one of the most affected aspects due to the pandemic. It is essential to develop health literacy as individuals need to have the capacity to obtain, process, and understand basic health information to make appropriate health decisions. Insuring that older adults have Internet and technology access may reduce health inequities and improve health-care delivery. Patient access to technology needs to be considered within the broader context of structural inequities and policies to address them. This is frequently accompanied by other types of inequities that make information access an important social determinant of health. These topics deserve further research.

Sullivan (2019) warns us that evaluating superficial aspects of websites such as the design as indicators of reliability, overlooks one of the most serious problems of the contemporary fake news problem: the ability of sites to mimic reputable sources and appear authoritative. Also, many websites are still inaccessible for people with motor, hearing, cognitive, visual, and other impairments. We should evaluate online content following patterns and diverse criteria (see the list of resources at the end of the chapter). Multilevel approaches are needed to increase equitable technology availability and use. Information specialists play important roles in the development of information services for underserved and unserved populations.

We should be able to support our communities while we all learn together, and from each other. When we register and acknowledge these projects, we get closer to that state when we make sense of a situation. Unfortunately, with climate change and its consequences, more pandemics can occur. We need to apply the lessons learned from this pandemic, including quick adaptability and rapid change. We also need keep working to obtain technology equity and provide information access, a basic human right.

REFERENCES

Aaronson, E. & Spencer, A. (2021). Informing information professionals: A case report on creating a shared site of pandemic resources. *Journal of Hospital Librarianship, 21*(2), 198–201. https://doi.org/10.1080/15323269.2021.1899789

Ali, M. Y., & Bhatti, R. (2020). COVID-19 (Coronavirus) pandemic: Information sources channels for the public health awareness. *Asia-Pacific Journal of Public Health, 32*(4), 168–169. https://doi.org/10.1177/1010539520927261

Ali, M, Y., & Gatiti, P. (2020). The COVID-19 (coronavirus) pandemic: Reflections on the roles of librarians and information professionals. *Health Information and Libraries Journal, 37*, 158–162. https://doi.org/10.1111/hir.12307

Allee, N. J., Friedman, C. P., Flynn, A. J., Masters, C., Donovan, K., Ferraro, J., Patel, R., & Rubin, J. C. (2021). Partnership development of the COVID-19 front door: A best evidence resource. *Journal of the Medical Library Association: JMLA, 109*(4), 680–683. https://doi.org/10.5195/jmla.2021.1353

American Library Association (ALA). (1996). *Access to electronic information, services, and networks: An interpretation of the library bill of rights.* www.ala.org/advocacy/intfreedom/librarybill/interpretations/digital

American Library Association (ALA). (2000). ACRL standards: Information literacy competency standards for higher education. *College & Research Libraries News, 61*(3), 207–215. https://doi.org/10.5860/crln.61.3.207

American Library Association (ALA). (2006, June 30). *Library Bill of Rights.* www.ala.org/advocacy/intfreedom/librarybill

American Library Association (ALA). (2007, July 30). *Interpretations of the Library Bill of Rights.* www.ala.org/advocacy/intfreedom/librarybill/interpretations

American Library Association (ALA). (2022). *Digital literacy. ALA Literacy Clearinghouse.* https://literacy.ala.org/digital-literacy.

Amoah, P. A., Leung, A. Y. M., Parial, L. L., Poon, A. C. Y., Tong, H. H.-Y., Ng, W.-I., Li, X., Wong, E. M. L., Kor, P. P. K., & Molassiotis, A. (2021). Digital health literacy and health-related well-being amid the COVID-19 pandemic: The role of socioeconomic status among university students in Hong Kong and Macao. *Asia-Pacific Journal of Public Health, 33*(5), 613–616. https://doi.org/10.1177/10105395211012230

Bin Naeem, S. B., & Bhatti, R. (2020). The COVID-19 "Infodemic": A new front for information professionals. Health Information and Libraries Journal, 37(3), 233–239.

Bin Naeem, S. B., & Kamel Boulos, M. N. (2021). COVID-19 misinformation online and health literacy: A brief overview. *International Journal of Environmental Research and Public Health, 18*(15), 8091. MDPI AG. http://dx.doi.org/10.3390/ijerph18158091

Bishop, C. A. (2012). *Access to information as a human right.* El Paso: LFB Scholarly Publishing LLC.

Charney, R. L., Spencer, A., & Tao, D. (2021). A novel partnership between physicians and medical librarians during the COVID-19 pandemic. *Medical Reference Services Quarterly, 40*(1), 48–55. https://doi.org/10.1080/02763869.2021.1873617

Chartered Institute of Library and Information Professionals-CLIP. (2018). The Library and Information Association. Information Literacy Group. *Definition of Information Literacy*. https://infolit.org.uk/ILdefinitionCILIP2018.pdf

Choi, N. G., DiNitto, D. M., Marti, C. N., & Choi, B. Y. (2022). Telehealth use among older adults during COVID-19: Associations with sociodemographic and health characteristics, technology device ownership, and technology learning. *Journal of Applied Gerontology: The Official Journal of the Southern Gerontological Society, 41*(3), 600–609. https://doi.org/10.1177/07334648211047347

Cochrane. (2021). *Cochrane on WHO call for action on managing the infodemic*. www.cochrane.org/news/cochrane-signs-who-call-action-managing-infodemic

Covington, C. (2021, July 28). Media literacy can be crucial for making informed pandemic decisions. *Texas Standard*. www.texasstandard.org/stories/media-literacy-can-be-crucial-for-making-informed-pandemic-related-health-decisions

Dawkins, A. (2017, March 8). *Intellectual Freedom Blog*. Office for Intellectual Freedom of the American Library Association. www.oif.ala.org/oif/access-information-universal-human-right

DeRosa, A. P., Jedlicka, C., Mages, K. C., and Stribling, J. C. (2021). Crossing the Brooklyn Bridge: A health literacy training partnership before and during COVID-19. *Journal of the Medical Library Association: JMLA, 109*(1), 90–96. https://doi.org/10.5195/jmla.2021.1014

Faraj, S., Renno, W., & Bhardwaj, A. (2021). Unto the breach: What the COVID-19 pandemic exposes about digitalization. *Information and Organization, 31*(1). https://doi.org/10.1016/j.infoandorg.2021.100337

Gai, N., Aoyama, K., Faraoni, D., Goldenberg, N. M., Levin, D. N., Maynes, J. T., McVey, M. J., Munshey, F., Siddiqui, A., Switzer, T., & Steinberg, B. E. (2021). General medical publications during COVID-19 show increased dissemination despite lower validation. *PLoS One, 16*(2), e0246427. https://doi.org/10.1371/journal.pone.0246427

Guldin, R., Noga-Styron, K., & Britto, S. (2021). Media consumption and news literacy habits during the COVID-19 pandemic. *International Journal of Critical Media Literacy, 3*(1), 43–71. https://doi.org/10.1163/25900110-03030003

Hill, L. (2006). *Georeferencing: The geographic associations of information*. Cambridge, MA: MIT Press.

Howes, L., Ferrell, L., Pettys, G., & Roloff, A. (2021). Adapting to remote library services during COVID-19. *Medical Reference Services Quarterly, 40*(1), 35–47. https://doi.org/10.1080/02763869.2021.1873616

Ibacache, K., Rybin K. A., & Vance, E. (2021). Emergency remote library instruction and tech tools: A matter of equity during a pandemic. *Information Technology & Libraries, 40*(2), 1–30. https://doi.org/10.6017/ital.v40i2.1275

Information Access. (2022). *Computer Sciences Encyclopedia*. www.encyclopedia.com/computing/news-wires-white-papers-and-books/information-access.

Ioannidis, J. Salholz-Hillel, M. Boyack, K.W., & Baas, J. (2021). The rapid, massive growth of COVID-19 authors in the scientific literature. *Royal Society Open Science, 8*(9). https://doi.org/10.1098/rsos.210389

Kamel Boulos, M. N. & Geraghty, E. M. (2020). Geographical tracking and mapping of coronavirus disease COVID-19/severe acute respiratory syndrome coronavirus 2 (SARS- CoV-2) epidemic and associated events around the world: How 21st century GIS technologies are supporting the global fight against outbreaks and epidemics. *International Journal of Health Geographics, 19*(1), 1–12. https://doi.org/10.1186/s12942-020-00202-8

Lama, Y., Davidoff, A. J., Vanderpool, R. C., & Jensen, R. E. (2022). Telehealth availability and use of related technologies among Medicare-enrolled cancer survivors: Cross-sectional findings from the onset of the COVID-19 Pandemic. *Journal of Medical Internet Research, 24*(1), e34616. https://doi.org/10.2196/34616

Lewis, R. (2021). *Technology, media literacy, and the human subject: A posthuman approach.* Cambridge, UK: Open Book Publishers.

Lindsay, J. M., Petersen, D., Grabeel, K. L., Quesenberry, A. C., Pujol, A., & Earl, M. (2021). Mind like water: flexibly adapting to serve patrons in the era of COVID-19. *Medical Reference Services Quarterly, 40*(1), 56–66. https://doi.org/10.1080/02763869.2021.1873622

Lucey, C. R., Davis, J. A., and Green, M. M. (2022). We have no choice but to transform: The future of medical education after the COVID-19 Pandemic. *Academic Medicine, 97*, S71–S81. https://doi.org/10.1097/ACM.0000000000004526

Ma, J., Stahl, L., & Knotts, E. (2018). Emerging roles of health information professionals for library and information science curriculum development: A scoping review. *Journal of the Medical Library Association: JMLA, 106*(4), 432–444. https://doi.org/10.5195/jmla.2018.354

Martínez-Alcalá, C., Rosales-Lagarde, A., Pérez-Pérez, Y., Lopez-Noguerola, J. Bautista-Díaz, M., & Agis-Juarez., R (2021). The effects of Covid-19 on the digital literacy of the elderly: Norms for digital inclusion. *Frontiers in Education, 6.* https://doi.org/10.3389/feduc.2021.716025

Mazure, E. S., Colburn, J. L., Wallace, E., Justice, E., Shaw, S., & Stigleman, S. (2021). Librarian contributions to a regional response in the COVID-19 pandemic in Western North Carolina. *Medical Reference Services Quarterly, 40*(1),79–89. https://doi.org/10.1080/02763869.2021.1873626

Mi, M., Zhang, Yingting, W., Lin, & Wu, W. (2020). Four health science librarians experiences: How they responded to the COVID-19 pandemic crisis. *College & Research Libraries News, 81*(7), 330–334. https://doi.org/10.5860/crln.81.7.330

Mohn, E. (2020). Internet of Things. *Salem Press Encyclopedia of Science.*

National Library of Medicine. (1980). Mass media. *Medical Subject Headings.* National Center for Biotechnology Information. www.ncbi.nlm.nih.gov/mesh/68008402.

National Library of Medicine. (1999). Education, Distance. *Medical Subject Headings.* National Center for Biotechnology Information. www.ncbi.nlm.nih.gov/mesh/68020375

National Library of Medicine. (2020). COVID-19. *Medical Subject Headings.* National Center for Biotechnology Information. www.ncbi.nlm.nih.gov/mesh/?term=covid-19

National Telecommunications and Information Administration (NTIA). (1995). *Falling through the net: A survey of the "have nots" in rural and urban America.* www.ntia.doc.gov/ntiahome/fallingthru.html

Orcutt, R., Campbell, L., Gervits, M., Opar, B., & Edwards, K. (2021). COVID-19 pandemic: Architecture librarians respond. *Art Documentation: Bulletin of the Art Libraries Society of North America, 40*(1), 123–140. https://doi.org/10.1086/714593

Paakkari, L., & Okan, O. (2020). COVID-19: health literacy is an underestimated problem. *The Lancet. Public Health, 5*(5), e249–e250. https://doi.org/10.1016/S2468-2667(20)30086-4

Rahaman, T. (2021). An introduction to telehealth and COVID-19 innovations—A primer for librarians. *Medical Reference Services Quarterly, 40*(1), 122–129. https://doi.org/10.1080/02763869.2021.1873647

Rainer G., Jaffe, M., Newman, R., & Schwarz, N. (2021). *The psychology of fake news: Accepting, sharing, and correcting misinformation.* London: Routledge.

Rajabifard, A., Paez, D., & Foliente, G. (2021). *COVID-19 pandemic, geospatial information, and community tesilience: Global applications and lessons.* Boca Raton, FL: CRC Press.

Rajasekhar, S., Makesh, D., & Jaishree, S. (2021). Assessing media literacy levels among audience in seeking and processing health information during the COVID-19 pandemic. *Media Watch, 12*(1), 93–108. https://doi.org/10.15655/mw/2021/v12i1/205461

Renu, N. (2021). Technological advancement in the era of COVID-19. *SAGE Open Medicine, 9*, 20503121211000912. https://doi.org/10.1177/20503121211000912

Richard, S. L. (2021). *Technology, media literacy, and the human subject: a posthuman approach.* Cambridge, UK: Open Book Publishers.

Sahoo, S., & Pandey, S. (2020). Growth analysis of global scientific research on COVID-19 Pandemic: A scientometrics analysis. *Library Philosophy & Practice,* 1–11.

Schmidt, F., Dröge-Rothaar, A., & Rienow, A. (2021). Development of a Web GIS for small-scale detection and analysis of COVID-19 (SARS-CoV-2) cases based on volunteered geographic information for the city of Cologne, Germany, in July/August 2020. *International Journal of Health Geographics, 20*(1), 1–24. https://doi.org/10.1186/s12942-021-00290-0

Schweitzer, E. J. (2015, November 23). Digital divide. *Encyclopedia Britannica.* www.britannica.com/topic/digital-divide

Siegal, R., Cooper, H., Capers, T., Kilmer, R. P., Cook, J. R., & Garo, L. (2021). Using geographic information systems to inform the public health response to COVID-19 and structural racism: The role of place-based initiatives. *Journal of Community Psychology.* https://doi.org/10.1002/jcop.22771

UNESCO. (2003). R*ecommendation concerning the promotion and use of multilingualism and universal access to cyberspace.* http://portal.unesco.org/en/ev.php-URL_ID=17717&URL_DO=DO_TOPIC&URL_SECTION=201.html

UNESCO. (2021). *Promoting open access to information for all and supporting multilingualism.* https://en.unesco.org/ci-programme/open-access

United Nations. (2021). *Universal Declaration of Human Rights.* www.un.org/en/about-us/universal-declaration-of-human-rights?msclkid=2ab15941b6b111ecb9dea7ed44c58559

Walker. P. (2021). The library's role in countering infodemics. *Journal of the Medical Library Association, 109*(1). https://doi.org/10.5195/jmla.2021.1044

Wells, J., Grant, R., Chang, J., & Kayyali, R. (2021). Evaluating the usability and acceptability of a geographical information system (GIS) prototype to visualise socio-economic and public health data. *BMC Public Health, 21*(1), 2151. https://doi.org/10.1186/s12889-021-12072-1

Winchester, C. L., & Salji, M. (2006). Writing a literature review. *Journal of Clinical Urology, 9*(5), 308–312. https://doi.org/10.1177/2051415816650133

World Health Organization. (2022a). *Infodemics.* www.who.int/health-topics/infodemic#tab=tab_1

World Health Organization. (2022b). *Social Determinants of Health.* www.who.int/health-t//opics/social-determinants-of-health#tab=tab_1

Zarocostas, J. (2020). How to fight an infodemic. *Lancet, 395*(10225), 676. https://doi.org/10.1016/S0140-6736(20)30461-X

INFORMATION RESOURCES: INFORMATION LITERACY GUIDELINES AND FRAMEWORKS

ACRL Information Literacy Framework. www.ala.org/acrl/standards/ilframework
CDC. COVID data tracker. https://covid.cdc.gov/covid-data-tracker/#global-counts-rates
CDC. COVID-19. www.cdc.gov/coronavirus/2019-ncov/communication/index.html
Coronavirus Misinformation Watch-Canada. https://covid19misinfo.org.

The CRAAP Test Worksheet (southcentral.edu). https://southcentral.edu/webdocs/library/CRAAP%20Test%20Worksheet.pdf
Fairleigh Dickinson University Guideline on How to Recognize Fake News. https://fdu.libguides.com/fakenews
Fake News and Information Literacy Guidelines by University of Oregon. https://researchguides.uoregon.edu/fakenews
How to Identify a Fake News, Library Guide—University of Washington. https://guides.lib.uw.edu/research/news/fake-news
Indiana University East fake news library guides. https://iue.libguides.com/fakenews
International Federation of Library Association's (IFLA) guideline in eight simple steps on "how to spot fake news." www.ifla.org/publications/node/11174
Media Bias/Fact Check rates various news organizations of "factual reporting." https://mediabiasfactcheck.com
Medical Library Association COVID-19 Literature Searches.
www.mlanet.org/p/cm/ld/fid=1713
National Command and Control Center of Pakistan. http://covid.gov.pk
News Guard Coronavirus Misinformation Tracking Center. www.newsguardtech.com
UNESCO "MIL CLICKS" campaign (Media & Information Literacy (MIL), Critical Thinking and Creativity, Literacy, Intercultural, Citizenship, Knowledge, Sustainability (CLICKS). https://en.unesco.org/MILCLICKS
News Literacy Resources. www.commonsense.org/education/articles/news-literacy-resources-for-classrooms
University of Michigan Library's Fake News, Lies, and Propaganda: How to Sort Fact from Fiction. https://sites.google.com/umich.edu/library-fake-news/home
UNESCO and Athabasca University Media and Information Literacy Course. http://elab.lms.athabascau.ca

3 Spatial Pattern of COVID-19 Positivity Rates in Indonesia during Local Restriction Phase

A Case Study of Jakarta and Surabaya City

Saut Sagala, Danang Azhari, Fathia Lutfiananda, and Indah Salsabiela

CONTENTS

3.1 INTRODUCTION

The emergence and spread of COVID-19 have highlighted the lack of understanding of pandemic response management from the global to the local level. The complexity of this pandemic has undoubtedly raised many questions, specifically on the impact of a pandemic on highly populous areas like cities. While cities are more likely to be well prepared to respond to the COVID-19 pandemic, they tend to be more exposed to COVID-19 transmission due to the proximity among citizens and challenges in

DOI: 10.1201/9781003227106-3

implementing the social-distancing protocol (OECD, 2020). Accordingly, the metropolitan and bigger cities have also been the epicenter of pandemic emergence globally, exhibiting higher incidence and cases during the long-term course of the pandemic (Adhikari et al., 2020; Ribeiro et al., 2020).

Indonesia, the fourth most populous country in the world, was predicted to suffer significantly due to the COVID-19 pandemic (Barron, 2020). Thus, the metropolitan cities of Indonesia, such as Jakarta and Surabaya, have been the epicenter of the COVID-19 pandemic in the country. Jakarta and Surabaya, respectively, had the first- and the second-highest positive cases of COVID-19 since the emergence of COVID-19 in Indonesia (COVID-19 Response Task Force, 2022). These cities have also implemented several measures to handle COVID-19 through various policies related to large-scale social restrictions (*Pembatasan Sosial Berskala Besar*/PSBB) and restrictions on community activities (*Pemberlakuan Pembatasan Kegiatan Masyarakat*/PPKM). As these cities pose high-density characteristics, there is a need to investigate the spatial pattern of COVID-19-positive cases. Nonetheless, the lack of understanding of the spatial pattern of COVID-19 cases in these two cities as a response measure remains. This understanding can be the basis for formulating a more localized approach to manage the COVID-19 pandemic and enhance the capacity to respond to the current pandemic and future risks.

Previous research used the spatial approach in understanding COVID-19. For instance, Shariati et al. (2020) provided a global context of high- and low-risk clusters of COVID-19. In Indonesia, Eryando et al. (2020) and Ratnasari and Dewi (2021) also explained the risk of COVID-19 cases in Indonesia at the Provincial level. However, there is still a lack of discussion on high- and low-risk clusters of COVID-19 in metropolitan cities for implementing appropriate COVID-19 policy responses.

This chapter explores and applies geographic information systems (GIS) data to understand the dynamic changes in COVID-19 cases in Jakarta and Surabaya with 765 local restriction policies. Spatial autocorrelation analysis and other neighborhood statistics are applied using spatial processing software. Temporal COVID-19 confirmed case data were Jakarta Smart City (2021) and Surabaya (East Java Government, 2021). The assessment was done for both Jakarta and Surabaya City.

Despite Jakarta being assessed at the provincial level, while Surabaya was assessed at the city level, it was relevant to compare both cities. Jakarta and Surabaya are the first- and the second most-populated areas on Java Island (Mardiansjah & Rahayu, 2019). This study focused on five administrative cities in Jakarta and Surabaya as the capital of East Java Province, where the population and economic activities are concentrated. Both cities are well-established and have integrated public transportation. Regarding COVID-19 cases, DKI Jakarta and Surabaya have experienced surging waves (i.e., positive cases) several times. This study focused on the second wave of COVID-19 cases in Indonesia, which started in June 2021. An assessment of confirmed COVID-19 cases was then conducted for up to four months until September 2021 to get the temporal analysis of active case progression, along with the applied local restriction policies, which are usually updated each week.

To assess the spatial risk and pattern of confirmed active COVID-19 cases in both study areas, spatial autocorrelation and hot spot analysis of the concentrated active cases were applied. Areas that are most likely to show clustering of the virus transmission were also calculated in this study. The spatial analytical assessment was done at the district level of each city. The information was then enriched with each district's population data, obtained from the DKI Jakarta Central Bureau of Statistics (BPS) (2022) and the Surabaya Central Bureau of Statistics (BPS) (2022). The population data are utilized to calculate the cumulative incidence rate (CIR), which is expected to give a proportionate calculation of active COVID-19 cases compared to population density in a district.

The autocorrelation analysis was done by utilizing the spatial autocorrelation (Global Moran's I) tool in the spatial toolbox in ArcMap. The toolbox calculates feature location (the location of active COVID-19 cases) and attributes values (the CIR of COVID-19) to determine the COVID-19 distribution pattern in an area. The patterns are categorized as clustered, dispersed, or random. The hot spot analysis (Getis-Ord Gi*) was applied to calculate significant clusters of high values (i.e., hot spots) and low values (i.e., cold spots) statistically. The hot spots indicate where the active COVID-19 cases are highly concentrated spatially. The hot spot and cold spot analyses delivered three levels of confidence.

Furthermore, to determine the highly transmittable area of COVID-19 cases, cluster and outlier analysis (local Anselin Moran's I) was utilized. The analysis determines a particular area with its surrounding values, indicating the potential transferability or risk of COVID-19 cases. The analysis resulted in five classes: high-high, low-low, high-low, low-high, and insignificant. The area determined as a high-high area indicates a high incidence of COVID-19 cases, and its surrounding area is considered to have a high clustering risk. Meanwhile, the low-low area indicates low cases, and its surrounding area is considered to have a low clustering risk. The case of Jakarta and Surabaya as two of the most populous cities and the highest number of positive cases in Indonesia can capture a concrete practice on dynamic mapping of the COVID-19 progression and formulate evidence-based recommendations to be applied to another city with a similar context.

3.2 POLICY AND SPATIAL PATTERN

In the policy process, policy formulation is necessary to determine the following measures. In the early development of policy studies, the formulation is associated with a system theory model (Hadna, 2021). The application of theories is the first step in determining elements of an analytical framework to answer particular assumptions about causal relations and stimulate new research questions. Public actors are the main actors who usually decide policies; however, it is likely that private actors may be involved in the decision-making and even implementation, which creates a more complex process (Eising, 2013).

Three policy models were proposed to develop the COVID-19 policy in Indonesia: policy cycle, multiple streams approaches (MSA), and an advocacy coalition framework (ACF). A policy cycle is considered an ideal one, a sequential process where policy is developed logically in response to a perceived problem (Lasswell, 1956; Bridgman & Davis, 2003). However, the policy cycle does not fit Indonesia's actual policy-making practice. Blomkamp et al. (2017) highlighted that some stages in the policy cycle, such as consultation and evaluation, were not practically evident. The policy process did not operate consecutively on policy analysis, decision-making, and coordination (Hadna, 2021).

Another theory is the MSA, which better suits the establishment of a COVID-19 policy in Indonesia than the policy cycle does the MSA was first conceptualized by Kingdon and Stano (1984). It considers how three separate issues are integrated with policy making: public sentiment, policy, and political streams (Eising, 2013; Blomkamp et al., 2017). A problem stream is a perceived problem shared by the public and Policymakers because it has generally disrupted life (Eising, 2013). A policy stream is a solution offered through evidence-based discussions among the community, experts, and other stakeholders to respond to problems (Eising, 2013). Political streams consist of many factors, such as changes in national conditions, changes in officials and parliament members, and campaigns under pressure to be carried out by interest groups, including political parties.

Another study by McBryde et al. (2020) indicated that modeling COVID-19 influenced policy-making at all outbreak stages. Imperial College London's (2020) modeling study analyzed the potential of COVID-19 causing widespread infection across the United Kingdom and the United States if a mitigation rather than a suppression strategy was pursued. Consistent model findings of high infection rates and mortality collectively resulted in many countries grasping the seriousness of the pandemic (McBryde et al., 2020). Pribadi et al. (2021) also raised the pivotal role of spatial analysis, modeling, and the need to include the spatiotemporal dynamics of transmission risk in developing the COVID-19 response policy. While there is a city's pandemic curve and a decrease in the intensity of cases due to policy implementations, the spatial transmission has persisted.

3.3 SPATIAL PATTERN OF COVID-19 PANDEMIC IN DKI JAKARTA AND SURABAYA CITY

Jakarta has been recorded as the city with the highest number of confirmed cases per day since Indonesia's first day of COVID-19 (Aisyah et al., 2020). The onset of the second wave of COVID-19 in DKI Jakarta arrived earlier than in Surabaya City. In the first week of June 2021, the total number of active COVID-19 cases in DKI Jakarta was 356,217. East Jakarta had the highest number of active cases in DKI Jakarta throughout the observed period (June–September 2021), as shown in Figure 3.1. This condition seems relevant because East Jakarta has the widest area among other cities in DKI Jakarta. The CIR was calculated to assess the density of active cases in this region. The CIR can help estimate the risk of an individual developing or experiencing symptoms of a disease during a specified duration. Based on the calculated CIR per city, Central Jakarta is the region with the most highly dense COVID-19 cases per 10,000 people, which can be seen in Figure 3.2.

During this first week of June, DKI Jakarta and the national government applied the PPKM-Micro policy to control the spread of COVID-19 cases. The escalation of active cases multiplied around twice each week until the fifth week of June 2021 (June

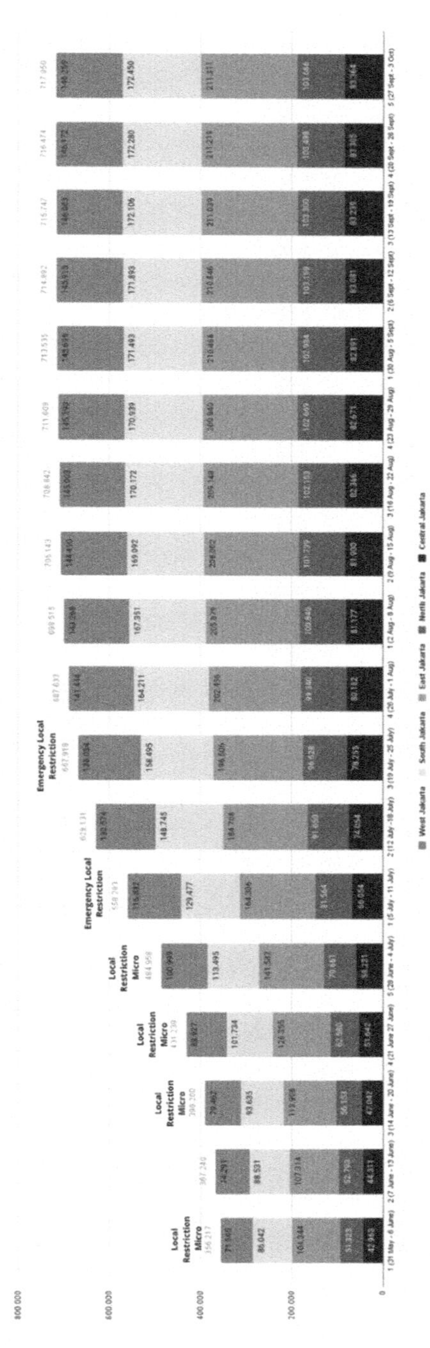

FIGURE 3.1 COVID-19 active cases per city in DKI Jakarta.

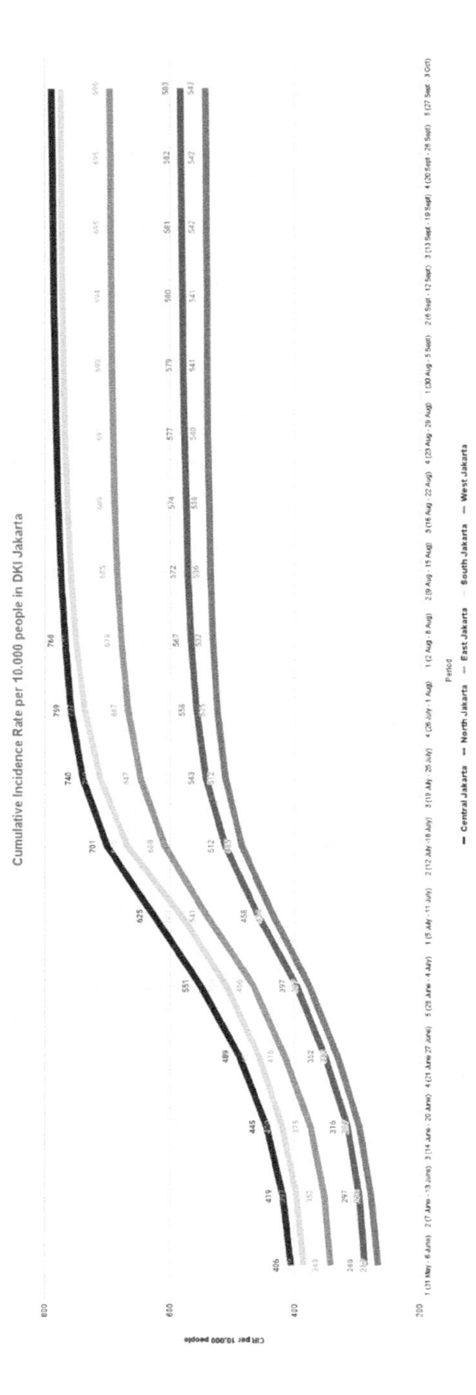

FIGURE 3.2 CIR per city in DKI Jakarta.

28–July 4). Over the observed period, the COVID-19 cases in DKI Jakarta increased gradually from the first week of June until the first week of August 2021 (August 2–8). The progression of COVID-19 cases over the rest of the weeks seemed stabilized, although there was still an escalation in the number of confirmed cases. The stabilized condition is also seen in the CIR number, where each city in DKI Jakarta experienced minor to no changes in the CIR. This condition can be assumed as no significant progression of new cases, or that the number of people who recovered from COVID-19 is almost equivalent to the number of new cases.

The number of active COVID-19 cases in Surabaya City remained stable from the first week of June (May 31–June 6) until early July 2021. The onset of the COVID-19 second wave in Surabaya City arrived in the third week of July 2021 (July 19–July 25). The number kept rising until the following week when the government changed the policy to PPKM Level 4. The COVID-19 cases in Surabaya City kept increasing until the third week of August (August 16–22), reaching 63,831 active cases (see Figure 3.3). The trend seems to stabilize for the rest of the observed period, where the increasing number of cases was around fewer than 1,000 per week.

East Surabaya has the highest number of active COVID-19 cases, followed by South Surabaya. Based on the CIR number assessment, South Surabaya has a higher density of active cases than East Surabaya at the beginning of the observed period. During the increased period of active cases, the second and third week of July (July 12–18, July 19–25), East Surabaya turned into the region with the highest density of active COVID-19 cases. Both areas remained in a tight position for the rest of the period, as shown in Figure 3.4. The trend only showed a sudden change from the second week of July to the third week of July and happened in each area in Surabaya City. The significant change during this period might have been caused by massive travel activities among provinces triggered by the Eid Al-Fitr celebration and that happened at the end of May 2021. The wave, which started to intensify in the second week of June, might suggest the cases were imported infections from people who just came back to work in Surabaya from other cities. Furthermore, all confirmed cases were referred to the local quarantine center, which is managed by the government (Kurniawan, 2021).

To further assess where the cases are concentrated, a hot spots analysis was done on the CIR number of each district in both study areas as illustrated in Figure 3.5. In the first week of June 2021 (May 31–June 6), the hot spots in DKI Jakarta were mainly concentrated in 12 districts around the central part of Jakarta, with the highest confidence of hot spots located in Menteng and Johar Baru District, Central Jakarta. This finding means the area has relatively higher accumulated cases than its surroundings. At least 425 people have active COVID-19 cases among 10,000 during the first week of June 2021 in Menteng Districts. Although the highest CIR during this period was found in Cempaka Putih District by 560 cases per 10,000 people, a hot spots analysis might show different results. The reason is that the calculation considers the cases' spatial density, which means if both areas have the same CIR, the one with the broader area will have less dense cases.

At least 12 districts were considered hot spots during this period. The following week, in the third week of June 2021 (June 14–20), the hot spots decreased to nine districts, with Menteng and Johar Baru remaining as hot spot districts with the highest

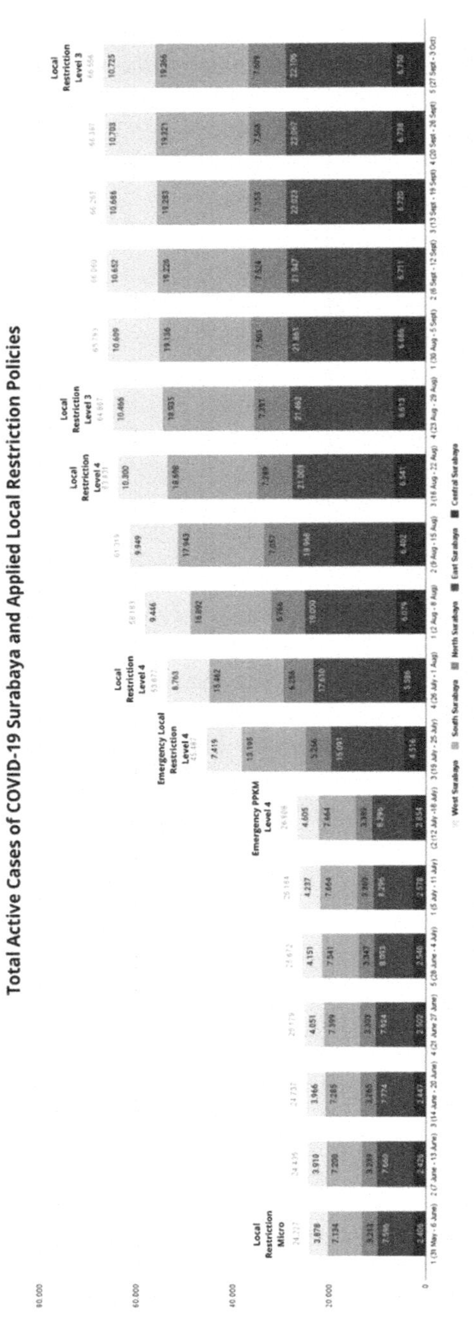

FIGURE 3.3 COVID-19 Active cases per region in Surabaya City.

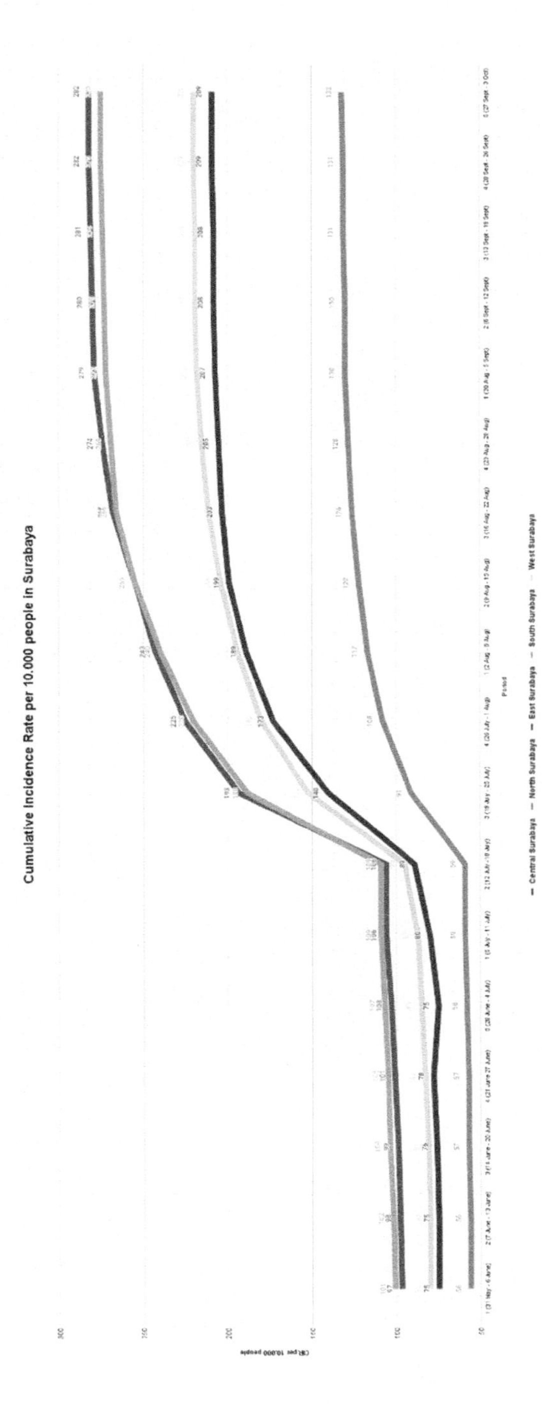

FIGURE 3.4 CIR per region in Surabaya City.

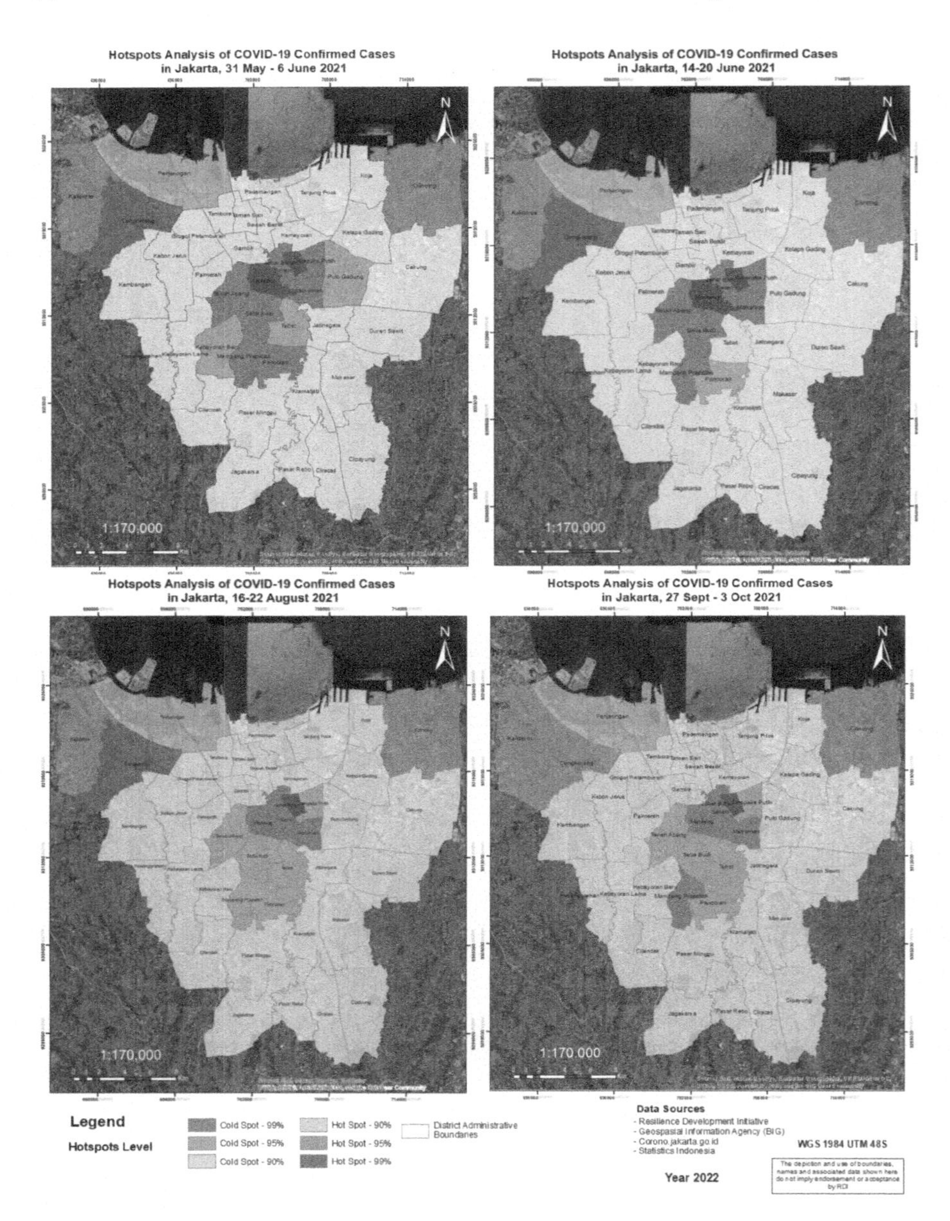

FIGURE 3.5 Hot Spots and Cold Spots areas in DKI Jakarta.

confidence level. A subtle change happened in the third week of August 2021 (August 16–20) when ten districts were hot spot areas. At the end of the observed period, a slight change happened in Mampang Prapatan District, South Jakarta, where its hot spots' confidence level increased to 95%. Hot spots are generally concentrated in the central area of DKI Jakarta, and cold spots are mainly identified in the northern area. Continuous changes in several hot spot districts indicate a swift progression of confirmed cases that were not local in a particular district. This condition also suggests that transmission cases across districts are relatively common relatively common.

In the first week of June 2021 (May 31–June 6), Surabaya's hot spot areas were all located in South Surabaya (see Figure 3.6). There were five districts categorized

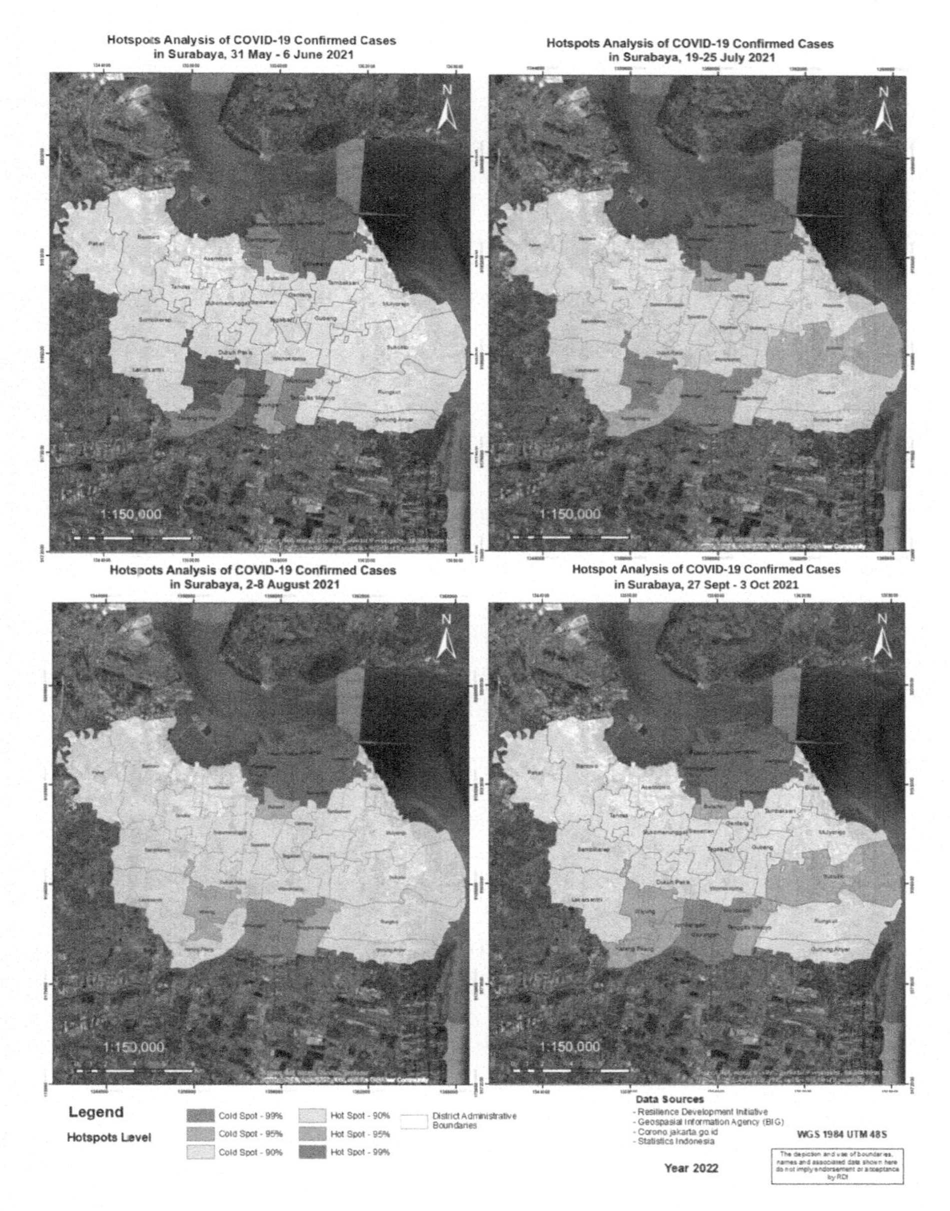

FIGURE 3.6 Hot Spots and Cold Spots areas in Surabaya City.

as hot spot areas. Wiyung and Jambangan Districts were districts with the highest hot spot level of confidence. The CIR of Wiyung was 128 cases per 10,000 people, and Jambangan was 119 cases. Wiyung District also remained a hot spot area during the observed period (May 31–October 3, 2021) with an average CIR (236 cases per 10,000 people). The result does not seem to align with the CIR of Surabaya City, where East Surabaya exhibits the highest CIR. Due to higher population density, South Surabaya has a greater possibility of turning into a hot spot than East Surabaya districts.

A significant change happened in the third week of July 2021 (July 19–25) when Sukolilo, a district in East Surabaya, was also categorized as a hot spot with a 90% confidence level. In total, six districts were considered hot spot areas. During this week, no district was categorized as a hot spot with the highest confidence level (99% confidence level). In the first week of August 2021 (August 2–8), the hot spots decreased to five districts located in South Surabaya. Trenggilis Mejoyo became a hot spot area with a 90% confidence level, while Karang Pilang District was no longer a hot spot area. Wiyung decreased its level of hot spot confidence to 90%. In the fifth week of September 2021 (September 27–October 3), the hot spot area increased to seven districts. Sukolilo, East Surabaya, became a hot spot in the third week of July. Karang Pilang also turned back into a hot spot area, and Trenggilis Mejoyo remained the same. On average, the CIR of Jambangan, Gayungan, and Wonocolo from June to September 2021 are 223, 224, and 205 cases per 10,000 people.

The local Anselin analysis provides insight into data at the local or district level to give a clue to the coarser area clustering statistic. As shown in Figure 3.7, at the beginning of the observed period, the first week of June 2021 (May 31–June 6), seven districts were categorized high-high areas, which indicated an increased risk of COVID-19 transmission in the area and its surrounding areas, or it might also suggest an area with high-density cases. The vaccination campaign also could be started in these areas to minimize the potential outbreak. Two districts of Central Jakarta (Tanah Abang and Johar Baru) are considered in the low-high category, indicating an area with lower transmission risk but surrounded by high-risk areas. Cengkareng in West Jakarta was categorized in low-low condition, which means a cluster area with a low risk of transmission. In the third week of July 2021 (July 19–25), a subtle change happened to Kemayoran, Central Jakarta, which turned into a high-high cluster. In total, six districts were considered high-risk transmission areas during this period. During the fourth week of August 2021 (August 23–29), the high-high risk areas decreased to five districts (Pulo Gadung, Matraman, Menteng, Tebet, and Mampang Prapatan). This condition remained the same until the fifth week of September 2021 (September 27–October 3). Tanah Abang and Johar Baru remained in low-high states and Cengkareng in low-low conditions for as long as the observed period.

Six districts in Surabaya were categorized as high-high areas, or areas with a high risk of transmission cases in the first week of June 2021 (May 31–June 6), as seen in Figure 3.8. Mainly the northern region of Surabaya was in the low-low category (Krembangan, Pabean Cantian, Semampir, Kenjeran), and one of them is in Central Surabaya (Simokerto). In the third week of July 2021 (July 19–25), the

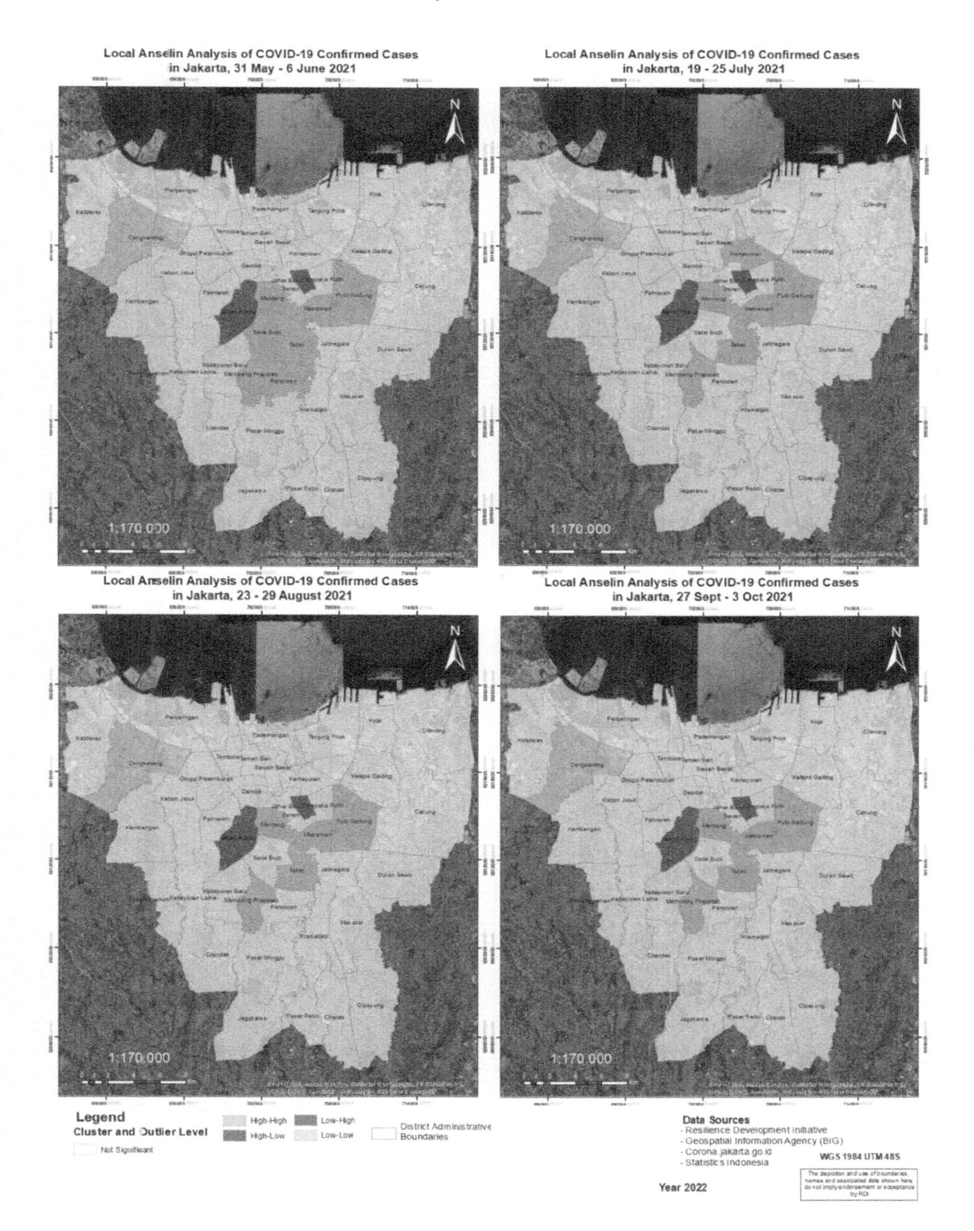

FIGURE 3.7 Local Anselin analysis in DKI Jakarta.

high-high area decreased to only four districts, three in South Surabaya (Jambangan, Gayungan, Wonocolo), and one in East Surabaya (Sukolilo). Karang Pilang, South Surabaya, turned into a low-high area, where the area is considered low-risk transmission cases but surrounded by high-risk areas, which suggests Karang Pilang in this period had considerably low cases but was surrounded by neighborhoods with

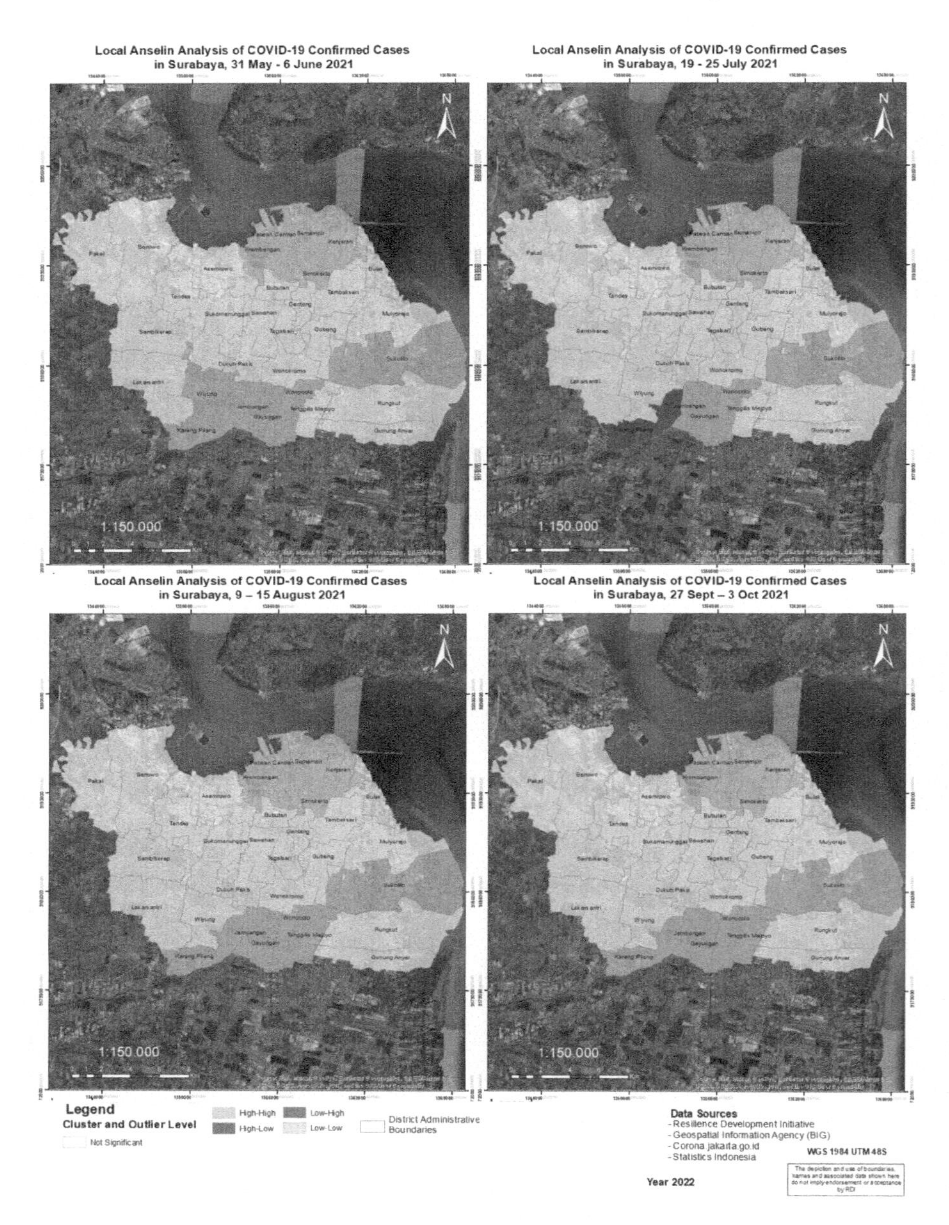

FIGURE 3.8 Local Anselin analysis in Surabaya City.

high cases. The high-high area increased to six districts during the second week of August (August 9–15), when Trenggilis Mejoyo, East Surabaya, turned into a high-risk transmission area. This condition remained the same until the fifth week of September 2021 (September 27–October 3). The low-low areas did not change during the study period.

3.4 LESSONS LEARNED FROM COVID-19 POLICY RESPONSE IN JAKARTA AND SURABAYA

Jakarta and Surabaya began the restriction of public activities and facility use during the first period of large-scale social restriction (PSBB) on April 9, 2020, after Jakarta formed a COVID-19 response team on March 2, 2020. Surabaya started transitioning into a "new normal" phase from June 8 to June 22, 2020, to revive the economic situation. During this phase, economic activities returned while health protocols were still observed. Since then, daily cases have escalated, which led East Java to surpass Jakarta cases in June 2020 (Purwanto et al., 2021). The significant changes in this regulation phase are shown below in the recapitulation of PPKM-Micro and PPKM-Emergency Level 3 and Level 4 in Table 3.1.

The PPKM-Micro policy was applied in all provinces in Indonesia, according to Ministry of Home Affairs Instructions No.12/202 and No.13/2021. The regulation was implemented and monitored in the neighborhood (*rukun tetangga*/RT) and the hamlet (*rukun warga*/RW). The trend in active cases per province was also applied to determine the zoning. The zoning was categorized into four main zones: red zones, yellow zones, orange zones, and green zones, suggesting that the red zone area had the highest number of active cases and the green zones the lower number of cases. The policy only focused on the active cases in the area. Within the government and society, there is a greater feeling of urgency to alter and tighten measures by implementing the PPKM Emergency, which was much stricter and more extensive. However, there was a lack of the spatiotemporal dynamics of transmission risk in developing this policy. In addition, while the MSA conceptualized by Kingdon (1984) stated the need for a balanced stream in policy-making, Indonesia's policy response to COVID-19 has also leaned more toward the policy stream but without community involvement. Hence, there is a gap in the policy to fulfill the public's needs while responding to the current pandemic.

3.5 THE PROGRESSION OF COVID-19 SPATIAL PATTERNS AND LOCAL RESTRICTION POLICY

Assessing the total number of confirmed COVID-19 cases and CIR in both areas shows how COVID-19 progressed differently. The COVID-19 cases in DKI Jakarta show a gradual increase during the second wave of the pandemic. A gradual progression should be the basis for local authorities to take appropriate responses by distributing health services and medical logistics to the most impacted areas. Several areas in mid-Jakarta are considered the most impacted areas. The results of hot spot analysis show a continuously changing outcome through the observed period, where some districts, such as Menteng, Johor Baru, and its surroundings, were identified as hot spot areas with a high level of confidence. This indicates that the majority of the cases are detected in mid-Jakarta. The data were then uploaded to the local and national government data centers when a case was confirmed. The local government will then be in charge of contact tracing and following up on the patient's progress. Therefore, residential addresses become essential and primary information for registering cases. The hot spots, which are mainly shown in mid Jakarta also indicate that the high

TABLE 3.1
Summary of PPKM-Micro and PPKM Emergency Level 3 and Level 4 in Jakarta and Surabaya

Policy Details	PPKM-Micro (June1–July 2, 2021)	PPKM-Emergency Level 4 (July 3–August 23, 2021)	PPKM-Emergency Level 3 (August 24–October 4, 2021)
Mobility	Mobility limitations	70% capacity for public transportation; domestic and international travels must have at least first vaccination dose and PCR test result	70% capacity for public transportation; domestic and international travels must have at least first vaccination dose, and COVID-19 test
Public facilities	Capacity limit 50% for mall visitors and other public facilities; restaurants are open for 50% visitors and takeout	Restaurants only accept takeout; malls and markets are closed except to access restaurants and supermarkets; must follow strict protocols	Malls and markets have limited operational time with 25%–50% capacity; outdoor restaurants/cafes are allowed to have dine-in with time limits and indoors are only available for takeout
Non-essential sectors	Working from office (WFO) with limited capacity and strict protocols and home (WFH)	Work from Home (WFH)	Work from Home (WFH)
Essential and critical sectors	WFO with limitations and strict protocols	WFO with limitations and strict protocols	WFO for essential sectors (25%–50%, depending on the industry); critical sectors are open with strict protocols
Education sectors	Online for zones with high risk, offline with limitations, and strict protocols for other zones	Full online	Online and offline 50%, except for special schools and preschools
Religious sectors	50% for zones with lower risks	Temporarily closed	Capacity limit 50% and must follow strict protocols
Other activities (social-cultural gatherings)	25% capacity limits	Temporarily restricted	Outdoor exercise and activities are allowed with strict protocols

population and internal mobility still became influencing factors in COVID-19 progression despite the application of PPKM. In addition, local governments need to strengthen health facility capability and capacity in these districts by localizing their approaches to local restrictions since a total lockdown is not applicable.

Strengthening the accessibility of local health services such as *Pusat Kesehatan Masyarakat* or the first level of the public health center is essential. During this period, the national government will have already started the vaccination campaign, and several confirmed cases might not experience severe symptoms of COVID-19, which would be suggested to take self-quarantine. The local health centers become pivotal in tracing the close contact of the newly confirmed cases and ensuring that the patients are still receiving essential health assistance despite self-quarantine.

The hot spot, which is mainly identified in mid-Jakarta, also suggests a high transmission risk in the surrounding areas. This finding also aligns with the local Anselin analysis result. The high-high areas in Jakarta are also mainly located in the center of Jakarta. Districts such as Menteng, Matraman, Tebet, Pulo Gadung, and Mampang Prapatan show a high incidence and high risk of transmitting cases based on the local Anselin's result. This result means that the local government needs to strengthen health facilities and policies taken in the identified hot spot areas and implement mitigation actions in the high transmission risk areas. As a result of Anselin's analysis, the high-high area could potentially cause an outbreak cluster if not handled wisely. Therefore, areas with high-high categories need to strengthen their local policies, and use social distancing and tracing for potential cases to ensure constraints in virus transmission. Meanwhile, in low-high areas, the areas in the low category need to minimize their activity with the high neighborhood category to prevent anticipated virus transmission.

Based on Global Moran's Index Analysis, the active COVID-19 cases distribution pattern in DKI Jakarta, as shown in Figure 3.9, exhibited a clustered pattern in the first week of June 2021 (May 31–June 6). This pattern represents the distribution of the cases based on an analysis of all districts in DKI Jakarta or a global provincial pattern. During this time, the government applied PPKM-Micro to restrain the transmission of COVID-19 cases. The condition remained the same in the second week of June 2021 (June 7–13). The clustered pattern became less significant in the third week of June (June 14–20). The government still applied the PPKM-Micro policy for the next two weeks. The clustered pattern then lessened in the fourth week of June, where the distribution of COVID-19 cases was categorized as random but bounced back to the clustered pattern in the fifth week of June 2021 (June 28–July 4). The government started to apply the Emergency PPKM policy in the first week of July 2021 (July 5–11), and the COVID-19 distribution pattern was random until the end of the observed period. The rapid change in the distribution pattern might be caused by the swift progression of confirmed cases in each district. DKI Jakarta, the capital city of Indonesia, is supported by comprehensive screening facilities, allowing better coverage and response to cases than other cities or provinces. The rapid change in the pattern might also be caused by high mobility among DKI Jakarta citizens, which makes virus transmission grow rapidly and randomly despite the application of local policies.

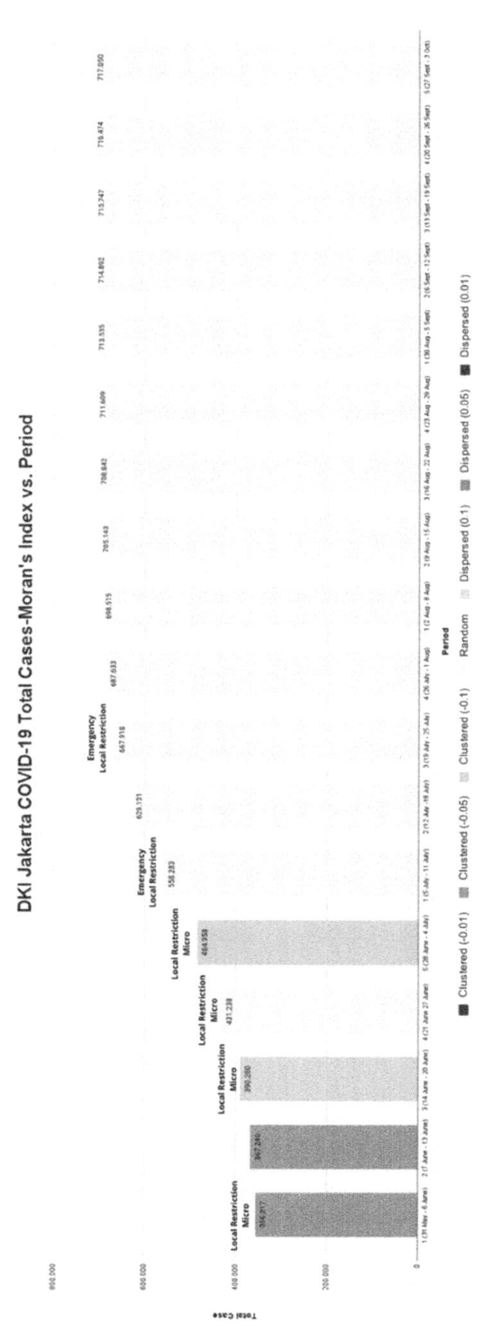

FIGURE 3.9 COVID-19 active cases pattern in DKI Jakarta.

Moran's Index results seem relevant to the previous hot spots and local Anselin analyses, which show continuous changes in the hot spots and high-risk areas in DKI Jakarta throughout the study period. The application of the Emergency PPKM policy seems able to restrain the progression of the COVID-19 cases from turning into a clustered condition in DKI Jakarta even though the number of active cases kept rising and seemed stabilized only around after the first week of August 2021 (August 2–8). This condition might be caused by the availability of integrated public transportation, which connects each city in DKI Jakarta. Several actions have been taken to mitigate the transmission of COVID-19 cases through public transportation on each applied local restriction policy. These include reducing public transport occupation by 50% (during the PPKM-Micro) and up to 70% (PPKM Emergency).

The distribution pattern of COVID-19 cases in Surabaya City remained clustered, with the highest P-value delete by around −0.01 for the rest of the observed period. These Moran's Index results are relevant to the spatial analysis of hot spots and local Anselin of Surabaya. The hot spot areas in Surabaya City are concentrated in the southern region, and the cold spots are mainly located in the northern region, with minimum changes from June to September 2021. This condition also correlated with the local Anselin results, where the high-risk transmission of COVID-19 cases was mainly located in the southern region. Several applied local restriction policies could not constrain the progression of COVID-19 cases in Surabaya City into a clustered pattern.

Furthermore, the Karang Pilang local authorities, as a low-high area based on Anselin analysis in the third week of July 2021, needed to limit their citizens' interactions or activities with the neighboring districts with high categories such as Jambangan and Gayungan. As the Anselin analysis is also highly clustered, the Surabaya authorities can use this analysis as a basis for decision-making to focus their medical assistance allocation to the high-high areas or by regulating support in low-low areas. This support can only be done when the local restriction in the low-low area has already ensured that the measures taken can constrain or mitigate the possibility of emerging cases.

Active cases also increased suddenly despite the application of the Emergency PPKM Level 4 policy in the second week of July 2021 (July 12–18), as seen in Figure 3.10. The clustered pattern of COVID-19 cases in Surabaya is relevant to the location of Surabaya Industrial Estate Rungkut, which is a major employer in the southern region of Surabaya. The availability of public transportation, which is not as integrated as in DKI Jakarta, can also be the reason why the COVID-19 distribution pattern in Surabaya City tends to be clustered. The public transportation in Surabaya City is mainly supported by the Suroboyo Bus system only, which operates from 6.00 a.m. to 10:00 p.m. Furthermore, the population in Surabaya City mostly relies on private vehicles and the Suroboyo Bus is mainly accessed by students (Sulistyowati & Muazansyah, 2019; Sunirno, Halim, & Setiawan, 2019). The number of Suroboyo Bus commuters also decreased since most Surabaya City students participated in fully online education during the Emergency PPKM Level 4 policy implementation. As the cases were highly clustered and concentrated in the southern region, the local government of South Surabaya needs to take adaptive and more localized approaches such

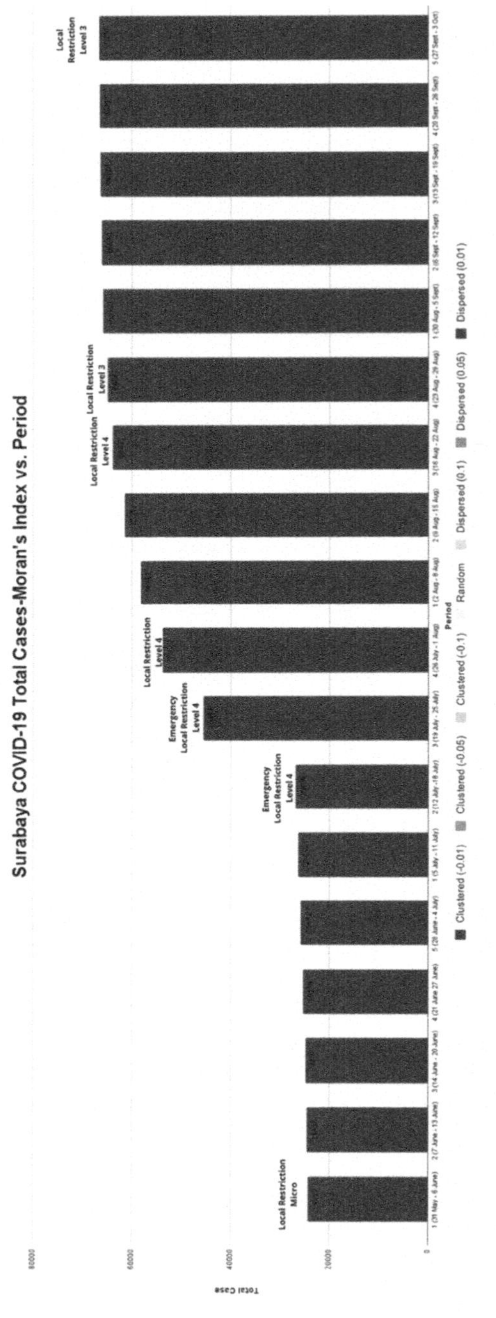

FIGURE 3.10 COVID-19 active cases pattern in Surabaya City.

as extending measures from the PPKM-Micro, -Emergency Level 4, and -Emergency Level 3, which were instructed by national authorities.

As mentioned in the previous section, a total lockdown policy is not possible to conduct in developing countries such as Indonesia; such policy should be made based on economic and public health considerations. Localized policies, such as PPKM, are expected to overcome those considerations. Unfortunately, the current local policy is solely taken based on the progression of total cases instead of the spatial interactions of the communities and the vector of disease, making it ineffective in constraining the transmission among districts or regions. Policy makers should consider mitigation responses based on the spatial modeling of case transmission. As mentioned by Pribadi et al. (2021), every finer geographical location has its own pandemic curve. A spatial and temporal assessment should be considered to address this condition. This condition highlights the importance of a spatial approach in addressing aerosol disease transmission such as COVID-19.

3.6 CONCLUSION

The progression of cases in Jakarta during the second wave grew gradually, while in Surabaya City, the cases spread rapidly from the second week of July to the third week. This condition indicated that the hot spots of COVID-19 cases are highly related to the prevalent human interaction. The cases in DKI Jakarta showed a high concentration of cases in Central Jakarta, while in Surabaya, they appeared in South Surabaya. The results align with the geographical condition of both study areas where the main economic activities are concentrated. Local Anselin analysis gave insight into particular locations with high-risk transmission or that should be targeted to strengthen their prevention plans.

The overall COVID-19 case pattern in DKI Jakarta is a random distribution, which may be the result of better testing surveillance and high interaction or mobility among districts or cities supported by an integrated public transportation system. In contrast, the distribution pattern in Surabaya City is strongly clustered with hot spots in the southern region. This pattern can be used as a basis to develop more adaptive local policies. DKI Jakarta needs to strengthen its coordination with other cities in providing medical assistance. All cities must practice equal prevention and protection since restricting mobility seems too difficult to conduct. On the other hand, Surabaya City can focus its medical resources on the southern region.

There is a need for more localized COVID-19 measures by strengthening the involvement of local actors to take adaptive actions in mitigating cases of transmission among districts or communities. Furthermore, spatial modeling of case transmission by socioeconomic status is needed since lower socioeconomic populations cannot conduct remote work from home. High population density layered with lower socioeconomic status might stimulate higher transmission of cases since these individuals may not have access to basic necessities in COVID-19 prevention such as sanitation, frequently changed masks, and social distancing even among household members.

ACKNOWLEDGMENT

This research was a part of a project entitled "Analysis of COVID-19 Policy Governance at the National and Regional Level" funded by BP-PTNBH Kemenristek of Indonesia / BRIN 2020 research funding 2/E1/KP/PTNBH/2021, and Riset Unggulan ITB entitled "The Impact of COVID-19 Pandemic on Urban Health Determinant during the Early Stage of the Outbreak in Urban Area in Indonesia" awarded to Dr. Saut Sagala.

REFERENCES

Adhikari, S., Pantaleo, N. P., Feldman, J. M., Ogedegbe, O., Thorpe, L., & Troxel, A. B. (2020). Assessment of community-level disparities in coronavirus disease 2019 (COVID-19) infections and deaths in large US metropolitan areas. *JAMA Network Open, 3*(7), e2016938. https://doi.org/10.1001/jamanetworkopen.2020.16938

Aisyah, D. N., Aryanti Mayadewi, C., Diva, H., Kozlakidis, Z., & Adisasmito, W. (2020). A spatial-temporal description of the SARS-CoV-2 infections in Indonesia during the first six months of outbreak. *PloS One, 15*(12), e0243703.

Barron, L. (2020, February 28). A silent epidemic? Experts fear the coronavirus is spreading undetected in Southeast Asia. *Time* magazine. https://time.com/5792180/southeast-asia-undetected-coronavirus/

Blomkamp, E., Sholikin, M. N., Nursyamsi, F., Jenny M Lewis, & Tessa Toumbourou. (2017, June). Understanding policymaking in Indonesia: In search of a policy cycle. PSHK. 1–45. www.pshk.or.id/wp-content/uploads/2017/08/Understanding-Policy-Making-in-Indonesia-PSHK.pdf

Bridgman, P., & Davis, G. (2003). What use is a policy cycle? Plenty, if the aim is clear. *Australian Journal of Public Administration, 62*(3), 98–102.

COVID-19 Response Task Force. (2022). COVID-19 condition per city level in Indonesia. Covid19.Go.Id . https://covid19.go.id/situasi

DKI Jakarta Central Bureau of Statistics (BPS). (2022). DKI Jakarta Province in Figures 2022. 31000.2203. DKI Jakarta.

East Java Government. (2021). East Java Alert COVID-19. https://infocovid19.jatimprov.go.id/.

Eising, R. (2013). *Theories of policy formulation*. Bonn: ZEI, Center for European Integration Studies.

Eryando, T., Sipahutar, T., & Rahardiantoro, S. (2020). The risk distribution of COVID-19 in Indonesia: A spatial analysis. *Asia Pacific Journal of Public Health, 32*(8), 450–52. https://doi.org/10.1177/1010539520962940

Hadna, A. H. (2021). Policy formulation during pandemic COVID-19: A new evidence of multiple streams theory from Yogyakarta, Indonesia. *Journal of Public Administration and Governance, 11*(3), 3655–3655.

Jakarta Smart City. (2021). COVID-19 case in Jakarta. https://corona.jakarta.go.id/id.

Kingdon, J. W., & Stano, E. (1984). *Agendas, alternatives, and public policies*. Vol. 45. Boston: Little, Brown.

Kurniawan, D. (2021, June 27). Gedung Parkir RSUD Dr. Soetomo Jadi Ruang Isolasi Pasien Covid-19. *Liputan6*. https://surabaya.liputan6.com/read/4592485/gedung-parkir-rsud-dr-soetomo-jadi-ruang-isolasi-pasien-covid-19

Lasswell, H. D. (1956). The political science of science: An inquiry into the possible reconciliation of mastery and freedom. *American Political Science Review, 50*(4), 961–979.

Mardiansjah, F. H., & Rahayu, P. (2019). Urbanisasi dan pertumbuhan kota-kota di Indonesia: Suatu perbandingan antar-wilayah makro Indonesia. *Jurnal Pengembangan Kota, 7*(1), 91–110.

McBryde, E. S., Meehan, M. T., Adegboye, O. A., Adekunle, A. I., Caldwell, J. M., Pak, A., Rojas, D. P., Williams, B. M., & Trauer, J. M.. (2020). Role of modelling in COVID-19 policy development. *Paediatric Respiratory Reviews*, *35*, 57–60.

Organization for Economic Cooperation and Development (OECD). (2020, July 23). OECD policy responses to coronavirus (COVID-19): Cities policy responses. *OECD*. www.oecd.org/coronavirus/policy-responses/cities-policy-responses-fd1053ff/

Pribadi, D. O., Saifullah, K., Syah Putra, A., Nurdin, M., & Rustiadi, E. (2021). Spatial analysis of COVID-19 outbreak to assess the effectiveness of social restriction policy in dealing with the pandemic in Jakarta. *Spatial and Spatio-Temporal Epidemiology, 39*, 100454.

Purwanto, P., Utaya, S., Handoyo, B., Bachri, S., Astuti, I. S., Utomo, K. S. B., & Aldianto, Y. E. (2021). Spatiotemporal analysis of COVID-19 spread with emerging hotspot analysis and space–time cube models in East Java, Indonesia. *ISPRS International Journal of Geo-Information, 10*(3), 133. https://doi.org/10.3390/ijgi10030133

Ratnasari, N. R. P. & Dewi, V. R. (2021). Spatio-temporal model for predicting Covid19 cases in Indonesia. *Seminar Nasional Official Statistics, 2020*(1), 196–209. https://doi.org/10.34123/semnasoffstat.v2020i1.723

Ribeiro, H. V., Sunahara, A. S., Sutton, S., Perc, M., & Hanley, Q. S. (2020). City size and the spreading of COVID-19 in Brazil. *PloS One, 15*(9), e0239699. https://doi.org/10.1371/journal.pone.0239699

Shariati, M., Mesgari, T., Kasraee, M., & Jahangiri-rad, M. (2020). Spatiotemporal analysis and hotspots detection of COVID-19 using geographic information system (March and April, 2020). *Journal of Environmental Health Science and Engineering, 18*(2), 1499–1507. https://doi.org/10.1007/s40201-020-00565-x

Sulistyowati, A., & Muazansyah, I. (2019). Optimalisasi pengelolaan dan pelayanan transportasi umum (studi pada "Suroboyo Bus" di Surabaya). In *Iapa Proceedings Conference* (pp. 152–165). ISSN 2686-6250. https://doi.org/10.30589/proceedings.2018.189.

Sunirno, F. C., Halim, K. C., & Setiawan, R. (2019). Karakteristik pengguna Suroboyo Bus. *Jurnal Dimensi Pratama Teknik Sipil, 8*(2), 136–413.

Surabaya Central Bureau of Statistics (BPS). (2022). Surabaya Municipality in Figures 2022. 35780.2202. Surabaya.

4 Investigating Spatial Relationships of Age and Comorbidities on COVID-19 Spread Using Geographically Weighted Regression

Naomi W. Lazarus

CONTENTS

4.1 INTRODUCTION

COVID-19 is a respiratory illness caused by the SARS-CoV-2 virus. The virus originated in Wuhan, China, in late 2019 and quickly spread to become a global pandemic. The first recorded coronavirus case in the United States was reported in January 2020. Common symptoms of COVID-19 infection include fever, cough, and fatigue. Patients with advanced stages of the virus exhibit more serious symptoms including difficulty breathing and chest pain (CDC, 2021a). As of June 30, 2021, the total number of coronavirus cases reported in the United States exceeded 33.4 million and total deaths were just over 600,000 (CDC, 2021b). Globally, the virus has been responsible for approximately 183 million infections and four million deaths as of early July 2021 (WHO, 2021).

The Centers for Disease Control and Prevention (CDC) reports that older adults and persons with preexisting medical conditions are at a higher risk of contracting COVID-19 and dying from the virus. Adults aged 65–74 years are six times more

DOI: 10.1201/9781003227106-4

likely to be hospitalized due to the virus than persons aged 18–29. Hospitalizations among the 85+ age cohort increase dramatically to 15 times higher than the 18–29 comparison group. Among those hospitalized, adults aged 65–74 are 95 times more likely to die from the virus compared to persons aged 18–29. Fatalities due to COVID-19 increase to 600 times in the 85+ age category in relation to the comparison group (CDC, 2020a). The CDC references several underlying medical conditions, such as cancer, diabetes, and heart disease, that are likely to increase the risk of contracting COVID-19, but specifies that there is not enough evidence to qualify causal or direct links to these risk factors (CDC, 2020b).

The chapter outlines the process and outcomes of a county-level assessment of COVID-19 in relation to age demographics and comorbidities of the exposed population in the United States. Geographically weighted regression (GWR) was used to explore how these risk factors are related to the incidence rate and death-case ratio of COVID-19 at the county level in the contiguous United States. GWR is a localized regression model that captures the spatial dependence in the relationships between the dependent and explanatory variables. This method produces individual regression coefficients for each spatial unit in the study (Mayfield et al., 2018; Ndiath et al., 2015). The purpose of this chapter is to discuss data management and methodological considerations related to the GWR model as it is applied to investigating the dynamics between coronavirus, age, and comorbidities. The study was part of a year-long project begun in October 2020 and funded by the Geospatial Fellowship program of the University of Illinois Urbana-Champaign and the National Science Foundation's Computer and Information Science and Engineering program.

4.2 ECOLOGICAL AND EPIDEMIOLOGICAL APPROACHES TO DISEASE RISK

Epidemiological research has recognized the importance of space to examine the public health impacts of infectious diseases at different geographical scales (Meade & Emch, 2010; Glass, 2000; Jones & Moon, 1993). Geographical space has often functioned as a mechanism to aggregate disease incidence, but is now increasingly viewed as a participating factor to analyze cause and effect of disease outcomes (Jones & Moon, 1993). Human-environmental interactions related to disease epidemics are explored using the *ecological* approach that derives its roots in the subfields of medical and hazards geography. Medical geography has adopted the ecological approach to specifically examine the spatial patterns of disease with a focus on cultural-environmental factors that determine the level of human exposure (Glass, 2000). These relationships are captured in the *triangle of human ecology* as described by Meade and Emch (2010). The framework consists of three components: habitat, population, and behavior. The habitat represents the physical and social environments in which human populations carry out their lives. The population is characterized primarily by its demographic and biological characteristics, namely age, gender, immunity, and nutritional status. Behavior is a function of social constructs set up by the human population that dictate their responses to events. Technology and education are important social constructs that determine behavior (Meade & Emch, 2010). The ecological approach in hazards geography recognizes that the impact of hazard events is a combination of the physical characteristics of the event and the characteristics of the population that occupies hazard-prone areas (Hewitt & Burton, 1971). It adopts a similar argument as put forth by Meade and Emch (2010) stating that the ordinary day-to-day activities undertaken by the population contribute to environmental changes and these changes translate to the timing and intensity of hazard events and people's experience with them (Hewitt & Burton, 1971). Drawing from these theoretical underpinnings, the ecological approach helps explain spatial and social dimensions of disease risk (Keesing, Holt, & Ostfeld, 2006; Glass, 2000; Jones & Moon, 1993).

Risk of exposure to disease is driven by prevailing vulnerabilities that are inherent in the environment and in the exposed population. The conceptual underpinnings of vulnerability have been extensively studied in hazards research and can be applied to disease risk. Physical vulnerability is generated by location and environmental factors that can adversely impact the exposed population (Adger, 2006; Montz, Tobin, & Hagelman, 2017). Social vulnerability is tied to the demographic and socioeconomic characteristics of the population that increase the likelihood of adverse outcomes resulting from exposure to hazards (Wisner et al., 2004; Cutter, 1996). The two dimensions of vulnerability frequently play out simultaneously to determine the risk level of the community. In a case study examining the vulnerability of populations in Georgetown, South Carolina, Cutter, Mitchell, and Scott (2000) classified physical vulnerability by demarcating hazardous zones that were exposed to natural and anthropogenic hazards, such as floods, hurricanes, chemical spills, and traffic incidents. Social vulnerability was addressed by developing an index that accounted for several demographic and socioeconomic variables which characterized the population in the county. These variables included the female population, the minority population, the young and the elderly population, housing value, and number of mobile homes. In a study examining childhood exposure to lead in Chicago, physical vulnerability was tied to the location of high-risk populations in the West Side and South Side of Chicago (Oyana & Margai, 2010). The study cited age, gender, household income, and housing age as significant factors in determining the social vulnerability of the population to lead exposure.

Epidemiological research has pointed to environmental and social factors that determine risk of exposure to disease. These dynamics are investigated extensively in the spread of vector-borne diseases. The World Health Organization (WHO) has put together several indices to monitor the presence of dengue hot spots that is tied to the physical location of vector breeding sites. The Breteau Index, for example, takes into account the number of water-holding containers per household to monitor the spread of the virus in densely populated areas (WHO, 2003). This measure is important because the vector responsible for the spread of dengue breeds in standing water that has been accumulating for more than seven days. A number of indoor and outdoor repositories are identified where such water collects to create conducive breeding grounds for mosquitos. These include water storage tanks, refrigerator and a/c water trays; plastic containers, tree holes, bamboo stumps, discarded tires, and coconut shells (WHO, 2003). The breeding pattern of mosquitos in the tropics is driven by the monsoon season, and this in turn, promotes the accumulation of standing water. A study on the dengue virus in Sri Lanka found that precipitation, together with temperature and humidity were significant drivers of the prevalence of the vector particularly in vegetated areas in the interior of the island (Weeraratne, 2013). When considering the social vulnerability of populations to vector-borne diseases, demographic characteristics like age and gender play a prominent role in diagnosing infections. Kanakaratne et al. (2009) investigated how age is related to the number of dengue cases and how this has varied over time. Previously, the disease disproportionately impacted children, but recent epidemics in South Asia have shown that viral loads are increasing in young adults. Urbanization, population density, and poverty are other factors that determine the prevalence of vector-borne diseases and increase the social vulnerability of communities exposed to such viruses (Sutherst, 2004).

The global pandemic of COVID-19 has revealed that the virus is undeterred by physical–environmental factors. The temporality and the transmissibility of the virus have provided medical professionals with insights into identifying high-risk populations based on age and underlying conditions. A study conducted by the CDC found that 78% of deaths due to COVID-19 that occurred between May and August 2020 were among patients aged 65 and older (Boehmer et al., 2020). Adults aged 65–74 years are six times more likely to be hospitalized due to the virus than persons aged 18–29. Among those hospitalized, adults aged 65–74 are 95 times more likely to die from the virus compared to persons aged 18–29 (CDC, 2020a). Mueller, McNamara, and Sinclair (2020) point to the decline of the immune system among older adults as being responsible for the trajectory of COVID-19 among this age cohort. A study out of China found that the median age of hospitalized coronavirus patients who were put on a ventilator was 69 years (Chen et al., 2020). While the research on investigating the relationship between underlying conditions and coronavirus is still evolving, emerging national and international studies show that comorbidities increase the risk of infection and death from COVID-19. A review of studies related to underlying conditions has found that hypertension, cardiovascular disease, obesity, and chronic obstructive pulmonary disease were among the top-level comorbidities present in hospitalized patients due to COVID-19 (Javanmardi et al., 2020; Sanyaolu et al., 2020). A comparative analysis of coronavirus prevalence in India and England revealed that diabetes, hypertension, and obesity were the prominent underlying conditions among those infected with the virus (Novosad et al., 2020).

The ecological approach to epidemiological studies has highlighted the importance of utilizing appropriate measures of morbidity and mortality to monitor public health outcomes. In the case of COVID-19, raw case numbers and death counts have been the primary method of disseminating information to the public as is evidenced by the dashboards maintained and shared by the WHO and the John's Hopkins University (WHO, 2021; JHU, 2021). While raw case and death counts are useful to monitor trends in individual localities, Pearce et al. (2020) caution that these indicators do not account for population size and are poor measures of comparison across states and countries. The CDC proposes several measures of morbidity and mortality that address the overall population, geographical scale, and time. Incidence proportion and incidence rate are measures of morbidity that focus on *new* case counts in a given time period in relation to the total population in a locality (CDC, 2012). Point prevalence and period prevalence monitor *cumulative* case counts as a ratio of the total population. In addition to the crude death rate, the CDC provides several options to monitor mortality associated with a specific disease or condition. Cause-specific death rate, proportionate mortality, and death-to-case ratio are all measures of mortality that account for number of deaths attributed to a specific health condition during a given time period, taken as a ratio of the population and multiplied by 100, 1,000, or 100,000 (CDC, 2012). In the case of COVID-19, Pearce et al. (2020) recommend the use of the population attack rate as a measure of morbidity, which is similar to the incidence rate proposed by the CDC. The attack rate is defined as a measure of the frequency of new cases among the contacts of known patients (CDC, 2012). The infection fatality rate is suggested by the authors of the same study as a measure of mortality, which is similar to the death-to-case ratio proposed by the CDC (Pearce et al., 2020; CDC, 2012).

The study of the spread of disease and their public health outcomes has demonstrated the value and challenges of spatial and/or temporal methodological approaches. In a case study examining the transmission of cholera in Bangladesh, Koelle et al. (2005) adopted a model that combined intrinsic factors like population immunity and extrinsic factors like climate variability to assess the temporal trends in disease outbreaks among the exposed population. The findings of the study revealed a temporal trend in the number of infected people based on the level of immunity. Major outbreaks of cholera occurred when the host population's immunity was very low after sufficient time had passed after the previous significant outbreak of the disease. With regard to extrinsic factors, a negative correlation was observed between low-frequency transmission rates and regional rainfall patterns and river discharge indicating that climate variability alone did not explain the incidence of cholera. Thomson et al. (2005) utilized logistic regression to examine the relationship between rainfall patterns and the incidence of malaria in Botswana. The study found that, while rainfall was a significant predictor of malaria incidence, the relationships between the variables were nonlinear. In some years where rainfall was significantly higher, the incidence of malaria was low, indicating nonlinearity due to flushing of vector breeding sites. In another study, multivariate Poisson regression was used by Siriyasatien et al. (2016) to predict the incidence of dengue in Thailand. The model results showed similarities between the actual and predicted dengue cases, identifying seasonality (a categorical predictor) and female mosquito infection rate as significant parameters. A limitation of the study was that it assumed a linear relationship between the dependent variable and the predictors, whereas other studies have shown that the relationships between vector-borne diseases and environmental and population variables are nonlinear based on type of virus strain and other environmental factors (Thomson et al., 2005; Tam et al., 2013).

Geographically weighted regression (GWR) has emerged as an important spatial analytical tool in epidemiological research in recent years (Mayfield et al., 2018; Brunton et al., 2017; Ndiath et al., 2015). GWR is a technique that accounts for spatial dependence as articulated in Tobler's first law of geography (Charlton & Fotheringham, 2009). Mayfield et al. (2018) used a logistic GWR model to evaluate the spatial relationships between incidence of leptospirosis (a zoonotic disease) and a combination of environmental and sociodemographic variables that included rainfall, proximity to water bodies, and poverty. The results included a map of predicted probabilities that displayed hot spots of the disease in the study area located in Fiji.

GWR was used in a study in Senegal to examine how prevalence rates of malaria were related to temperature, rainfall, and housing characteristics among other factors (Ndiath et al., 2015). The GWR model in this case explained 82% of the variability in malaria, and local coefficient maps showed similar spatial trends in household size and usage of mosquito nets as they related to prevalence rates. In another case study related to bovine tuberculosis in the United Kingdom, Brunton et al. (2017) used GWR in combination with principal component analysis to identify influential predictors and to observe regional variations in the occurrence of the disease in England and Wales. Prevailing epidemiological and ecological studies on disease epidemics have explored a wide variety of theoretical and methodological approaches as illustrated in this section. These studies provide opportunities for the application of different approaches to examine the evolution of COVID-19.

4.3 APPLICATION OF GEOGRAPHICALLY WEIGHTED REGRESSION

The study investigated the effects of age and comorbidities in relation to the spread of COVID-19 during two peak periods for the contiguous United States. The first peak occurred between 03/01/20 and 04/30/20 and the second, between 06/01/20 and 07/31/20. These time frames were selected based on daily nationwide trends of COVID-19 cases during the early stages of the pandemic (CDC, 2021b). The first step in the data management process was to configure the data to reflect comparable indicators on incidence, mortality, and morbidity. Current data and research on the coronavirus have focused on raw case numbers and on identifying hot spots related to positive cases. While this approach is useful to monitor the effect of mitigation strategies, it does not account for the geographical variation in population demographics across the country (Pearce et al., 2020). These criticisms are addressed in the study by the use of incidence rates and death-case ratios as measures of morbidity and mortality to examine the spatial variation of the virus across counties. Incidence rate and death-case ratio function as the dependent variables in the GWR model. Incidence rates are calculated for each peak period using the following formula:

$$\text{Incidence Rate} = \frac{\textit{Number of New Cases during a specified time period}}{\textit{Total County Population}} \times 100{,}000$$

The death-case ratio is used to measure mortality and is defined as follows:

$$\text{Death to case Ratio} = \frac{\textit{Number of New Deaths during a specified time period}}{\textit{Number of New Cases during that same time period}} \times 100$$

The above equations are derived from CDC recommendations associated with monitoring disease risk (CDC, 2012, p. 188). Raw case and death counts were

TABLE 4.1
Explanatory Variables Used in the GWR Analysis

Variable Code	Variable Definition	Source
PCT_50to74	Percent population aged 5 –74	American Community Survey 2018 5-year estimates data.census.gov
PCT_over75	Percent population aged 75 and above	American Community Survey 2018 5-year estimates data.census.gov
DIAB_PCT	Percent population diagnosed with diabetes	Centers for Disease Control and Prevention https://gis.cdc.gov/grasp/diabetes/DiabetesAtlas.html#
CARDIO_MR	Heart disease mortality rate—number of deaths per 100,000 of population	Centers for Disease Control and Prevention https://wonder.cdc.gov/
OBESE_PCT	Percent adult population diagnosed with obesity	Centers for Disease Control and Prevention https://gis.cdc.gov/grasp/diabetes/DiabetesAtlas.html#

obtained from USAFacts.org (https://usafacts.org/visualizations/coronavirus-covid-19-spread-map/).

The independent or explanatory variables are listed in Table 4.1. High-risk age cohorts related to coronavirus are identified as percent population aged 50–74 and above 75. Variables representing underlying conditions include percent population diagnosed with diabetes, percent population diagnosed with obesity, and heart disease mortality rate.

The data collection process revealed some discrepancies related to availability and consistency in terms of the time period. The age cohort variables are derived from the 2018 American Community Survey five-year estimates. Obesity and diabetes prevalence are based on 2017 estimates and heart disease mortality rates are based on 2018 estimates. Comorbidity trends revealed marginal changes in recent years. Diabetes and obesity prevalence rates changed on average by 0.22% from 2010 to 2015. Heart disease mortality rates varied by 0.003% from 2007 to 2017. Given these marginal changes in comorbidity, it was expected that the limitations outlined above would have a minimal effect on the analysis results.

The analysis phase of the project involves executing the GWR model to examine the spatial relationships between coronavirus, age, and comorbidities. Spatial dependence recognizes that counties in close proximity are likely to display similar characteristics and trends than those that are distant (Charlton & Fotheringham, 2009). GWR is useful to conduct an exploratory analysis to identify these spatial relationships. The generalized GWR formula is presented as follows:

$$Y_i = b_{i0} + \sum_{j=1}^{m} b_{ij} x_{ij} + \varepsilon_i$$

where Y_i is the dependent variable in the geographical location, *i*; b_0 is the intercept associated with that location; b_{ij} is the coefficient; and x_{ij} is the observed value of the independent variable *j* in location *i* in a set of *m* number of variables. The error term, ε, is the residual associated with location *i*. The model results include individual coefficients for each spatial feature or location (Fotheringham, Brunsdon, & Charlton, 2002; Rogerson, 2015). Python code for Jupyter notebooks and ArcGIS software developed by the Environmental Systems Research Institute (ESRI) were used to conduct the analysis.

The framework of the GWR model is determined by the model type, the bandwidth, and the weighting method. Histograms generated for the dependent variables reveal that coronavirus incidence rates and death-case ratios are positively skewed. In this scenario, a logistic regression model would be appropriate for the GWR where the dependent variable is classified as a binary variable (recording a value of 1 or zero) to capture the significant variations in the attribute across the study area (ESRI, 2021). A binary model, however, would have the effect of reducing the dependent variables to two possible outcomes and fail to capture the variations in the incidence and mortality patterns of the virus. An alternative method would be the continuous (Gaussian) model type, which is typically applied to data that are normally distributed. In this case, transforming the dependent variable would be a solution to minimize skewness in the data. Log transformation and square root transformation are possible candidates, but both these strategies contain shortcomings. The log10 transformation was selected for the analysis as it is viewed as a more robust method to address skewed data (Manikandan, 2010; Bland, 2000). One of the shortcomings of log transformation is that it does not accommodate zero values. Several counties recorded zero coronavirus incidence rates and death-case rates during the two peak periods under investigation. In order to implement the transformation, counties with zero values for the dependent variables were removed, which resulted in the number of observations to vary for each of the peak periods related to incidence and death-case ratio. Table 4.2 lists the data sets that were used for each iteration of the GWR analysis. Given that this was an exploratory study, and was not meant to be used for future prediction of coronavirus cases and deaths, the Gaussian model type was used for the GWR analysis with log10 transformed dependent variables. This method assumes that the dependent variable is normally distributed, which is accomplished by the transformation (ESRI, 2021).

The bandwidth (or kernel) defines the size and extent of the neighborhood of influence for each individual feature. Bandwidths can be defined by the number of

TABLE 4.2
Size of Data Sets Used in the GWR Analysis

Peak Period	Dependent Variable	Number of Observations
Peak 1: 03/01/20–04/30/20	Incidence Rate	N = 2,807
Peak 2: 06/01/20–07/31/20	Incidence Rate	N = 3,061
Peak 1: 03/01/20–04/30/20	Death-Case Ratio	N = 1,445
Peak 2: 06/01/20–07/31/20	Death-Case Ratio	N = 1,964

individual neighbors or by a distance band. When the neighborhood is determined by a fixed number of nearest neighbors, the size of the neighborhood will vary based on the density of features. This is referred to as the adaptive kernel type that defines smaller neighborhood sizes where observations are dense, and larger neighborhoods where they are dispersed. The distance band method or fixed kernel type adjusts the number of observations per neighborhood so that the size of the neighborhood remains the same throughout the study area. It results in a greater number of observations per neighborhood where features are dense and in fewer observations per neighborhood where features are dispersed (ESRI, 2021). The distance band method can result in bias in favor of densely populated regions and underrepresentation of areas where features are dispersed. For this study, the adaptive kernel type is used as it is better able to adjust to varying spatial patterns and sizes of features throughout the study area (Oshan et al., 2019). A golden section search method is used to define the size of the neighborhood where the lowest Akaike Information Criterion (AIC) score determines the number of neighboring features that participate in the analysis based on Euclidean distance (ESRI, 2021; Nakaya et al., 2005). The size of the neighborhood varied relative to the size of the data set for each iteration as outlined in Table 4.2. The bandwidths for the GWR analysis with incidence rates were 102 neighbors for peak 1 and 111 neighbors for peak 2. The bandwidths pertaining to death-case ratios were 210 neighbors for peak 1 and 137 neighbors for peak 2.

The weighting method is a critical feature in GWR as it accounts for spatial dependence by estimating the distance decay effect of neighboring features. The Gaussian weighting method is used to assign greater weights for features that are in close proximity and smaller weights to those that are distant (Nakaya et al., 2005). It ensures that each feature in the neighborhood retains some influence over the regression point regardless of its distance from that point (ESRI, 2021; Oshan et al., 2019). This method is deemed appropriate due to the transmissibility of COVID-19 that shows higher incidence of the virus occurring in clusters relative to population density and mobility. The Gaussian weighting method would apply greater weights to counties in and around these high clusters and smaller weights to areas that are located away from these clusters.

4.4 SPATIALTEMPORAL PATTERNS OF COVID-19 IN RESPONSE TO AGE AND COMORBIDITIES

Four iterations of the GWR analysis were run using log transformed incidence rates and log transformed death-case ratios for peak 1 and peak 2 as the dependent variables. The results of the ordinary least squares regression (OLS) are presented in Table 4.3. The coefficients are listed alongside their respective *p* values in brackets. Comorbidities related to diabetes and heart disease generated significant positive coefficients for coronavirus incidence rates. The age cohort variables (50–74 and above 75) and heart disease mortality rates recorded significant positive coefficients for death-case ratios. The coefficient values represent the log transformed dependent variable, and therefore, the b coefficient values were back-transformed to provide meaningful interpretations. In both time periods, the age cohort predictors generated negative coefficients for coronavirus incidence rates and positive coefficients for

TABLE 4.3
B Coefficients—OLS Regression

Variable	Incidence Rates (Peak 1)	Incidence Rates (Peak 2)	Death-Case Ratio (Peak 1)	Death-Case Ratio (Peak 2)
PCT_50to74	−0.002 *(0.352)*	−0.032 *(0.000)*	0.0097 *(0.002)*	0.011 *(0.000)*
PCT_over75	−0.044 *(0.000)*	−0.019 *(0.000)*	0.017 *(0.028)*	0.0249 *(0.000)*
DIAB_PCT	0.025 *(0.006)*	0.0088 *(0.000)*	−0.004 *(0.216)*	−0.0018 *(0.550)*
CARDIO_MR	−0.0001 *(0.171)*	0.0004 *(0.000)*	0.0007 *(0.000)*	0.0002 *(0.084)*
OBESE_PCT	−0.018 *(0.357)*	0.002 *(0.137)*	−0.0015 *(0.462)*	−0.0028 *(0.134)*

TABLE 4.4
Model Fit Statistics—OLS and GWR

	Adjusted R2		AIC	
Model	*OLS*	*GWR*	*OLS*	*GWR*
Incidence Rates (Peak 1)	0.053	0.316	4149.47	3277.51
Incidence Rates (Peak 2)	0.229	0.578	2903.70	1102.17
Death-Case Ratio (Peak 1)	0.111	0.166	1019.74	935.77
Death-Case Ratio (Peak 2)	0.070	0.247	1922.28	1526.64

death-case ratios. The b coefficient for percent population above 75 related to incidence rates in peak 1 was −0.0448 (significant at $p < .001$). The back-transformed value of 0.901 ($10^{-0.0448}$) is subtracted from 1.00 and multiplied by 100 to provide the percentage change in incidence rates (0.901 − 1.00 * 100 = −9.8%). In this example, the b coefficient reveals that when percent population above the age of 75 increases by one unit, incidence rates decline by 9.8% in peak period 1. The b coefficient for percent population above 75 related to death-case ratios in peak 2 was 0.0249 (significant at $p < .001$). The back-transformed value subtracted by 1.00 is 0.059 ($10^{0.0249}$ − 1.00). The b coefficient reveals that when percent population above the age of 75 increases by one unit, death-case ratios increase by 6% in peak period 2. Coefficients related to diabetes revealed a marginal increase in coronavirus incidence rates of 2% in peak 1 and an increase of 6% in peak 2. Coefficients associated with obesity were not significant across all models.

Model fit statistics of the OLS and GWR models are presented in Table 4.4. The GWR model pertaining to incidence rates in peak 2 recorded an adjusted R^2 value of 0.578, indicating that the model explains approximately 58% of the variation in coronavirus incidence rates during the second peak period between June and July 2020. A review of the AIC values shows that this model had the second-lowest value of 1102.17. The GWR model related to death-case ratios in peak 1 recorded the lowest AIC value of 935.77; however, this model also recorded a very low adjusted R^2 (0.166). As referenced in Table 4.2, several counties recorded zero coronavirus deaths

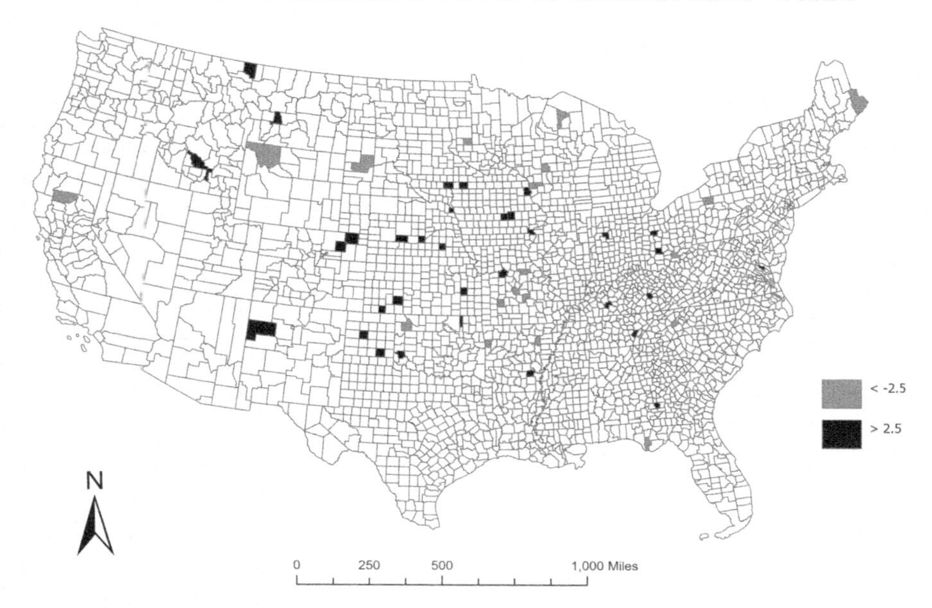

FIGURE 4.1 Standardized residual map of COVID-19 incidence rates for peak period 1 (03/01/20 – 04/30/20). GWR Model.

in peak 1 that resulted in the lowest number of observations compared to the other models (n = 1445). The model dealing with incidence rates in peak 2, as reported earlier, had the highest number of observations of 3,061.

The average log transformed incidence rate in peak 1 was 1.87, which, when back-transformed, returned a value of 74 new cases per 100,000 of the population. The standard deviation during this period was 0.52, which translates to 3.3 new cases per 100,000. The standardized residual maps of incidence rates are presented in Figures 4.1 and 4.2. The standardized residual map of incidence rates in peak 1 shows that a majority of counties recorded incidence rates within the range of 2.5 std. deviations (i.e., 8.5 per 100,000) of the mean (Figure 4.1). High-value outliers are dispersed throughout the Mountain West, Great Plains, and Midwest regions constituting 1% of counties that reported coronavirus cases during this period. The average log transformed incidence rate in peak 2 was 2.62, which, when back-transformed, returned a value of 417 new cases per 100,000 of the population. The standard deviation during this period was 0.44, which translates to 2.7 new cases per 100,000. High-value outliers greater than 2.5 standard deviations of the mean (or 6.75 per 100,000) intensify in peak 2 in the Southwest with marginal changes observed in the interior (Figure 4.2). Low-value outliers that are highlighted in gray (less than −2.5) are concentrated in New England and in the Mountain West region and these patterns are largely consistent during both peak periods.

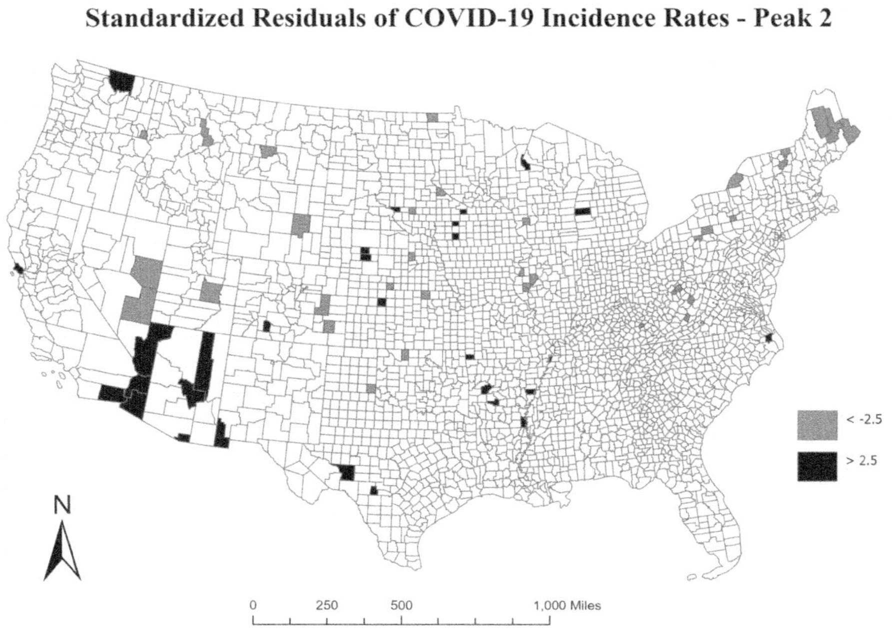

FIGURE 4.2 Standardized residual map of COVID-19 incidence rates for peak period 2 (06/01/20 – 07/31/20). GWR Model.

The average log transformed death-case ratio in peak 1 was 0.69, which, when back-transformed, returned a value of five new deaths per 100 of the population. The standard deviation during this period was 0.36, which translates to 2.29 new deaths per 100. The average death-case ratio dropped to 1.7 per 100 of the population in peak 2 and the standard deviation remained largely the same at 2.5 per 100. The standardized residual maps of death-case ratios are presented in Figures 4.3 and 4.4. Fifty-four percent of counties in peak 1 and 37% of counties in peak 2 did not report any COVID-19-related deaths. High-value outliers are identified as counties that recorded a death-case ratio of greater than six new deaths per 100 of the population of the mean in any given peak period. As illustrated in Figure 4.3, fewer high-value outliers are recorded in peak 1 compared to those related to incidence rates (Figure 4.1). High-value outliers are sparsely dispersed throughout the Great Plains and Midwest regions during peak 2 (Figure 4.4).

The series of maps that follow show the change in the back-transformed coefficients of the independent variables. The percent values represent the change in incidence rates and death-case ratios in response to a unit change in the predictor in question. The predictor associated with percent population above age 75 and percent population with diagnosed diabetes generated significant variations in the relationship with incidence rates. As illustrated in Figure 4.5, general clustering of increases in incidence rates in response to population above 75 is observed in the South, Southeast, Northeast, and Midwest regions during peak 1, 03/01/20 to 04/30/20. Six counties

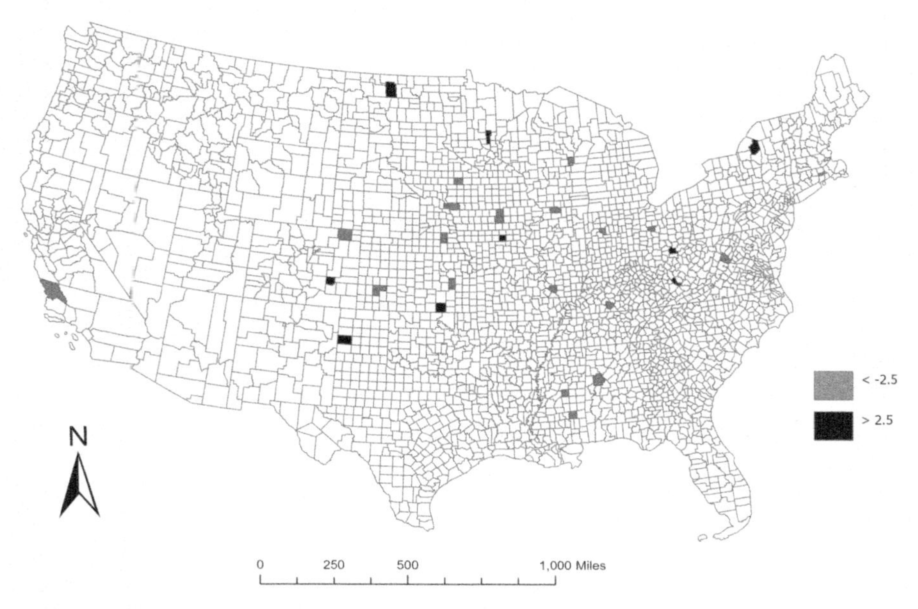

FIGURE 4.3 Standardized residual map of COVID-19 death-case ratios for peak period 1 (03/01/20 – 04/30/20). GWR Model.

FIGURE 4.4 Standardized residual map of COVID-19 death-case ratios for peak period 2 (06/01/20 – 07/31/20). GWR Model.

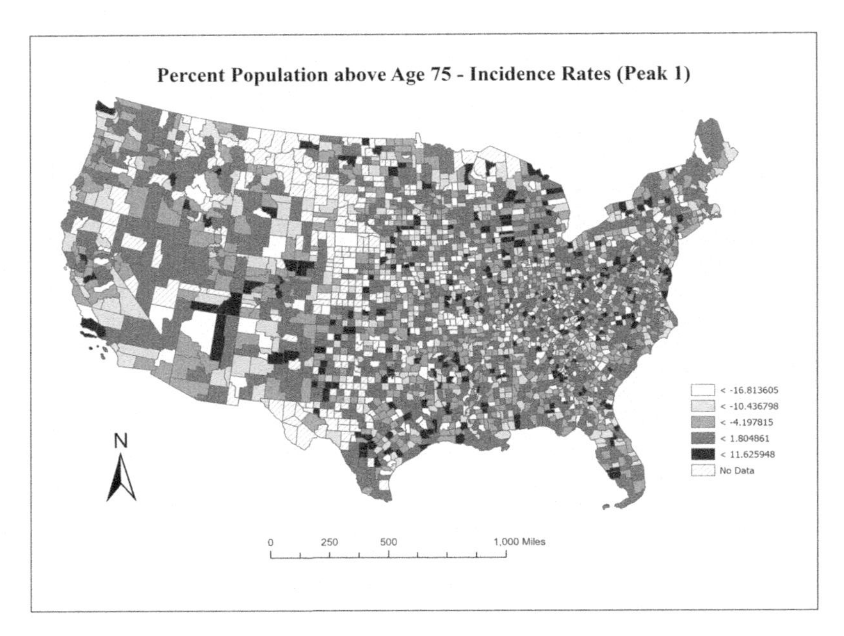

FIGURE 4.5 Map of b coefficients related to percent population above the age of 75 when dependent variable is COVID-19 incidence rates for peak period 1 (03/01/20 – 04/30/20).

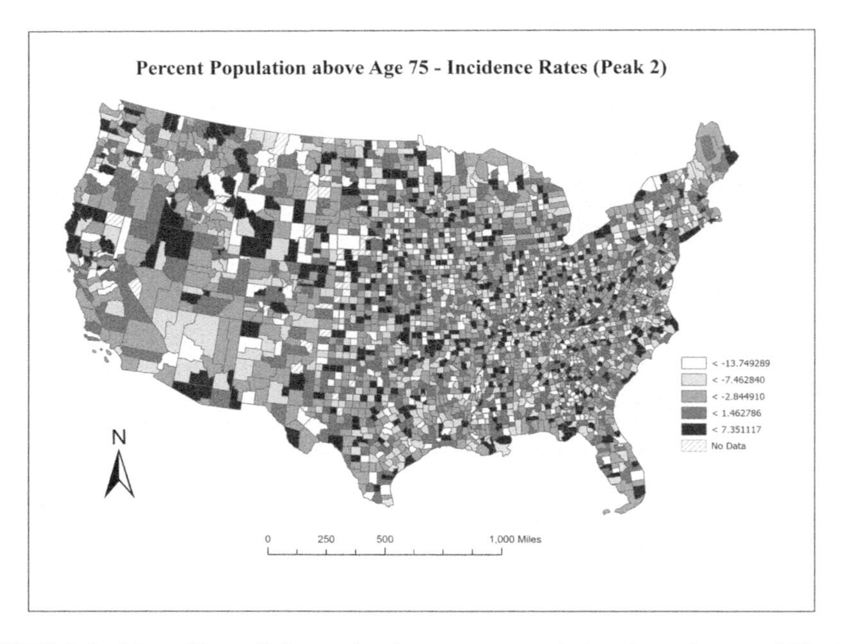

FIGURE 4.6 Map of b coefficients related to percent population above the age of 75 when dependent variable is COVID-19 incidence rates for peak period 2 (06/01/20 – 07/31/20).

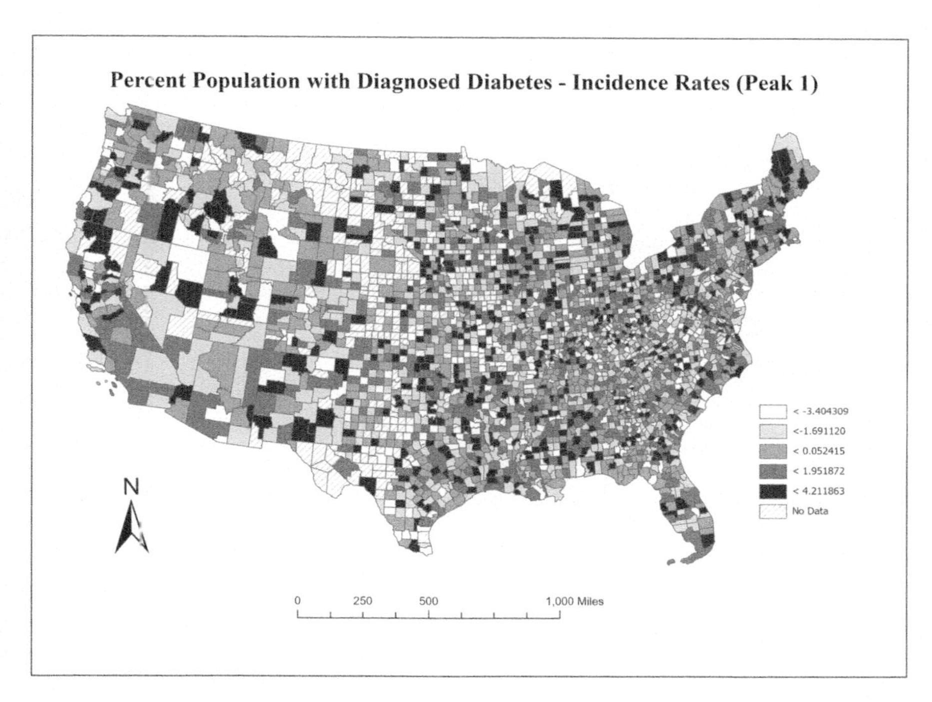

FIGURE 4.7 Map of b coefficients related to percent population with diagnosed diabetes when dependent variable is COVID-19 incidence rates for peak period 1 (03/01/20 – 04/30/20).

located in the South, mid-Atlantic, and in Minnesota recorded increases in incidence rates above 11% during this period. A similar spatial pattern of high clusters of incidence rates related to population above 75 was observed in peak 2 (06/01/20–07/31/20) with new clusters emerging in the Mountain and Western states (Figure 4.6). Seven counties that recorded an increase in incidence rates above 7% were dispersed across the states of Idaho, Colorado, Florida, West Virginia, New York, and Michigan. The range of increase in incidence rates in response to a unit increase in percent population with diabetes was between 1% and 4% in peak 1 (Figure 4.7). Increases in incidence rates above 2% were randomly distributed throughout the country with smaller clusters located in the Mountain West region. Increments in incidence rates in response to a unit increase in percent population with diabetes grew in peak 2 compared to peak 1. The range of increase recorded was between 2% and 10% in peak 2 (Figure 4.8). Three hundred four counties (10%) recorded rates above 5% in relation to diabetes during this period, displaying a similar spatial pattern of clusters as in peak 1. Eight counties recorded increases in rates above 10% with several located in the South and Midwest regions.

The predictor associated with percent population aged 50–74 generated mixed results in its relationship with coronavirus incidence rates. Moderate increases in incidence rates were observed in peak 1 between 1% and 6% in response to this predictor.

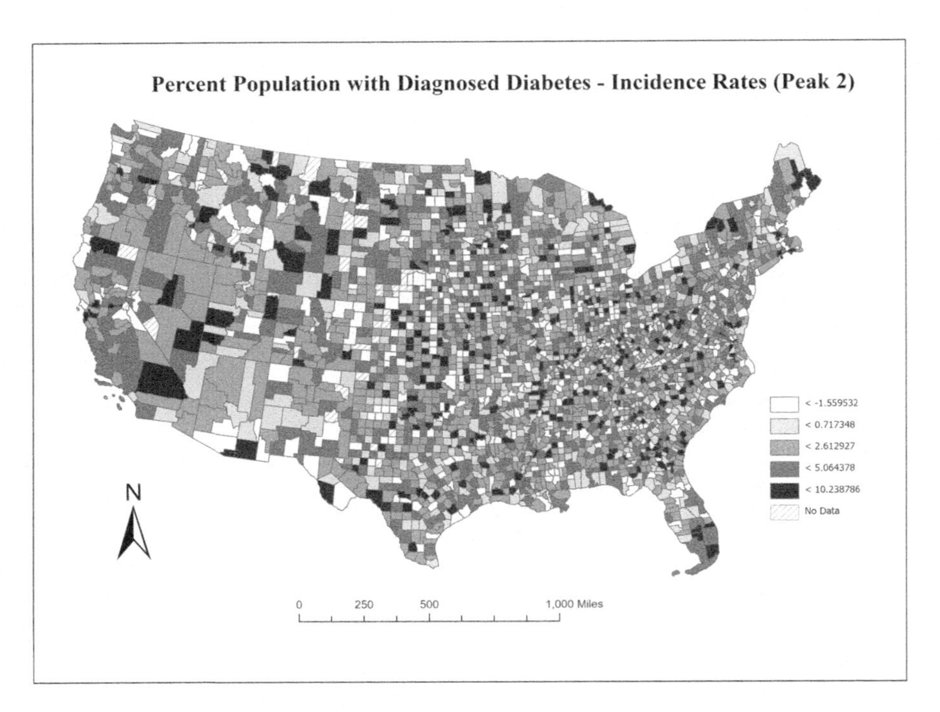

FIGURE 4.8 Map of b coefficients related to percent population with diagnosed diabetes when dependent variable is COVID-19 incidence rates for peak period 2 (06/01/20 – 07/31/20).

During peak 2, this age cohort had a negative effect, showing an overall decrease in incidence rates. Heart disease mortality rates had only a marginal effect on incidence rates in both peak periods. The overall increase in incidence rates in response to this predictor was less than 0.4% in peak 1 and less than 0.2% in peak 2. The coefficients related to obesity show a similar spatial pattern of increases in incidence rates across the contiguous United States. Counties with increases in incidence rates above 2% are randomly dispersed across the country in response to percent population with diagnosed obesity. It should be noted that this predictor was not significant at $p < .05$ in the global regression model.

The spatial pattern of variations in death-case ratios in response to each of the predictors is presented in a series of maps. The two predictors associated with age and percent population with diagnosed diabetes generated significant variations in the relationship with death-case ratios. As illustrated in Figure 4.9, 534 counties (37%) recorded an increase in death-case ratios above 3% per one unit increase in population aged 50–74 during peak 1. A majority of counties that recorded these increments are located in metropolitan areas in the Northeast, South, Southeast, and Midwest regions. The number of counties recording an increase in death-case ratios above 3% in response to percent population aged 50–74 dropped to 442 (23%) in peak 2 (Figure 4.10). The spatial pattern and overall range of increments in death-case

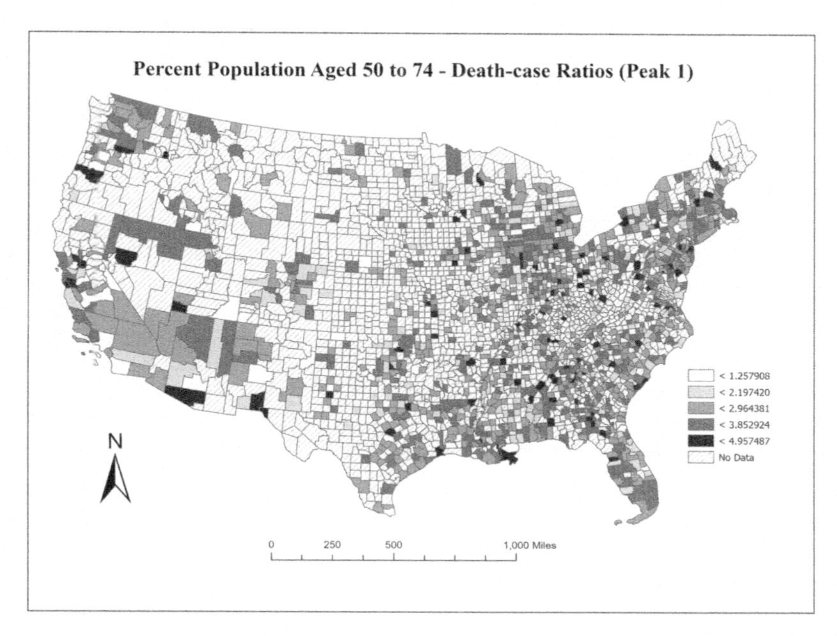

FIGURE 4.9 Map of b coefficients related to percent population aged 50 to 74 when dependent variable is COVID-19 death-case ratios for peak period 1 (03/01/20 – 04/30/20).

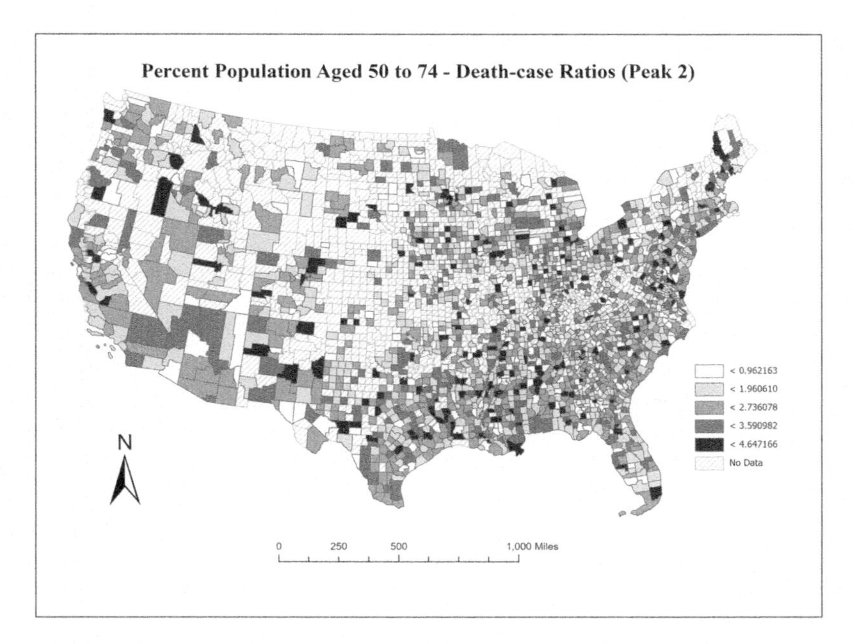

FIGURE 4.10 Map of b coefficients related to percent population aged 50 to 74 when dependent variable is COVID-19 death-case ratios for peak period 2 (06/01/20 – 07/31/20).

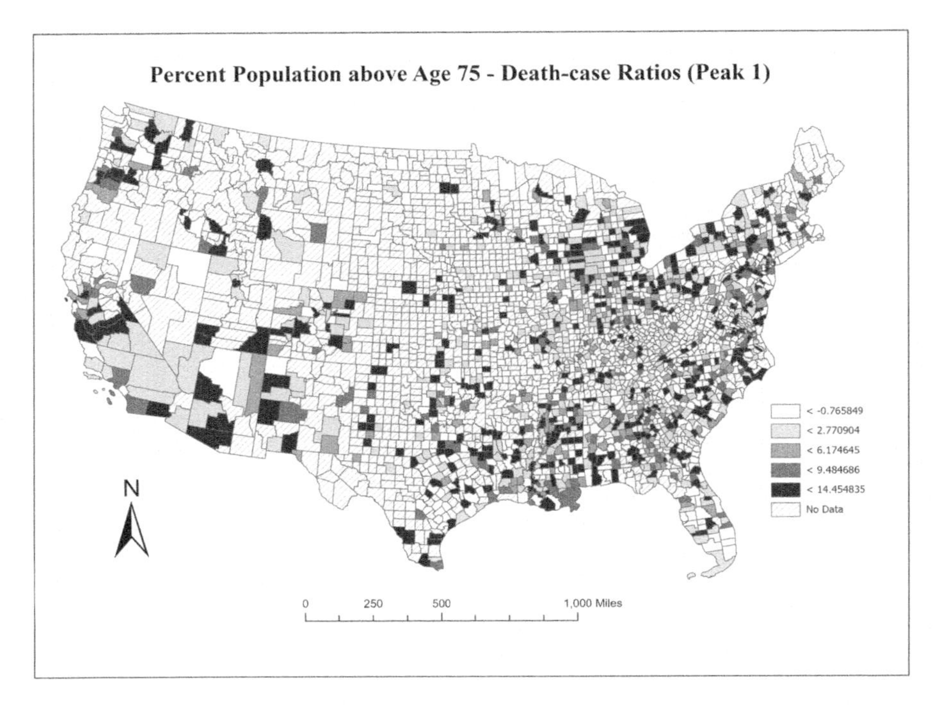

FIGURE 4.11 Map of b coefficients related to percent population above age 75 when dependent variable is COVID-19 death-case ratios for peak period 1 (03/01/20 – 04/30/20).

ratios were similar to those of peak 1. Increases in death-case ratio in response to percent population above aged 75 ranged from 2% to 14.5% in peak 1 (Figure 4.11). Counties that recorded increments in death-case ratios greater than 10% are located in the Midwest, Northeast, South, and Southeast. Some clusters were also observed in the Southwest and Pacific Northwest. The overall increase in death-case ratios in response to population above aged 75 was higher in peak 2, ranging from 4% to 23% compared to peak 1 (Figure 4.12). A similar spatial pattern was observed during this period as in peak 1 with clusters of increments greater than 10% intensifying in the Midwest, South, and Southeast.

Overall increases in death-case ratios in response to percent population with diabetes were marginal in peak 1 with the highest increase recorded as 2.6%. Forty-six counties (3%) recorded increases in death-case ratios above 2% in relation to this predictor and they were dispersed across the Eastern seaboard, the Midwest, and Southern states. Variations in death-case ratios in response to population with diabetes increased in peak 2 with the highest increase recorded to be 7%. A total of 376 counties (19%) recorded increases in death-case ratios above 3% in this period and were located in the same regions as in peak 1, with new clusters emerging in the Southeast and Southwest. As in the case of coronavirus incidence rates, heart disease mortality rates had only a marginal effect on death-case ratios in both peak periods. The overall increase in incidence rates in response to this predictor was less than

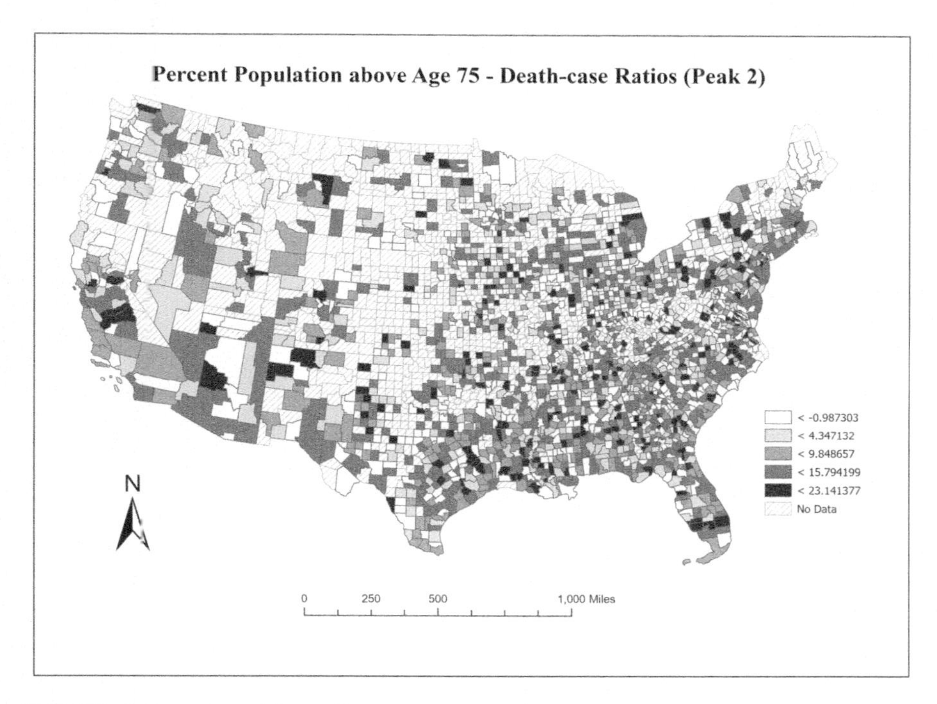

FIGURE 4.12 Map of b coefficients related to percent population above age 75 when dependent variable is COVID-19 death-case ratios for peak period 2 (06/01/20 – 07/31/20).

0.2% in peak 1 and peak 2. An overall increase of 3% in death-case ratios in response to obesity was observed in peak 2 compared to only a marginal increase of 0.7% in peak 1. Counties that recorded increments in death-case ratios greater than 2% in peak 2 in response to obesity were located mostly in the South and Southeast.

The analysis of the coefficients reveals that trends in coronavirus incidence rates and death-case ratios consolidated during the second peak period (06/01/20–07/31/20) as the virus became more widespread throughout the United States. The results of the OLS and GWR analyses also highlight that percent population above age 75 and percent population diagnosed with diabetes were correlated with outcomes of the virus compared to other age cohorts and comorbidities. The spatial patterns of the effects of percent population above age 75 and those with diagnosed diabetes were examined using a cluster-outlier analysis. High-value clusters of incidence rates in peak 2 in response to population above age 75 are located across rural counties in Kansas west of Kansas City as well as in northern Arkansas and West Virginia (Figure 4.13). High-low outliers are present in counties along the mid-Atlantic in Virginia and North Carolina. Other locations of high-low outliers are identified in central and northwestern Louisiana. Increases in death-case ratios in peak 2 in response to population above age 75 are clustered in counties around the metropolitan areas of Kansas City, Nashville, Knoxville, and Houston (Figure 4.14). Several high-value clusters were also present across Florida. High-low outliers are located

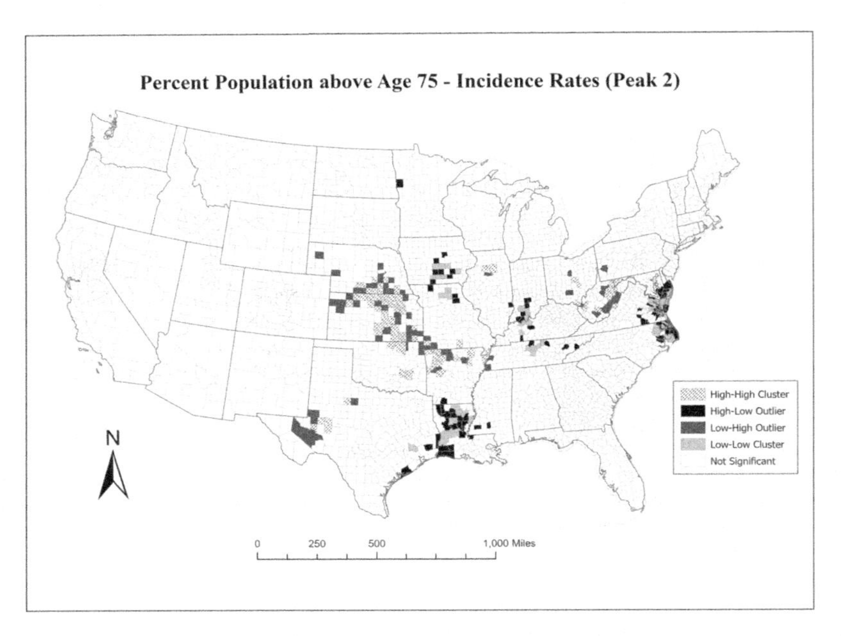

FIGURE 4.13 Cluster-outlier map related to percent population above age 75 when dependent variable is COVID-19 incidence rates for peak period 2 (06/01/20 – 07/31/20).

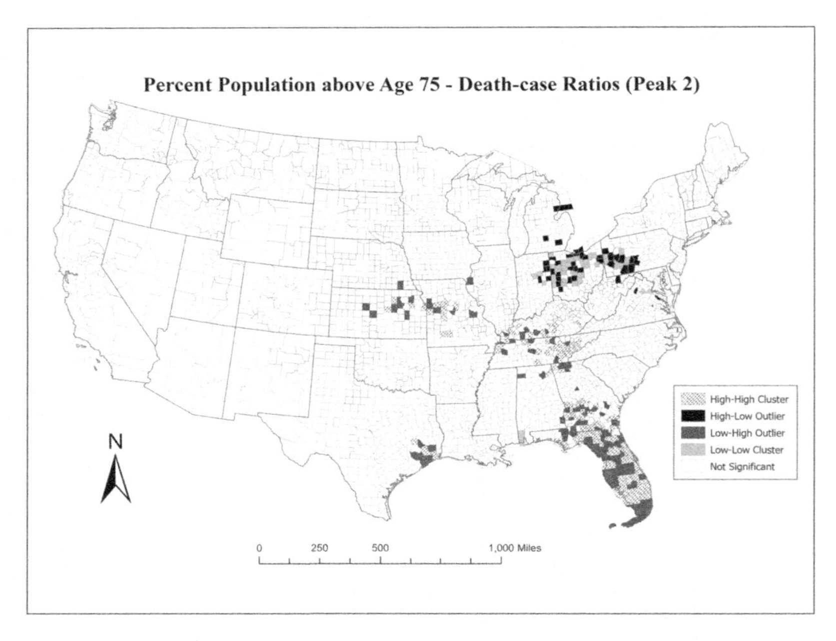

FIGURE 4.14 Cluster-outlier map related to percent population above age 75 when dependent variable is COVID-19 death-case ratios for peak period 2 (06/01/20 – 07/31/20).

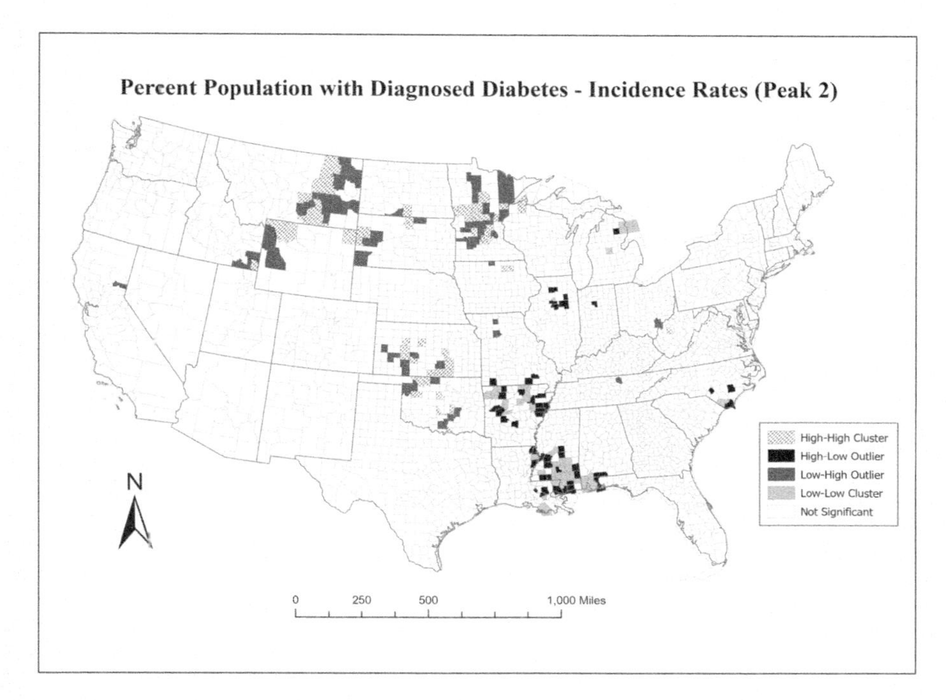

FIGURE 4.15 Cluster-outlier map related to percent population with diagnosed diabetes when dependent variable is COVID-19 incidence rates for peak period 2 (06/01/20 – 07/31/20).

in and around the metropolitan areas of Cleveland, Columbus, and Pittsburgh in the Midwest. Increasing incidence rates in peak 2 in response to population with diabetes are clustered across rural and urban counties in Oklahoma, eastern Montana, and Minnesota (Figure 4.15). High-low outliers are present in counties along the Gulf Coast in Louisiana, Mississippi, and Alabama. Other locations of high-low outliers are identified in Arkansas and North Carolina. Significant clustering of high death-case ratios in peak 2 in response to population with diabetes is observed in counties around the metropolitan areas of New York City and New Jersey and in rural counties located in Texas (Figure 4.16). Several high-low outliers are present in Oregon, Washington, Virginia, and Georgia amid counties that recorded consistently low death-case ratios during this period.

The results indicate that diabetes is a significant risk factor associated with coronavirus disease. The cluster-outlier maps were used to identify a sample set of counties associated with diabetes prevalence and coronavirus disease in peak 2. Table 4.5 provides information on variables associated with these counties that recorded high coronavirus incidence rates and death-case ratios in response to diabetes prevalence. Demographic and health indicators are drawn from the data set used in this study. Classification of counties as urban or rural is based on the US Census Bureau's definition of urbanized areas that consist of census tracts or census blocks with a combined

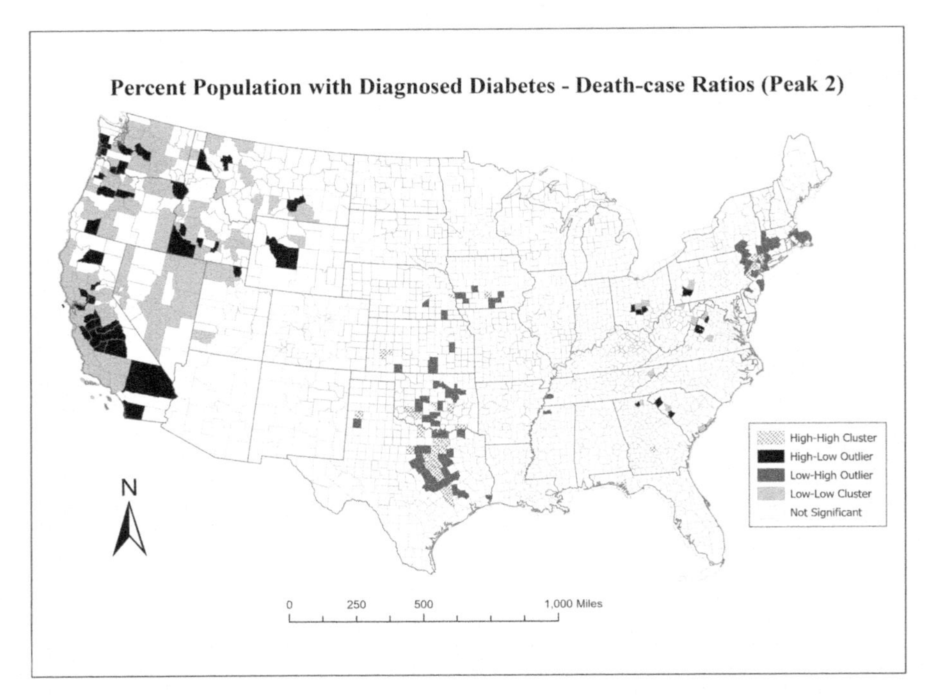

FIGURE 4.16 Cluster-outlier map related to percent population with diagnosed diabetes when dependent variable is COVID-19 death-case ratios for peak period 2 (06/01/20 – 07/31/20).

TABLE 4.5
County-level Analysis of Variables

County	Total Population	Urban / Rural	Incidence Rate	Death-case Ratio	Percent diagnosed with Diabetes	Percent above age 75	Percent Non-White
Big Horn, MT	13,319	R	**2,170**	3.1	16.3	4.4	66.7
McLain, OK	40,474	U/R	**687**	0.36	10.5	5.9	9.0
Carver, MN	105,089	U	**501**	0.19	7.7	4.5	5.5
Lee, AR	8,859	R	**9,721**	0.58	12.6	8.0	57.2
East Carroll Parish, LA	6,861	R	**4,868**	0.30	21.3	6.1	70.0
Wayne, MS	20,183	R	**2,571**	3.85	21.6	7.5	42.0
Hunterdon, NJ	124,371	U	108	**47.8**	5.7	6.9	8.0
Stephens, TX	9,366	R	267	**8.0**	11.8	8.7	6.3
Richmond, NY	476,143	U	258	**6.7**	9.8	6.4	23.7
Kittitas, WA	47,934	U/R	497	**7.1**	6.1	6.0	8.2
Wallowa, OR	7,208	R	236	**5.9**	10.2	12.0	2.6
Page, VA	23,902	R	439	**2.8**	15.6	8.4	3.0

population greater than or equal to 50,000 (USCB, 2010). Minority status is represented by the percentage of non-White population in each county. This indicator was derived from the 2018 American Community Survey (USCB, 2018). The first six counties listed in Table 4.5 recorded very high incidence rates compared to the national median of 417 cases per 100,000 in the second peak period. With the exception of Carver County, Minnesota, and McLain County, Oklahoma, the other locations recorded diabetes prevalence ranging from 12.6% to 21.6%, which was greater than the national average of 10.5%. Four out of the six counties are rural, with more than a third of the population identified as non-White. For example, Lee County, Arkansas, reported an incidence rate of 9,721 per 100,000 and a non-White population of 57.2%.

The rest of the counties listed in Table 4.5 are those that recorded high death-case ratios in response to diabetes prevalence. All six counties in this cluster recorded death-case ratios greater than the national median of 1.63% in peak 2. Unlike with incidence rates, diabetes prevalence had a marginal impact on coronavirus deaths in the counties listed in the table. Two out of the six counties—Page County, Virginia, and Stephens County, Texas—recorded a diabetes prevalence rate greater than the national average of 10.5%. Page County, Virginia, recorded the highest diabetes prevalence rate of 15.6%, but this county experienced the lowest coronavirus death-case ratio of 2.8% among the counties in that cluster. Hunterdon County, New Jersey, recorded the highest death-case ratio of 47.8% and a diabetes prevalence rate of 5.7%. While Hunterdon, New Jersey, and Richmond, New York, are characterized as urban counties, there are differences in the counties' minority status. Close to one-fourth of the population in Richmond, NewYork, is non-White, whereas this demographic accounts for only 8% of the population in Hunterdon, New Jersey. Based on the data from sampled counties highlighted in Table 4.5, there is no clear relationship between minority status and population size (urban/rural) with that of coronavirus deaths associated with diabetes prevalence.

4.5 DISCUSSION OF RISK FACTORS ASSOCIATED WITH COVID-19

The findings of this study reveal that older adults above the age of 75 and persons diagnosed with diabetes are at a higher risk of experiencing negative outcomes in relation to infection and death from COVID-19. These findings are supported by several epidemiological studies that have tracked the progression of COVID-19 during the early phases of the pandemic. Boehmer et al. (2020) reported that the incidence rate among people above age 80 in the United States was on average 328 per 100,000 of the population between May and July 2020. Another study conducted by the CDC found that 78% of deaths due to COVID-19 that occurred between May and August 2020 were among those aged 65 and older (Gold et al., 2020). Similar trends are noted by researchers investigating the global impact of age on COVID-19 incidence and mortality. Eighty percent of coronavirus deaths were reported among people over the age of 70 in Korea and Italy (Kang & Jung, 2020). Epidemiological studies cite the weakening of the immune system among the elderly to be the primary cause of infection and death from COVID-19 among this age cohort (Perrotta et al., 2020; Kang & Jung, 2020). Comorbidities related to hypertension, cardiovascular disease, obesity, and chronic obstructive pulmonary disease (COPD) were found to increase the risk of infection and death from coronavirus exposure (Javanmardi et al., 2020; Sanyaolu et al., 2020; Neumann-Podczaska, 2020). In a study conducted by Ejaz et al. (2020), it was observed that frequency of coronavirus infection increases by 58% among people with diabetes as a preexisting condition, while obesity increased the risk of severe illness from COVID-19 by 68%. As reported by the CDC, 10.2% of the US population (approximately 27 million) was diagnosed with diabetes based on 2018 estimates (CDC, 2020c). Regional disparities exist in these indicators with states in the South and Southeast recording above-average percentages of people diagnosed with diabetes ranging from 12% to 33% (CDC, 2020c). In the United States, the mortality rate of COVID-19 among persons with diabetes was the highest, at 58%, compared to rates in China and Italy (Ejaz et al., 2020). Comorbidities related to diabetes and obesity are likely to have significant implications in the diagnosis and treatment of coronavirus infections as the pandemic evolves with the emergence of new variants.

The coefficient maps and the cluster-outlier analysis revealed that, in general, population density and rural–urban distinctions determine spatial patterns of COVID-19 incidence and death-case ratios. High clusters of incidence rates in response to percent population above age 75 and those diagnosed with diabetes were observed mostly in rural counties in the South, Southeast, and Midwest. High-value death-case ratios, on the other hand, were recorded in and around major metropolitan areas across the country especially during the second peak period. While these trends provide a general overview of the geography of the virus, there are other factors like mobility, lockdowns, and other local restrictions that may explain the presence of high and low outliers related to coronavirus incidence and mortality in specific locations that are beyond the scope of this study.

GWR functions as a useful method to analyze local relationships between age, comorbidities, and coronavirus disease. It generates better model-fit statistics and parameter estimates compared to the OLS regression analysis. Limitations associated with GWR include assumptions related to model specification and local multicollinearity, which were addressed in this study. The Gaussian model type used in this case necessitated a log transformation of the dependent variable to address skewness. The resulting coefficients can be difficult to interpret due to this transformation. To address this issue, the *b* coefficients were back-transformed in order to accurately examine the effects of the predictors on incidence rates and death-case ratios. Local multicollinearity is affected by the size of the neighborhood (bandwidth) and the robustness of the model represented by the AIC value. In general, smaller neighborhood sizes tend to increase condition numbers, an indication that multicollinearity is increasing. On the other hand, larger bandwidths can minimize multicollinearity, while increasing AIC values as the model moves from the local to the global scale (ESRI, 2021). A major concern of local multicollinearity is that it produces unreliable coefficient estimates, but scholars have also pointed out that collinearity is not more pervasive in GWR than in other traditional regression models (Oshan et al., 2019). In this study, an effort was made to select the most efficient bandwidths for each iteration of the GWR based on the varying sample sizes for each peak period. Despite the efforts to specify a model that minimizes local multicollinearity, the condition numbers indicate this phenomenon still exists. As such, it is recommended that the results be adopted purely for exploring general relationships between age, comorbidities, and coronavirus trends and not be used for the purpose of predicting future trajectories of the virus.

4.6 CONCLUSION

The study discussed the process and outcomes of a county-level assessment of COVID-19 in relation to age demographics and comorbidities using GWR. Several challenges related to data management and model specification had to be addressed at the outset to carry out the analysis. These included cleaning up the data set by removing missing and zero values, implementing log transformation of the dependent variable, and defining the neighborhood sizes for each iteration of the model. Out of the five predictors used in the analysis, older adults above the age of 75 and persons diagnosed with diabetes produced significant results in terms of their effects on coronavirus incidence rates and death-case ratios. Data visualization of results highlighted regional disparities in the clustering of COVID-19 infections and deaths that are closely tied to population density and urban versus rural classification of counties.

A review of incidence rates in peak 2 revealed that counties with above-average diabetes prevalence experienced very high coronavirus case rates with clusters emerging in rural counties located in the South, Mountain West, and Midwest regions. Significant increases in death-case ratios in response to diabetes prevalence were recorded in urban counties located in the Northeast, Southwest, and Southeast regions. Heart disease mortality rates and obesity had only marginal effects on coronavirus incidence rates and death-case ratios in both peak periods. Small increases in incidence rates in response to heart disease mortality were observed throughout the country, and increments in death-case ratios in response to obesity were located mostly in the South and Southeast. The results provide a preliminary assessment of the spatial and temporal trends of the early stages of the pandemic and serve as a launching pad for future investigation into the vulnerabilities of exposed populations to the virus.

REFERENCES

Adger, W. N. (2006). Vulnerability. *Global Environmental Change, 16*(3), 268–281.

Bland, M. (2000). *An introduction to medical statistics* (3rd ed.). Oxford: Oxford University Press.

Boehmer, T. K., DeVies, J., Caruso, E., van Santen, K. L., Tang, S., Black, C. L., Hartnett, K. P., Kite-Powell, A., Dietz, S., Lozier, M., & Gundlapalli, A. V. 2020. Changing age distribution of the COVID-19 pandemic—United States, May–August 2020. *Morbidity and Mortality Weekly Report, 69*(39), 1404–1409.

Brunton, L. A., Alexander, N., Wint, W., Ashton, A., & Broughan, J. M. (2017). Using geographically weighted regression to explore the spatially heterogeneous spread of bovine tuberculosis in England and Wales. *Stochastic Environmental Research and Risk Assessment, 31*, 339–352. https://doi.org/10.1007/s00477-016-1320-9

Centers for Disease Control and Prevention (CDC). (2012). *Principles of epidemiology in public health practice: An introduction to applied epidemiology and biostatistics.* www.cdc.gov/csels/dsepd/ss1978/ss1978.pdf.

Centers for Disease Control and Prevention (CDC). (2020a). Coronavirus Disease 2019 (COVID- 19). www.cdc.gov/coronavirus/2019-ncov/covid-data/investigations- discovery/hospitalization-death-by-age.html

Centers for Disease Control and Prevention (CDC). (2020b). Coronavirus Disease 2019 (COVID- 19). www.cdc.gov/coronavirus/2019-ncov/need-extra-precautions/people-with- medical-conditions.html.

Centers for Disease Control and Prevention (CDC). (2020c). *National Diabetes Statistics Report, 2020*. US Department of Health and Human Services. www.cdc.gov/diabetes/pdfs/data/statistics/national-diabetes-statistics-report.pdf

Centers for Disease Control and Prevention (CDC). (2021a). Symptoms of COVID-19. www.cdc.gov/coronavirus/2019-ncov/symptoms-testing/symptoms.html

Centers for Disease Control and Prevention (CDC). (2021b). COVID Data Tracker Weekly Review. www.cdc.gov/coronavirus/2019-ncov/covid-data/covidview/index.html

Charlton, M., & Stewart Fotheringham, A. (2009). *Geographically weighted regression.* White Paper. Maynooth, Co Kildare, Ireland: National University of Ireland.

Chen, R., Liang, W., Jiang, M., Guan, W., Zhan, C., Wang, T., Tang, C., Sang, L., Liu, J., Ni, Z., Hu, Y., Liu, L., Shan, H., Lei, C., Peng, Y., Wei, L., Liu, Y., Hu, Y., Peng, P., Wang, J. et al. (2020). Risk factors of fatal outcome in hospitalized subjects with coronavirus disease 2019 from a nationwide analysis in China. *Chest Infections, 158*(1), 97–105.

Cutter, S. L. (1996). Vulnerability to environmental hazards. *Progress in Human Geography, 20*(4), 529–539.

Cutter, S. L., Mitchell, J. T., & M. S. Scott. 2000. Revealing the vulnerability of people and places: a case study of Georgetown county, South Carolina. *Annals of the Association of American Geographers, 90*(4), 713–737.

Ejaz, H., Alsrhani, A., Zafar, A., Javed, H., Junaid, K., Abdalla, A E., Abosalif, K O. A., Ahmed, Z., & Younas, S. (2020). COVID-19 and comorbidities: deleterious impact on infected patients. *Journal of Infection and Public Health, 13*, 1833–1839. https://doi.org/10.1016/j.jiph.2020.07.014

Environmental Systems Research Institute (ESRI). How Geographically Weighted Regression (GWR) Works. https://pro.arcgis.com/en/pro- app/latest/tool-reference/spatial-statistics/how-geographicallyweightedregression- works.htm

Fotheringham, A. S., Brunsdon, C., & Charlton, M. (2002). *Geographically weighted regression: The analysis of spatially varying relationships.* West Sussex, UK: John Wiley & Sons.

Glass, G. E. (2000). Update: Spatial aspects of epidemiology: The interface with medical geography. *Epidemiologic Reviews, 22*(1), 136–139.

Gold, J. A.W., Rossen, L. M., Ahmad, F. B., Sutton, P., Li, Z., Salvatore, P. P., Coyle, J. P., DeCuir, J., Baack, B. N., Durant, T. M., Dominguez, K. L., Henley, S. J., Annor, F. B., Fuld, J., Dee, D. L., Bhattarai, A., & Jackson, B. R. (2020). Race, ethnicity, and age trends in persons who died from COVID-19—United States, May–August 2020. *Morbidity and Mortality Weekly Report, 69*, 1517–1521. http://dx.doi.org/10.15585/mmwr.mm6942e1

Hewitt, K., & Burton, I. (1971). *The hazardousness of a place: A regional ecology of damaging events.* Toronto: University of Toronto Press.

Javanmardi, F., Keshavarzi, A., Akbari, A., Emami, A., & Pirbonyeh, N. (2020). Prevalence of underlying diseases in died cases of COVID-19: A systematic review of meta-analysis. *PLoS One, 15*(10), e0241265. https://doi.org/10.1371/journal.pone.0241265

John's Hopkins University and Medicine (JHU). (2021). COVID-19 Dashboard. Last modified September 24. https://coronavirus.jhu.edu/map.html.

Jones, K., & Moon, G. (1993). Medical geography: Taking space seriously. *Progress in Human Geography, 17*(4), 515–524.

Kanakaratne, N., Wahala, W. M. P. B., Messer, W. B., Tissera, H. A., Shahani, A., Abeysinghe, N., de Silva, A. M., & Gunasekera, M. (2009). Severe dengue epidemics in Sri Lanka, 2003–2006. *Emerging Infectious Diseases, 15*(2), 192–199. https://dx.doi.org/10.3201/eid1502.080926

Kang, S.-J., & Jung, S. I. (2020). Age-related morbidity and mortality among patients with COVID-19. *Infection and Chemotherapy, 52*(2), 154–164. https://doi.org/10.3947/ic.2020.52.2.154

Keesing, F., Holt, R. D., & Ostfeld, R. S. (2006). Effects of species diversity on disease risk. *Ecology Letters, 9*, 485–498. https://doi.org/10.1111/j.1461-0248.2006.00885.x

Koelle, K., Rodó, X., Pascual, M., Yunus, M., & Mostafa, G. (2005). Refractory periods and climate forcing in cholera dynamics. *Nature, 436*(4), 696–700. https://doi.org/:10.1038/nature03820

Manikandan, S. (2010). Preparing to analyze data. *Journal of Pharmacology & Pharmacotherapeutics, 1*(1), 64–65. https://doi.og/10.4103/0976-500X.64540

Mayfield, H. J., Lowry, J. H., Watson, C. H., Kama, M., Nilles, E. J., & Lau, C. L. (2018). Use of geographically weighted logistic regression to quantify spatial variation in the environmental and sociodemographic drivers of leptospirosis in Fiji: A modelling study. *Lancet Planet Health, 2*, e223–232.

Meade, M. S., & Emch, M. (2010). *Medical Geography* (3rd ed.). New York: Guilford Press.

Montz, B. E., Tobin, G. A., & Hagelman III, R. R. (2017). *Natural hazards: Explanation and integration* (2nd ed.). New York: Guilford Press.

Mueller, A. L., McNamara, M. S., & Sinclair, D. A. (2020). Why does COVID-19 disproportionately affect older people? *Aging, 12*(10), 9959–9981.

Nakaya, T., Fotheringham, A. S., Brunsdon, C., & Charlton, M. (2005). Geographically weighted Poisson regression for disease association mapping. *Statistics in Medicine, 24(17)*, 2695–2717.

Ndiath, M. M., Cisse, B., Ndiaye, J. L., Gomis, J. F., Bathiery, O., Dia, A. T., Gaye, O., & Faye, B. (2015). Application of geographically-weighted regression analysis to assess risk factors for malaria hotspots in Keur Soce health and demographic surveillance site. *Malaria Journal, 14*, 463–473. https://doi.org/10.1186/s12936-015-0976-9

Neumann-Podczaska, A., Al-Saad, S. R., Karbowski, L. M., Chojnicki, M., Tobis, S., & Wieczorowska-Tobis, K. (2020). COVID 19—clinical picture in the elderly population: A qualitative systematic review. *Aging and Disease, 11*(4), 988–1008. http://dx.doi.org/10.14336/AD.2020.0620

Novosad, P., Jain, R., Campion, A., & Asher, S. (2020). COVID-19 mortality effects of underlying health conditions in India: A modelling study. *BMJ Open, 10*, e043165. https://doi.org/:10.1136/bmjopen-2020-043165

Oshan, T. M., Li, Z., Kang, W., Wolf, L. J., & Fotheringham, A. S. (2019). MGWR: A Python implementation of multiscale geographically weighted regression for investigating process spatial heterogeneity and scale. *International Journal of Geo- Information, 8*, 269–299.

Oyana, T. J., & Margai, F. M. (2010). Spatial patterns and health disparities in pediatric lead exposure in Chicago: Characteristics and profiles of high-risk neighborhoods. *The Professional Geographer, 62*(1), 46–65. https://doi.org/10.1080/00330120903375894

Páez, A., Farber, S., & Wheeler, D. (2011). A simulation-based study of geographically weighted regression as a method for investigating spatially varying relationships. *Environment and Planning A, 43*(12), 2992–3010.

Pearce, N., Vandenbroucke, J. P., VanderWeele, T. J., & Greenland, S. (2020). Accurate statistics on COVID-19 are essential for policy guidance and decisions. *American Journal of Public Health, 110*(7): 949–951.

Perrotta, F., Corbi, G., Mazzeo, G., Boccia, M., Aronne, L., D'Agnano, V., Komici, K., Mazzarella, G., Parrella, R., & Bianco, A. (2020). COVID-19 and the elderly: Insights into pathogenesis and clinical decision-making. *Aging Clinical and Experimental Research*, 32, 1599–1608. https://doi.org/10.1007/s40520-020-01631-y.

Rogerson, P. A. (2015). *Statistical methods in geography: A student's guide* (4th ed.). London: SAGE.

Sanyaolu, A., Okorie, C., Marinkovic, A., Patidar, R., Younis, K., Desai, P., Hosein, Z., Padda, I., Mangat, J., & Altaf, M. 2020. Comorbidity and its impact on patients with COVID-19. *SN Comprehensive Clinical Medicine, 2*, 1069–1076. https://doi.org/10.1007/s42399-020-00363-4

Siriyasatien,., Phumee, A., Ongruk, P., Jampachaisri, K., & Kesorn, K. (2016). Analysis of significant factors for dengue fever incidence prediction. *BMC Bioinformatics, 17*, 166–174. https://doi.org/10.1186/s12859-016-1034-5

Sutherst, R. W. (2004). Global Change and Human Vulnerability to Vector-Borne Diseases. *Clinical Microbiology Reviews, 17*(1), 136–173.

Tam, C. C., Tissera, H., de Silva, A. M., De Silva, A. D., Margolis, H. S., & Amarasinge, A. (2013). Estimates of dengue force of infection in children in Colombo, Sri Lanka. *PLoS Neglected Tropical Diseases, 7*(6), e2259–e2265. https://doi.org/10.1371/journal.pntd.0002259

Thomson, M. C., Mason, S. J., Phindela, T.., & Connor, S. J. (2005). Use of rainfall and sea surface temperature monitoring for malaria early warning in Botswana. *American Journal of Tropical Medicine, 73*(1), 214–221.

United States Census Bureau (USCB). (2010). 2010 Census Urban and Rural Classification and Urban Area Criteria. Last modified October 8, 2021. www.census.gov/programs-surveys/geography/guidance/geo-areas/urban-rural/2010-urban-rural.html

United States Census Bureau (USCB). (2018). DP05 ACS Demographic and Housing Estimates. American Community Survey. www.census.gov/data/developers/data-sets/acs- 5year.2018.html

Weeraratne, T. C., Perera, M. D. B., Mansoor, M. A. C. M., & Parakrama Karunaratne, S.H.P. (2013). Prevalence and breeding habitats of the dengue vectors aedes aegypti and aedes albopictus (diptera: culicidae) in the semi-urban areas of two different climatic zones in Sri Lanka. *International Journal of Tropical Insect Science, 33*(4), 216–226. https://doi.org/10.1017/S174275841300026X

Wisner, B., Blaikie, P., Cannon, T., & Davis, I. (2004). *At risk: natural hazards, people's vulnerability and disasters* (2nd ed.). London: Routledge.

World Health Organization (WHO). (2003). *Guidelines for dengue surveillance and mosquito control* (2nd ed.). Manila, Philippines: WHO.

World Health Organization (WHO). (2021). WHO Coronavirus (COVID-19) Dashboard. Last modified September 23. https://covid19.who.int/

5 Spatiotemporal Patterns of COVID-19

A District-Level Analysis of Kerala, India, 2020–2021

Madhuri Sharma, Shimod Kunduparambil, and Rajesh Kumar Abhay

CONTENTS

5.1 INTRODUCTION

The whole world was challenged by the global pandemic of COVID-19 (Gouda et al., 2020), first reported on December 8, 2019, in the City of Wuhan in Hubei Province of China (Carrat et al., 2021). While the global community was under the tight grip of the novel coronavirus pandemic for a majority of 2020–2021, and even though we are now in 2023, the world seems to be dealing with newer variants of COVID-19 such as Delta, Omicron and now BA.2 Omicron. This unknown virus has posed incredible threats to mankind all over the world (Mondal & Ghosh, 2020), especially because of the very nature of this disease due to airborne carriers. It spreads fast from person-to-person contact and/or proximity of space, if people are not separated by significant distance. Across almost the entire the world, the short- and long-term impacts of this

DOI: 10.1201/9781003227106-5

virus were so deeply felt that it caused the entire health infrastructure in the richest countries almost to crumble, and had heartbreaking damaging impacts felt especially in developing economies such as that of India. This disease took the entire world through a phase of economic, demographic, and societal changes that had never been seen in the past ten decades. Especially the first quarter of the year 2021 displayed the serious shape of COVID-19 infections in different parts of the world, along with the first published scholarly research (Table 5.1). As noted in this table, while the first virus was found in Wuhan, China, the instances of first cases of COVID-19 were noted during a wide period of December 2019 to June of 2020 in the ten countries of the world that showed the highest numbers of cases. While the whole world was under the grip of the pandemic at different levels, approximately three-fourths of the total cases were reported from 11 countries, namely, the United States, Brazil, India, Russia, Spain, Peru, Mexico, Chile, South Africa, Colombia, and Iran (Bag et al., 2020; Gouda et al., 2020; Palamim et al., 2020; Carrat et al., 2021; Fong et al., 2021; Arashi et al., 2021; Zinchenko, 2021).

COVID-19 has changed the normal functioning of society in almost every country in the world. It has had a multiplier effect on the different facets of life, and has been challenging our everyday capacity for managing public health and society at large (Liu, 2020; Morgan et al., 2021). These widespread cases of infections in almost every urban and rural space across the global community, and the fear of its spread through air and contagion/spatial proximity, have greatly disrupted everyday survival of the larger population. In terms of its impacts on economy, it has not only affected supply chains, the financial markets, and consumer demand at local, regional, national, and global scales, but also various economic sectors like tourism and transportation, and the global value-chain systems due to the globally integrated community that we are a part of (Mishra, 2020).

In terms of academic scholarship, while a large share of recent work has addressed a variety of issues pertaining to the social impacts of COVID-19, the pandemic has had

TABLE 5.1
Time Line of the Reported First Case of COVID-19 in Major Countries/States

S. No.	Country	First Reported Case
1.	China	December 8, 2019 (Carrat et al., 2021)
2.	France	December 27, 2019 (Carrat et al., 2021)
3.	South Korea	January 20, 2020 (Fong et al., 2021)
4.	USA	January 21, 2020 (Fong et al., 2021)
5	India/Kerala	January 30, 2020 (Gouda et al., 2020)
6.	UK	January 31, 2020 (Sartorius et al., 2021)
7.	Russia	January 2020 (Zinchenko et al., 2021)
8.	Egypt	February 14, 2020 (Arashi et al., 2021)
9.	Iran	February 19, 2020 (Fong et al., 2021)
10.	Brazil	June 5, 2020 (Palamim et al., 2020)

Source: Compiled from different sources by authors

short- and long-term impacts on almost every aspect of our social and economic lives. COVID-19 and the ensuing lockdowns led to enormous changes in the social lifestyle, impacting education (schools and colleges alike, but schools have experienced the greatest impacts by increasing the workloads on women by multiple folds), health infrastructure, and health-care concerns at short-term and long-term scales that have greatly affected the mental health of people belonging to all age groups (Zinchenko, 2021). Reverse migration of migrants from urban to rural areas, especially from the largest metropolises and urban areas of India, also have posed serious issues regarding the security of the livelihood of workers, a large share of whom are employed in informal economies of various sorts (Khan & Arokkiaraj, 2021). Besides enormous negative impacts, COVID-19 also had many positive outcomes, though. With reference to environmental sustainability, major cities in India, especially Delhi, experienced improvements in air quality due to strict restrictions imposed during the lockdown as the emissions of polluted air from industries and vehicles were limited to the minimum, leading to a decline in overall levels of water and air pollution (Rupani et al., 2020; Morgan et al., 2021).

While the entire world was engrossed in finding a variety of ways to research and fund a cure for this pandemic, the State of Kerala in India was going through a unique phase of COVID-19 time-series occurrence. The first COVID-19 case in India was reported as early as January 30, 2020, from Kerala. Since February of 2020, the news of COVID-19 cases gradually started accumulating from the global community, whereas in the state of Kerala, the cases kept rising without much effort being taken to contain it through measures. While the number of cases was negligible by February of 2020 in India, by March 15, it had expanded rapidly in the states of Tamil Nadu, Rajasthan, Uttar Pradesh, Haryana, Jammu, Maharashtra, Delhi, Punjab, and Karnataka (Gouda et al., 2020). Kerala, the first state in India to have attained 100% literacy in the early-to-late 1980s, had continued to rank one of the highest in terms of numbers of COVID-19 infection cases per 1000 people. In our review of the pertinent literature on this newly developing line of research (summarized in Table 5.2 below), we did not find any scholarly focus on district-scale analysis and the potential causes for such patterns in the State of Kerala. This chapter specifically maps and analyzes the spatial and temporal spread of COVID-19 infections across the 14 districts of Kerala from January 2020 to December 2021—a period of 24 months—and expands our understanding of the potential reasons for such high levels of infections and deaths. In doing so, our aim is to provide possible policy measures such that in the face of any newer strains of COVID-19 virus, or potential pandemic concerns, one could be better prepared to address such a mass disaster.

5.2 LITERATURE REVIEW

5.2.1 Brief Overview of COVID-19 Scholarship—the Global Context and the Indian Context

COVID-19-focused academic research has so far concentrated on the spatiotemporal dimensions of its spread, expansion, and regional disparities on a global scale. Chen et al. (2021), for example, analyzed the spatial distribution of COVID-19 infections over time in China, using province as the scale of analysis. The variables analyzed included the prevalence rate, recovery rate, and the mortality rates during the period of January 20 to March 12, 2021. Sartorius et al. (2021) used the methods of statistical modeling to predict the spatiotemporal spread of COVID-19 cases in cities in England. This study examined the associated deaths and the risk factors related to the spread of COVID-19. Silva et al. (2021) analyzed real-time forecasting of COVID-19 cases in Brazil for the period of May 25–27, 2020, whereas Palamim (2020) considered COVID-19 cases among the indigenous people of Brazil. The interesting factor in all these analyses is that given the very novel characteristics of the virus, not much was known except the number of infections and its geographic spread in the world. As such, much of this scholarship largely reported the numbers of cases and their spread in terms of temporal and spatial capture.

A few studies analyzed the spatiotemporal pattern of the COVID-19 outbreak in India at the national and state levels (Bag et al., 2020; Basu & Mazumdar, 2021; Bhunia et al., 2021; Gounda, 2020). India was chosen for study by these scholars because of its wide regional diversity along with many dimensions (Basu & Mazumdar, 2021), which made it possible to look for potential reasons for differential patterns. These studies involved the identification of epicenters with high incidence rates of COVID-19. The analysis showed that the spread of COVID-19 was trending toward the eastern parts of the country, starting from Mumbai and largely the western cities in the country, but that the concentration of cases was still clustered in the western part of the country (Bag et al., 2020). Numerous cities in the western part of the country comprise highly industrialized zones (e.g., Maharashtra, Gujarat, Karnataka), with many large and midsized metropolises, with regular transit of incoming tourists and visitors from various countries abroad. This could have contributed to the unchecked inflow of infected people. A study carried out by Bhunia et al. (2021) found that disease clusters were mostly concentrated in the central and western states of India, as indicated above. On the other hand, northeastern states—a region of the country that is also geographically isolated from the rest of the peninsula—experienced far lower rates of disease clusters. To confirm this pattern, indeed the study conducted by Gouda (2020) concluded that the onset of COVID-19 in a different parts of India occurred due to people with travel history from outside of India, largely the Middle Eastern countries, and that there had been an exponential growth of such cases during April–May 2020 in several cities of India (e.g., Mumbai, Bengaluru, and several cities in the State of Kerala).

Regarding regional disparity in COVID-19 infections, using state as the scale of analysis, Basu and Mazumdar (2021) established that COVID-19 cases were more prevalent in urbanized areas with higher development levels. Further, it was also found that poor people were less susceptible to COVID-19 infections in India, whereas these cases were high in developed countries like the United States and the United Kingdom. Though this was largely the empirical pattern, especially during the first wave of COVID-19, the exact reasons for such regional and global patterns were difficult to derive. This state-level regional disparity in COVID-19 infections in India was also observed by Mandi et al. (2020), which also prompted a direction for sequential lifting of the lockdown. Ray and Subramanian (2020) also noted similar types of regional disparity in infections, although their key objective was to provide an interim report on the Indian lockdown provoked by the COVID-19 pandemic.

Of the many studies conducted so far on the issue of COVID-19, very few have focused on a state-level analysis in India. Jayesh and Sreedharan (2020) analyzed the spread of COVID-19 cases in Kerala, and the measures adopted by the Kerala government to combat the pandemic. Chathukulam and Tharamangalam (2021) considered the Kerala model that was applied to manage the pandemic. This study examined the institutions, provision of health-care services, welfare and safety of people and public actions, and the like to examine their effectiveness in containing the pandemic. This study concluded that the Kerala model was still relevant and successful in coping with the COVID-19 disaster. Elias (2021) examined the system thinking process by analyzing the causal loop modeling in Kerala.

5.2.2 Empirical Coverage in COVID-19 Analysis

While a reasonable number of studies focused on COVID-19 examined its occurrence and its spatial distribution at various levels, our review of the literature suggests that the empirical coverage of the studies ranged from country level to state level to district level. Further, we also found that the country-level studies were far greater in number than those at the interstate or the intrastate level—presenting a finer scale of analysis. Our review of the literature summarized in this chapter shows a lopsided focus on country- or regional-state-level analyses rather than a finer scale, which gathers better perspectives of local geographies (see our compilation of the literature in Table 5.2). The reasons for this could be attributed to the difficulties involved in gathering COVID-19 data at the

TABLE 5.2
Empirical Coverage of the Studies Conducted on the Spatiotemporal Analysis COVID-19

S. No.	References	Study Area	Empirical Scale of Analysis
1.	Bhunia et al. (2020); Bag et al. (2020);	India	National/Country
	Fang et al. (2020);	Russia	
	Palamim et al. (2020);	Brazil	
	Arashi et al. (2021);	South Africa	
	Carrat et al. (2021);	France	
	Chen (2021);	China	
	Zinchenko (2021)	Russia	
2.	Gouda et al. (2020);	India	Interstate/Province
	Basu & Mazumder (2021)	India	
	Silva et al. (2021);	Brazil	
	Sartorius (2021)	UK	
3.	Jayesh & Sreedharan (2020);	Kerala	Individual State/ Interdistrict
	Elias (2021);		
	Chathukulam & Tharamangalam (2021)		

Source: Compiled from different sources by authors

level of districts and cities, especially when we are considering a developing economy such as that of India. The state/province-level data are easily available. However, fewer studies appear to be at the individual district level or the interdistrict level. To bridge up this gap in the existing literature focused on COVID-19 infections and the relevant attributes, this chapter makes a special contribution by conducting a time-series analysis of its inter-district spatial pattern for 24 months in Kerala.

5.2.3 Methodological Approaches to COVID-19 Analysis

Different types of methods have been adopted by different scholars to understand and explain the regional patterns and disparities in the spatiotemporal patterns of COVID-19 infections and their expansion. Various geographic information system (GIS) techniques and statistical predictive models have been employed by researchers to explain the spatial patterns and future temporal trends of COVID-19 cases, its distribution, expansion, and movement trends, along with predictive modeling of number of deaths, and the like. The benefits of such models have largely been about the precautions and the science behind the patterns such that preventive and control measures could be instituted at different scales to contain the virus.

The use of GIS techniques in analyzing the spatiotemporal pattern of the COVID-19 outbreak in India have been widely applied. Using GIS techniques, patterns of spatial clustering of COVID-19 infections, its hot spots, and spatial statistics have been identified. Moreover, the polynomial regression model has also been applied to predict the COVID-19 affected population and associated deaths (Bag et al., 2020). Different statistical algorithms have been employed to analyze the growth rates of COVID-19 cases, such as the exponential trends and logarithmic values (Gouda et al., 2020). Bhunia et al. (2021) used Voronoi statistics to identify the affected states from a series of polygons. Spatial patterns are explained by global spatial autocorrelation techniques. Further, a local spatial autocorrelation has also been applied using statistical methods (Arashi et al., 2021).

Mandi et al. (2020) constructed a multidimensional vulnerability index to capture the major reasons for infections, and how they could serve the purposes of a directive toward sequential lifting of the lockdown that was implemented all over India—across states and several districts on a phase-by-phase basis depending on the infection rates. Besides India, many other countries have also been studied by researchers who employed various techniques depending upon the quality and the scale of data available to them. Chen et al. (2021) applied spatial autocorrelation with Moran's-I; thereafter a simple correlation between population flow and the number of confirmed cases was applied to analyze the distribution characteristics of COVID-19 cases and influencing factors in China. Sartorius et al. (2021) employed the Bayesian hierarchical space-time SEIR model for COVID-19 transmission and deaths and their spatiotemporal variability on a weekly basis at a small scale in England. Silva et al. (2021) used linear regression modeling and artificial neural networks-based GIS techniques for monitoring real-time spatiotemporal forecasting in Brazil. Thus, while the above scholarship employed a wide range of sophisticated methods and tools to examine various aspects of COVID-19 parameters, the crux of methodological implementation largely depends also on the scale and the quality of data available for a scale/country. In this chapter, given the focus on interdistrict analysis of COVID-19 infections and related parameters, we employ GIS mapping tools along with simple statistical analysis to provide an interdistrict perspective on the temporal and spatial patterns of the disease. The following section on research design illustrates the steps we employed in conducting the gathered data and the methods that we employed in deriving the findings for the districts of Kerala, India.

5.3 RESEARCH DESIGN

5.3.1 Study Area

Kerala, known as "God's own country" and the "gateway of monsoon in India" is truly a land of eternal bliss and a tropical Eden with the fascinating beauty of its sun-bathed golden seashores edged with abundant coconut trees, the rocky terrain of the Western Ghats, straggling plantations and paddy fields, the cerulean lagoons and the bountiful rivers and mighty waterfalls, and the fascinating biodiversity of its flora and fauna. Kerala stands unique among Indian states with the status of the first state in the country to have achieved the 100% literacy rate. This state has also demonstrated a consistently higher level of human development, and access to highly improved health-care facilities. Despite lower per capita income, the state is comparable to those of many developed countries. Kerala, located on the southwestern tip of India, extends over 560 km (Figure 5.1), and given its physiographic features, it is divided into three distinct regions: hills and valleys, midland plains, and coastal region.

Over the last six decades, Kerala has consistently been a prominent outlier with regard to its health-care system and health outcomes compared to most of the other states in India. This can be attributed to the strong emphasis from the state government on public health care and the primary health-care system and investment in its infrastructure development. As per the NITI Aayog 2019-20, Kerala ranks first in the Health Index for the fourth time successively. Given the first instance of a COVID-19 case in Kerala, which subsequently kept on capturing its spatial spread, we decided to follow a protocol to conduct this analysis. While the State of Kerala gradually started gathering and providing data on COVID-19 cases to keep its larger population informed about the dangers of this potentially less studied virus, we also started looking for any relevant scholarly work or reports published by the government or private sources. Along with scholarly research and the granularity of available data in the State of Kerala, we started doing preliminary mapping analyses to get a grasp of the spatial patterns of the spread of this infection. When the maps started telling a story about its spread and outcomes, we began doing more focused research on solidifying these processes to get more substantive analyses. The flow chart (Figure 5.2) illustrates our steps involved in completing this research. The advent of this pandemic captured the attention of the entire nation, especially given the high rank and accolades the state had enjoyed with regard to its health-care management and practices. As such, this chapter is an illustration of the spatial and temporal analysis of the interdistrict level of COVID-19 infections and the related characteristics.

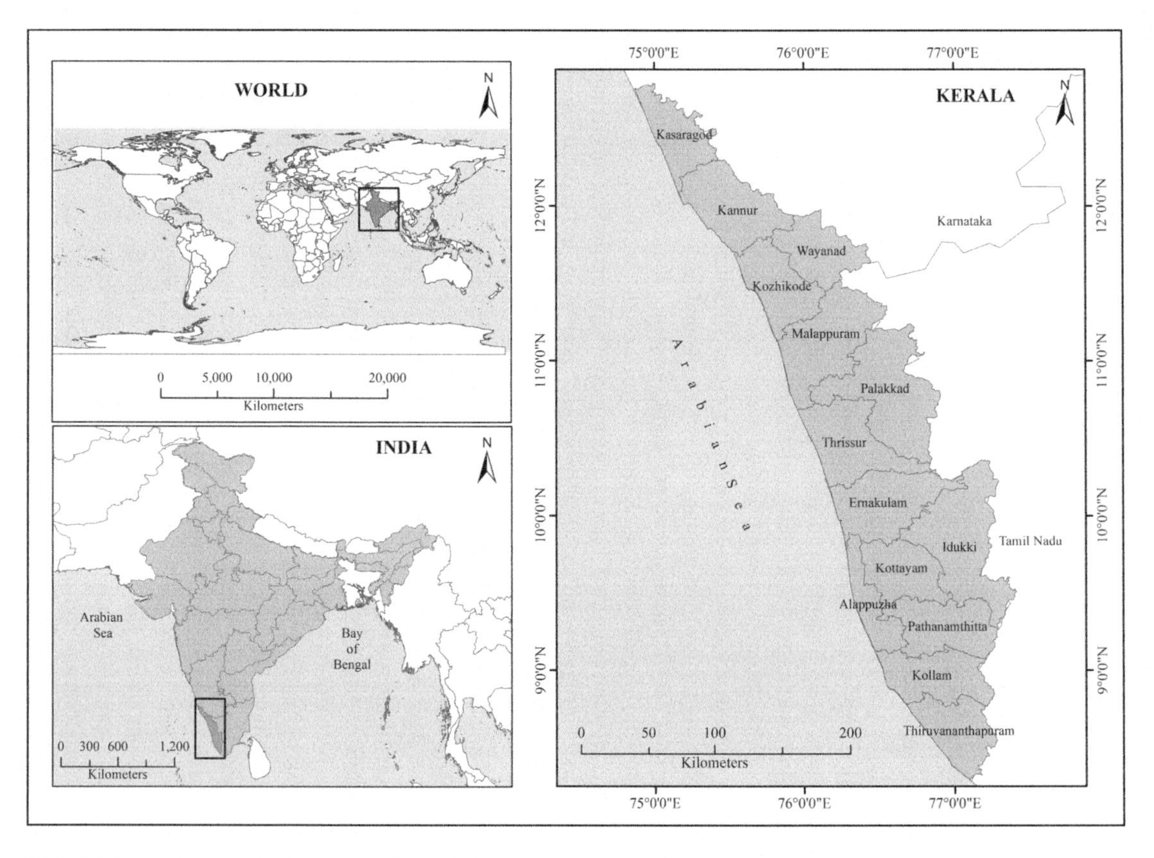

FIGURE 5.1 Geographical location of the study area.

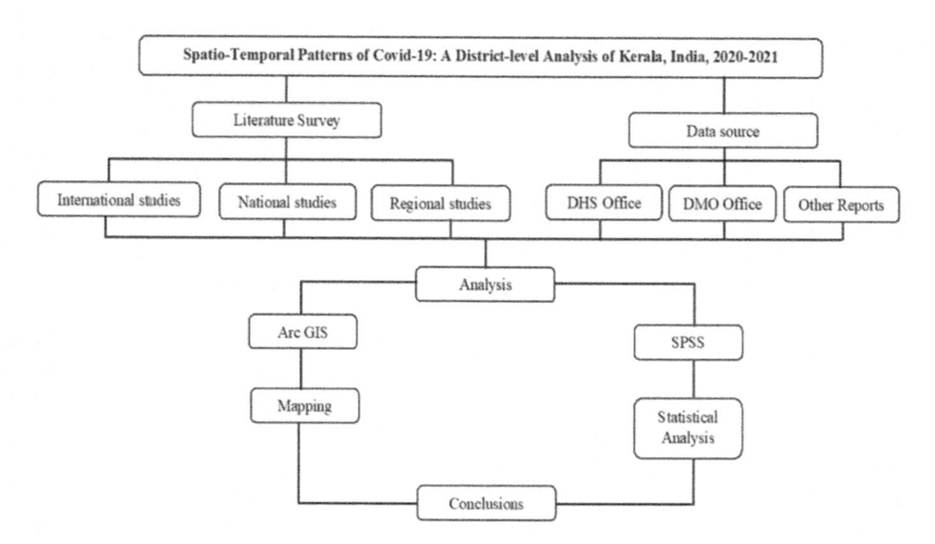

FIGURE 5.2 Flow chart of the research methodology.

5.3.2 Data Sources, Scale of Analysis, and Methodological Steps

This study is conducted using secondary data sources that were gathered from data published by the Ministry of Health and Family Welfare, the Government of Kerala, the District Medical Offices of all districts in the state, all other media reports, other official websites, and social media pages. In this analysis, we use weekly data across all 14 districts of Kerala to examine the temporal and spatial patterns of COVID-19 infections and cumulative deaths. To identify and understand the spatial pattern of COIVD-19 cases in the state, all the major parameters such as daily reported positive, negative, quarantine, and death cases were analyzed at the district scale at quarterly intervals. The spatial and cartographic analyses were conducted using ArcGIS software, and statistical analysis was carried out using SPSS software. Given the focus of this chapter, we employ simple mapping tools along with basic descriptive statistical analyses that provide insights into some important dimensions of COVID-19 infections in the state of Kerala.

In terms of the methodological approaches employed in this analysis, we first made district-level maps for select attributes of COVID-19 data for quarters, thus representing eight maps each for the selected variables. We felt this was the best method to capture the 24-month-long temporal variations that could tell the story of interdistrict variations. Thereafter, we also conducted graphical analysis of major patterns and changes in select variables to understand the reasons for those patterns. Many of these are represented in our maps and the bar graphs that tell the stories of months/durations when the COVID-19 infections and the related attributes had peaked during 24 months. This data is corroborated with a variety of secondary sources such as the local and national media and news channels, reports, government published data, and other academic work to make sense of what was happening in

Kerala and how one could explain those patterns. Some of these patterns, indeed, illustrated far greater knowledge on their spatial and temporal patterns. This could be studied at far greater depth by linking these data with travel data at interdistrict and intradistrict levels—something that is planned for future analysis.

5.4 ANALYSES AND FINDINGS

5.4.1 COVID-19 Infection, Hospitalization, and Deaths: Spatial Patterns of Quarterly Trends

We prepared district-scale maps for the major variables at quarterly intervals (March 2020, June 2020, September 2020, December 2020, March 2021, June 2021, September 2021, and December 2021) for the most important variables—total observations, positive cases under treatment, quarantined, hospitalized, and cumulative deaths. In the recorded monthly observations (Figure 5.3) from our data sources, we found that the minimum and maximum numbers of cases recorded in March 2020 were the first noticeably alarming numbers of 1,123 and 10,908 respectively, and the total number of cases noted in March 2020 was already 72,460—a time when the world was barely opening itself up to learning about this new virus for the very first time. This analysis also found two noticeable peaks: the first one during November–December of 2020, after which there was a downward trend, and the second one starting to peak again by May 2021, which also turned out to be the deadliest month in the State of Kerala.

Likewise, when examining the quarterly trends in the numbers of positive cases (Figure 5.4), cases being quarantined (Figure 5.5), hospitalized (Figure 5.6), and cumulative deaths (Figure 5.7), it is obvious that most districts showed very high numbers for different occurrences over the time period. Regarding the total observations, March 2020 was the first time, with 10,908 cases observed; this was exceeded by 154,308 cases observed in May 2021 (Table 5.3)—the summer months when almost all throughout India, people were under the deadliest grip of the pandemic. In terms of the numbers of cases in quarantine, once again May 2021 was the month, with 151,162 cases in quarantine. In terms of cumulative deaths, the first time when cumulative deaths were the highest was August 2020, with 7,770 deaths, followed by August 2021 onward until December 2021 when cumulative deaths exceeded 18,360 (August 2021), with December 2021 having 40,006 deaths. In terms of monthly-based numbers of cases for each of these, the district-level maps show that in terms of deaths also, Thiruvananthapuram had the worst-case scenario with the highest recorded deaths (Figure 5.7).

On a closer analysis of the district-scale spread of total observations, positive cases, in-quarantine, cumulative deaths, and the like, it is obvious that the geography of travel/migration, urban significance, and proximity to main corridors of urbanism were instrumental in these patterns. Despite having an incredible level of advancement in the sector of health care in the state, Kerala was one of the worst-affected states in the country. That was mainly due to the repatriation from different parts of the world as well as within the country. Various published reports and news channels showed that Thiruvananthapuram, Ernakulam, and Kozhikode were the most affected

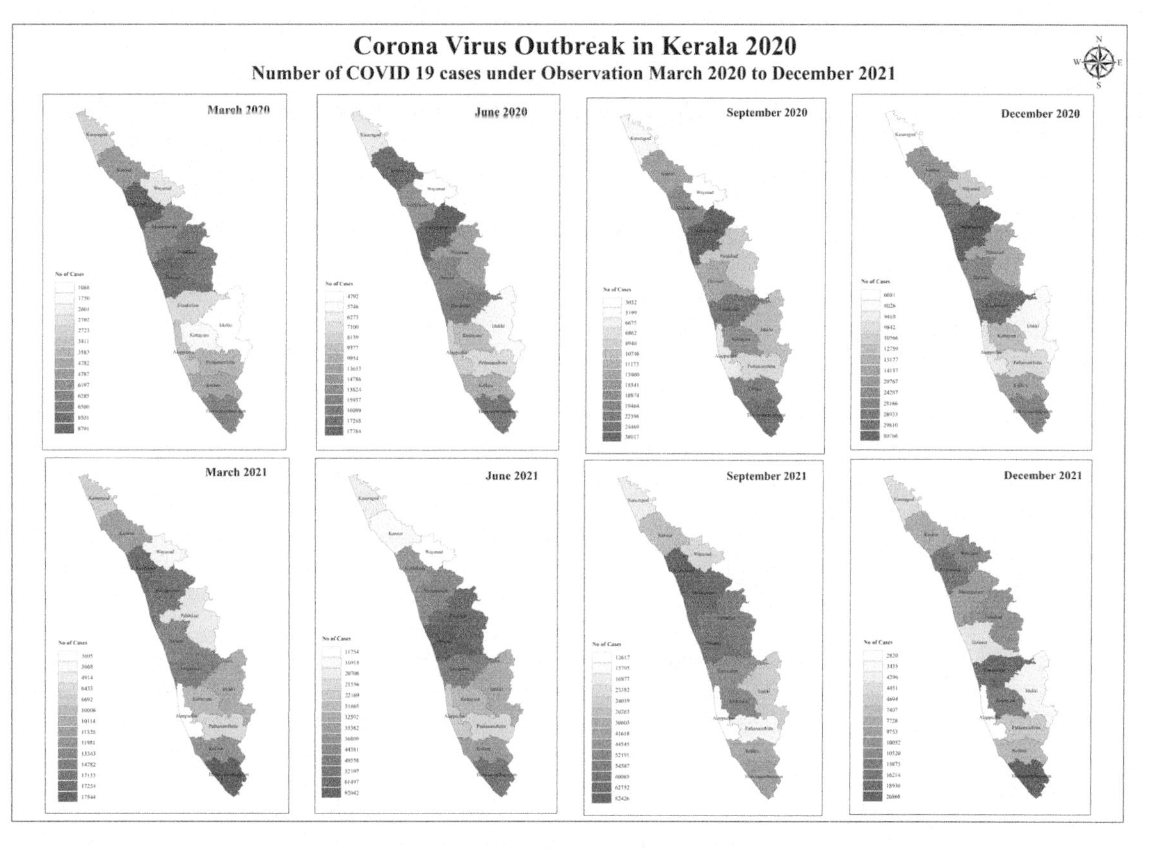

FIGURE 5.3 Interdistrict patterns for number of COVID-19 cases under observation, March 2020 to December 2021.

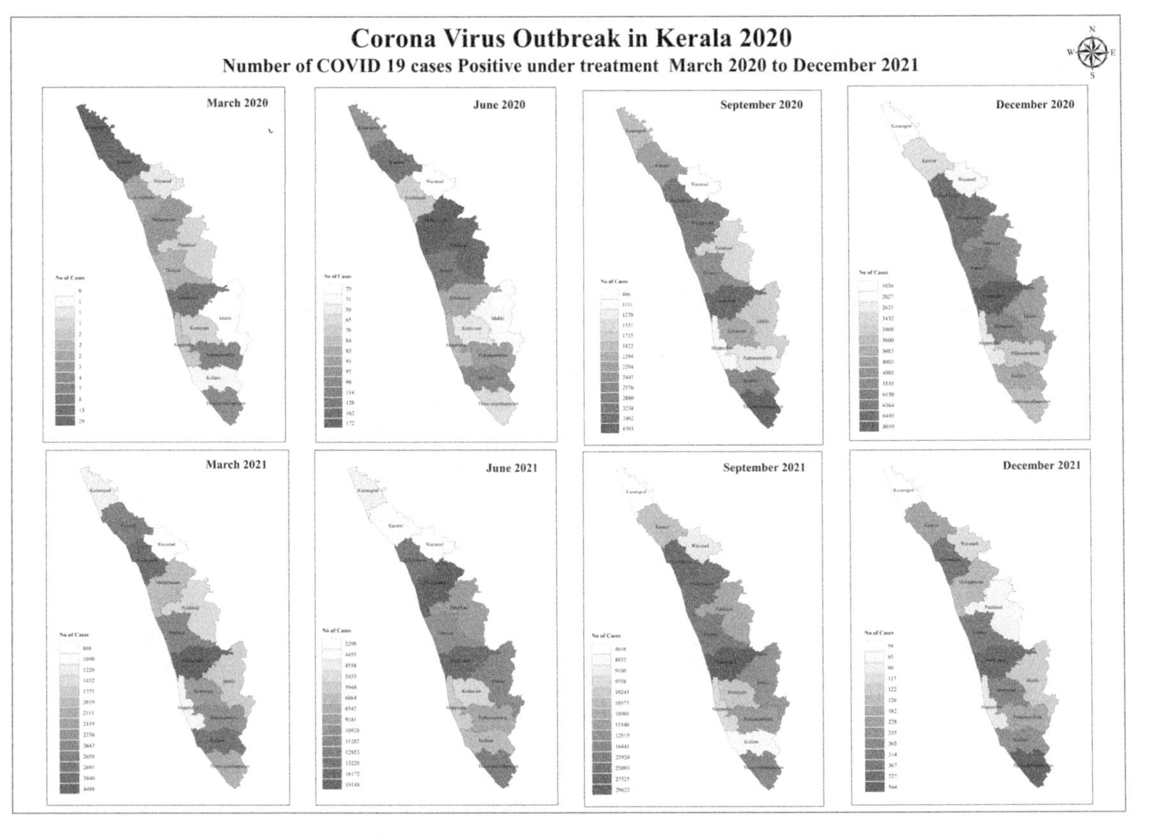

FIGURE 5.4 Interdistrict pattern for number of COVID-19 positive cases under treatment, March 2020 to December 2021.

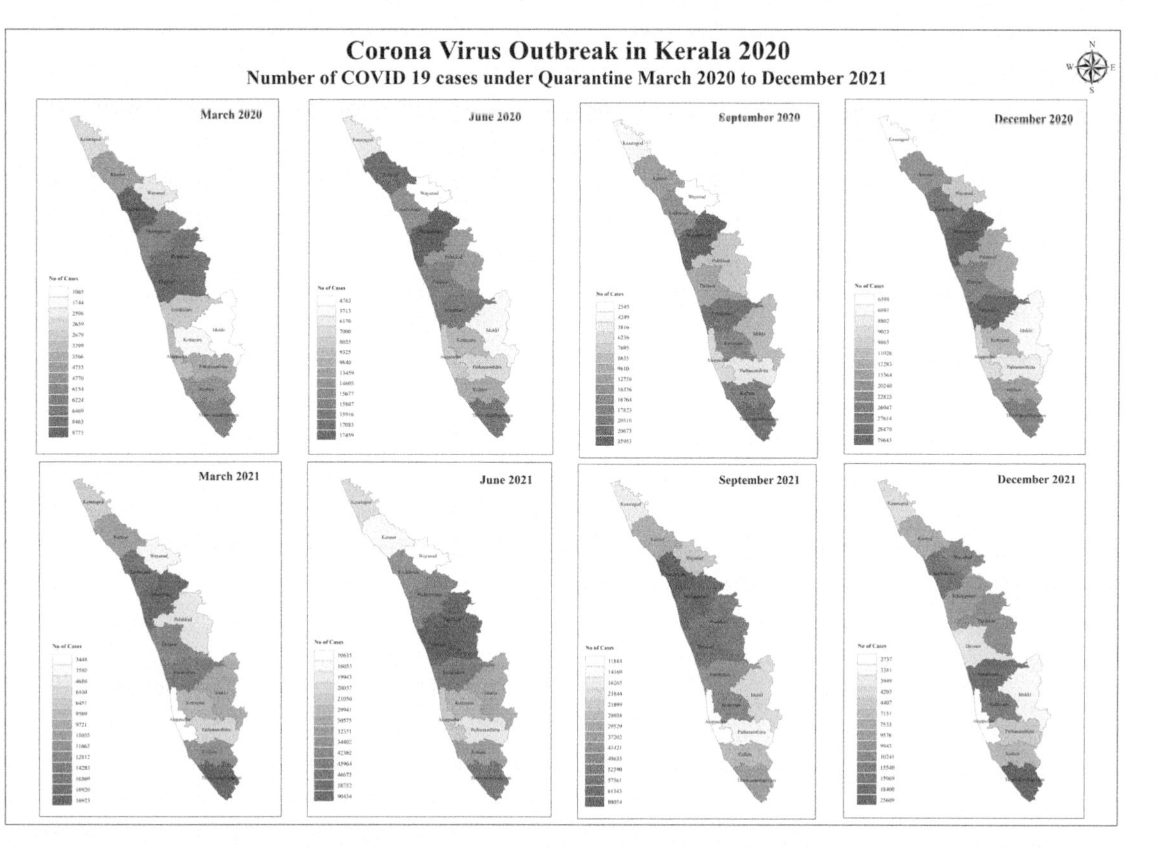

FIGURE 5.5 Interdistrict pattern for number of COVID-19 cases under quarantine, March 2020 to December 2021.

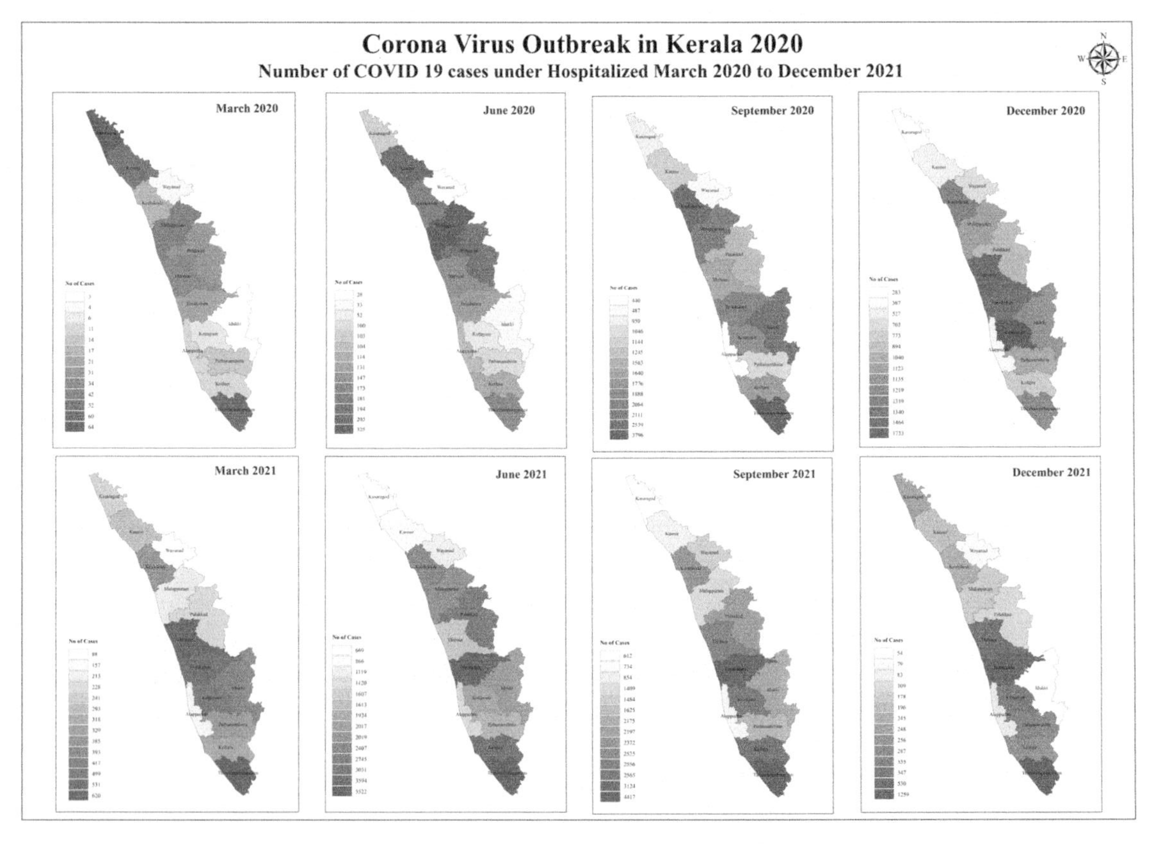

FIGURE 5.6 Interdistrict pattern for number of Covid-19 cases hospitalized, March 2020 to December 2021.

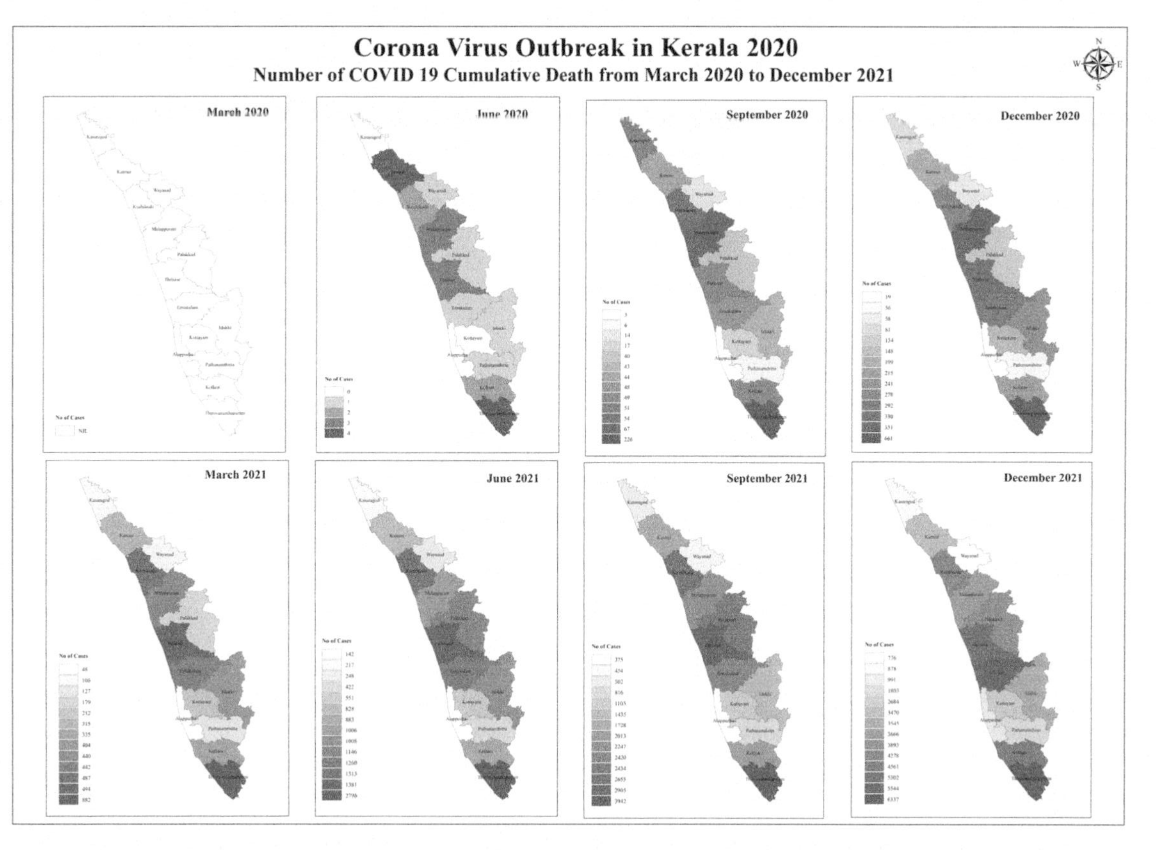

FIGURE 5.7 Interdistrict pattern for cumulative number of COVID-19 patients who died, March 2020 to December 2021.

TABLE 5.3
Total Numbers of Cases and Maximum Numbers of Cases for Major Variables by Month and Year

Months/ Year	Obsr-Max.	Quarantined-Max.	Hospitalized-Max.	In-Treatment-Max.	Cum-Deaths-Max.	Obser-Sum	Quarantined-Sum	Hospitalized-Sum	In-Treatment-Sum	Cum-Deaths-Sum
Jan'20	214	211	14	1	0	1471	1421	50	0	0
Feb'20	245	244	3	1	0	914	907	7	0	0
Mar'20	10908	10879	90	14	0	72460	71994	466	4	0
Apr'20	5203	5176	102	54	2	36667	36335	332	8	0
May'20	9176	9054	122	30	4	74398	73865	533	12	0
June'20	15732	15595	381	221	19	125307	123318	1989	97	1
July'20	42628	41885	856	939	36	184601	179612	4989	349	3
Aug'20	33763	32291	2792	3455	7770	155025	142291	12734	1007	555
Sep'20	39110	37244	3649	4927	380	203256	181764	21492	1991	27
Oct'20	48488	46478	4388	12405	924	273686	245261	28425	6554	66
Nov'20	69351	67974	2548	11372	1610	307828	286680	21148	5943	115
Dec'20	85994	84677	1794	7967	2329	314029	298929	15100	4386	166
Jan'21-1	66531	65640	1362	8867	3066	243828	231831	11997	4647	219
Jan'21	27198	25707	1491	11024	3675	215650	203898	11752	5160	263
Feb'21	26106	25224	914	9072	4135	217959	210293	7666	3671	295
Mar'21	16865	16320	545	3142	4524	127542	123751	3791	1734	323
Apr'21	64188	62228	2183	29708	5026	391463	374464	16999	12785	359
May'21	154308	151652	5998	50502	6965	988009	949300	38709	21882	498
June'21	79917	78460	5322	13287	11799	469522	441617	27905	7692	843
July'21	51808	50169	4221	19359	130	397164	372317	24847	8710	9
Aug'21	73701	71882	3888	31242	18360	490836	462416	28420	12857	1311
Sep'21	98188	95565	4645	35583	22255	621039	588784	32255	16975	1590
Oct'21	47010	44816	3159	17210	26008	371196	356899	14135	8332	1858
Nov'21	47451	45764	1687	11468	32982	247485	240859	6626	5267	2356
Dec'21	26162	25541	1269	6792	40066	153221	148515	3119	2862	10944

districts, and this occurred because Thiruvananthapuram is the administrative capital of the state, whereas Ernakulam is the commercial capital and Kozhikode is a prominent urban center in northern Kerala. Thiruvananthapuram is also known for its first information technology (IT) park in the country, which has 10 million sq. ft. of built-up area, with more than 460 companies providing employment to more than 63,000 professionals. Being the administrative center of the state, many people from different parts of the state as well as the country work in this office park at various capacities. Thiruvananthapuram district also has the highest density of population in the state records—1,509 persons per square km (2011 census). Ernakulam and Kozhikode were major urban centers of the state with access to the best of amenities needed to maintain a higher quality of life. Thus, in many ways, due to access to airports, seaports, container terminals, IT parks, large shopping malls, railway stations, and major highways, the large number of commuting population and huge number of migrant workers indeed collectively played a major role in the high numbers of COVID-19 cases in the state.

The summers of 2021 were the deadliest months for the whole country as news and Western media channels were all full of stories about deaths, cremations, and misery everywhere, and with the news media bashing the lack of control over healthcare distributional networks that brought lots of misery to almost every household. Households, rich and poor, were equally impacted by this pandemic, even though the poor had nowhere to go. There were images of low-income waged laborers migrating back to the countryside, near and far away, as many of these laborers engaged in informal economic activities were laid off due to the closure of the economy for prolonged periods of time. During such times, Kerala seems to have suffered significantly.

5.4.2 Descriptive Analysis of Major Variables across the Districts of Kerala

The spread of COVID-19 cases had already started showing its impact on the people of Kerala by February–March of 2020. In the recorded monthly observations from our reliable data sources, we found that the minimum and maximum numbers of cases recorded in March 2020 were the first noticeably alarming numbers of 1,123 and 10,908 respectively, and the total number of cases noted in March 2020 was already 72,460—a time when the world was barely opening itself up to hearing the name of this new virus for the very first time, and soon, in a few months, the whole world was under its dangerous grip. Thereafter, the numbers of observations of the disease started to increase rapidly in the State of Kerala, with total numbers exceeding almost 74,398 in May 2020, and 125,307 in June 2020, and very soon exceeding 200,000 by September 2020 and 300,000 by November 2020 (Figure 5.8).

While there was slight decline in total numbers of cases by March 2021, it started to peak once again in April 2021 (Figure 5.9), when the summers were beginning in most of India, with Kerala being the southernmost state and with relatively far higher temperatures. By May 2021, Kerala had a record total number of cases at 988,009, and once again the second-highest record of cases at 621,039 by September 2021,

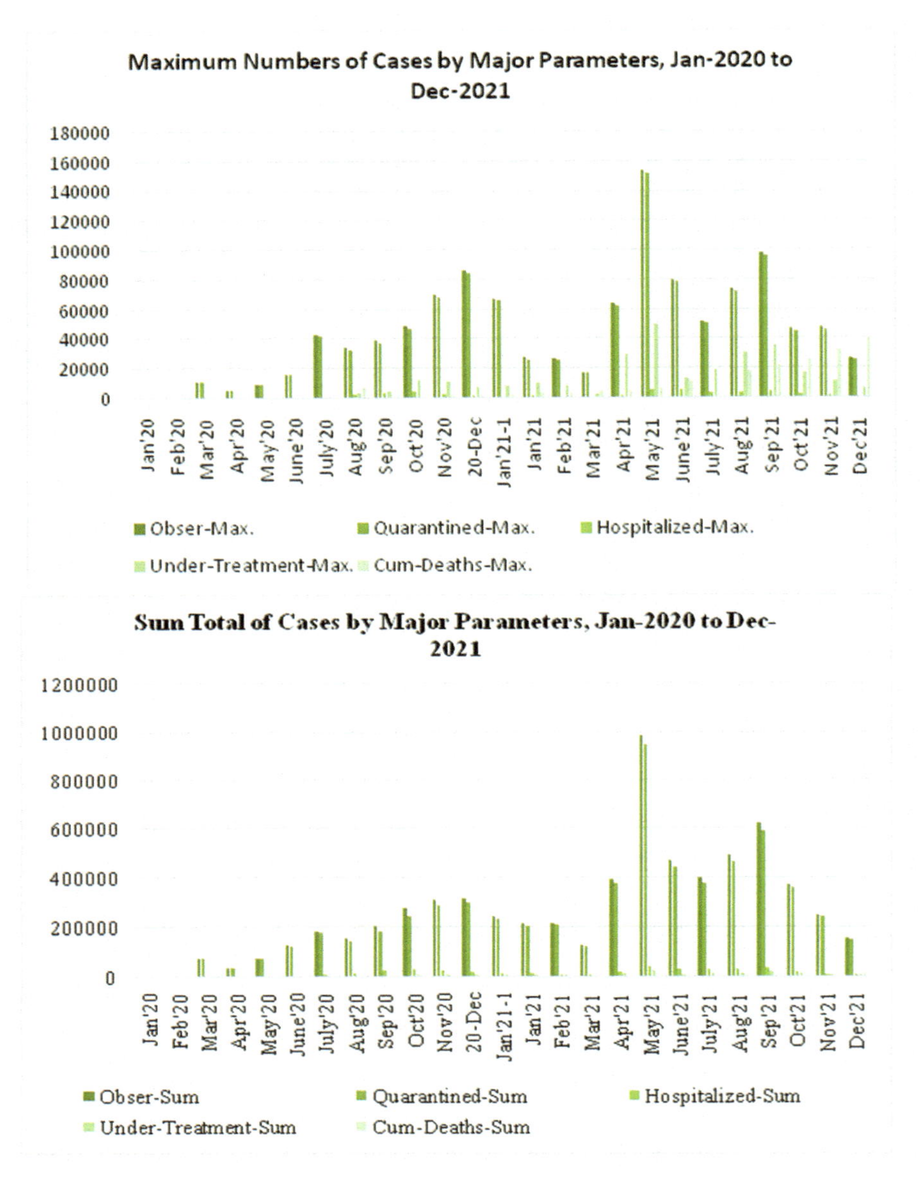

FIGURE 5.8 Monthly charts for major variables—maximum cases (upper) and sum total cases (Lower).

after which it started declining gradually, with a record observation of 153,221 in December 2021. This is also a time when, by the start of fall 2021, the vaccinations had started rolling in and the vaccination drive was soon catching up among the common masses across India, which helped slow the spread of the disease.

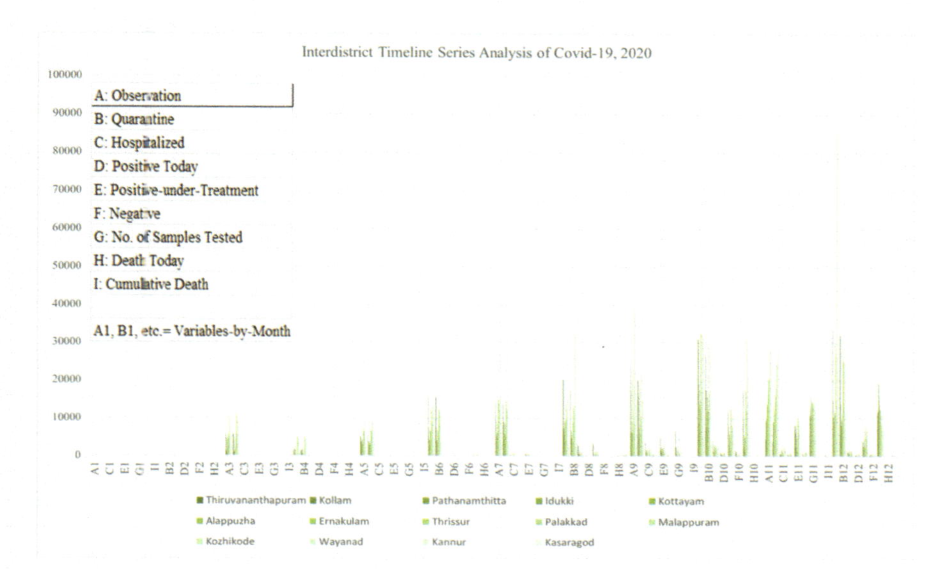

FIGURE 5.9 Interdistrict timeline series analysis of COVID-19 infections, 2020.

The charts in Figures 5.9 and 5.10 show that the numbers of cases and their subsequent effects had started peaking suddenly since July 2020, reaching unbearably enormous heights by December 2020. In particular, the variables count along every dimension—whether total observations, those quarantined, under treatment, monthly deaths, or cumulative deaths—all were alarmingly critical. While the numbers of samples that were sent for testing were quite high, often reaching a minimum of 17,685 to a maximum of 99,055 in the month of April 2021, the total numbers of cases that turned out as negative were very low, leading to a very high number of positive cases that needed to be quarantined (see Table 5.3, Figure 5.5, and Figure 5.6).

These high numbers of cases, however, slowed a little during January 2021 through March 2021, and after that these numbers once again started climbing rapidly in almost every district. The peaks in 2021 occurred again in April–May and August–September, the time when India as a nation was being criticized heavily all across the world for its inability to manage and control COVID-19 infections, with numerous Western and international news media flashing images of deaths, mass cremations being performed at crematoriums, and the like—projecting an enormously negative image of the country and its health-care system. People had dropped their guard against taking precautions, and the onset of marriage and festive seasons further exacerbated the problem. These infections and deaths gradually started declining after October–November of 2021.

In terms of infected people quarantining (Table 5.3 and Figure 5.5), the statistics suggested that the numbers of cases of those getting infected had almost reached a total of 123,318 people by June 2020, and the numbers of people quarantining kept on increasing rapidly over the next few months, with the maximum numbers peaking by

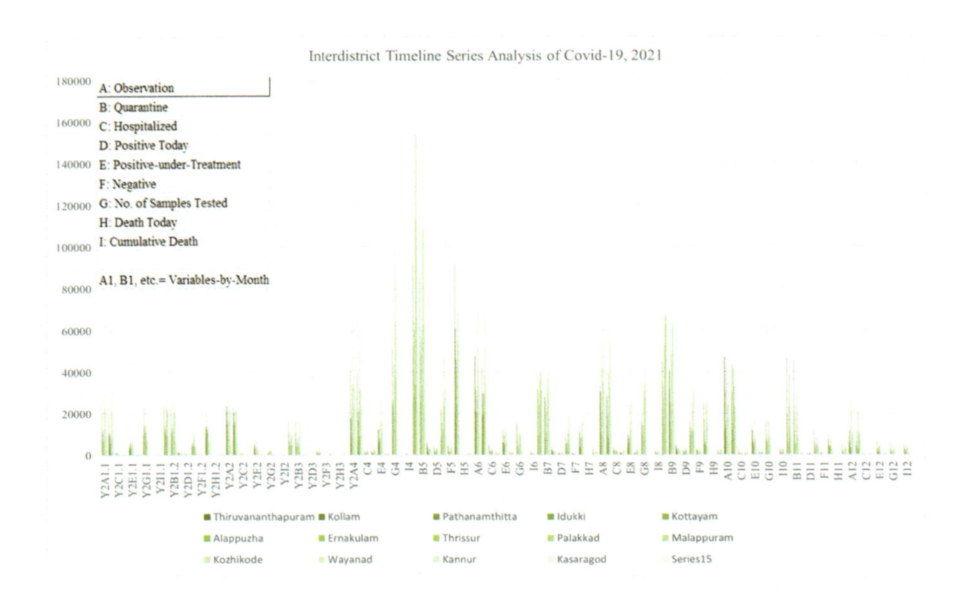

FIGURE 5.10 Interdistrict timeline series analysis of COVID-19 infections, 2021.

May 2021, at 949,300 (Figure 5.9). In terms of the total numbers of patients who had tested positive and were under treatment, the statistics started showing significantly alarming high numbers starting in March 2020—a time when the world was learning about the COVID-19 pandemic, its symptoms, and ways of dealing with its cure. In Kerala's Kollam district, 14 patients were already under treatment by March 2020, and thereafter the numbers kept on increasing with time in almost all the districts of Kerala, reaching to thousands of people. Starting August 2020 and onward, the numbers of positive cases-under-treatment had already surpassed 3,455, peaking at 50,502 in May 2021. During early spring to the summer of 2021, that is, during April–July 2021, the highest numbers of positive cases were in treatment, and totaled about 29,708; 50,502; 13,287; 19,359; 31,242; 35,583 respectively. By December 2021, the last month for which we collected this data, the maximum number of patients under treatment was still at 6,792.

In terms of total numbers of people hospitalized, our month-by-month analysis of these two years shows that the total number of hospitalizations had exceeded 1,989 by June 2020, and since then it remained very high, at almost 12,734 (August 2020) to 21,492 (September 2020) to 28,425 (October 2020); the second wave of this pandemic peaked again starting in April 2021 (16,999) and was 38,709 by May 2021; 27,905 in June 2021; and 32,255 in September 2021 (Figure 5.3); thereafter, the numbers started to decline gradually, with a total of 4,706 hospitalized in December 2021. This was also marked by a period when at least one or more types of vaccinations were becoming available in India.

When looking at district-by-district statistics for some of those months when the COVID-19 infections had peaked, we find that in May 2021, the cases noted had peaked to a dangerously high level (see Figure 5.2). Regarding the total numbers of observations recorded, the districts ranking the highest include Thrissur (154,308), followed by Ernakulum (113,174), Palakkad (111,515), Kozhikode (93,384), and Thiruvananthapuram (91,404). Concerning the positive cases under treatment in the same month, we found that the districts ranking the highest were Ernakulam (50,502), Malappuram (48,197), and Kozhikode (31,987), among others.

Regarding the cumulative number of deaths noted in this month that showed the highest peaking observations, were the districts of Thiruvananthapuram, Kozhikode (828), Thrissur (808), and Ernakulum (648). Overall, in the two-year period, however, the highest cumulative cases of deaths were noted in Thiruvananthapuram (5,611), Ernakulam (4,846), and Thrissur (4,531); others with alarmingly higher numbers of cumulatively deaths also included 4,050 (Kozhikode), 3,708 (Kollam), 3,230 (Palakkad), 2,994 (Malappuram), and 2,902 (Kannur); cumulative numbers of deaths in other districts were below 2000. In the previous year, 2020, the peak of cases was noted in December 2020, and a careful evaluation of data shows that the cumulative numbers of deaths were the highest in Thiruvananthapuram (567), followed by Thrissur (257), Malappuram (236), and Ernakulum (224). However, the cumulative deaths in 2021 were significantly higher.

5.5 CONCLUSIONS AND SIGNIFICANCE FOR FUTURE RESEARCH

In the two-year period being analyzed here, we found that the month-by-month analyses over 2020–2021 were marked by, sadly, very high numbers of deaths, which had indeed begun in August of 2020 itself—the first time the highest number of cumulative deaths exceeded 4,313 in the district of Kasaragod; the lowest cumulative death in this month was 37 in the district of Kottayam. While inspecting the total cumulative deaths thereafter for each month until December of 2021, we found that in almost every month, at least 134 deaths were noted, and not a single month passed without at least one death. By December 2021, cumulative deaths had reached 5,611 and the average number of deaths had reached 2,862 by December 2021. Without doubt, May 2021 turned out to be the deadliest month in terms of maximum numbers of cases as well as sum total. As can be noted, a total cumulative 10,944 deaths were already reported by December 2021, the last month when this data was collated for this analysis.

Our comprehensive analysis of the existing literature, news and media sources, visual analyses of COVID-19 cases, and the basic descriptive and graphical analysis based on available data suggests that having access to urban resources and local–global connectivity are critical in maintaining better quality of life. However, these factors, such as access to airports, railway stations, major hospitals, sea beaches, and large shopping malls, and a large floating population and the presence of large numbers of migrant workers who provided various types of formal and informal labor support also contributed to the wide spread of this pandemic across the districts in Kerala. In addition, Thiruvananthapuram, being the hub of the IT park and the state's administrative center, was a major source of employment for Kerala's population. Further, Ernakulam, as a major center of commerce and trade, is also the largest metropolitan region in the state, with a very high population density of 1,069 persons per square km. Kozhikode, being a major urban center in the northern part of the state, and with access to affordable health facilities, good transportation and communication facilities, IT parks, beaches, and large shopping centers and malls, and with a commuting population and migrant workers—all of these also played critical roles in the spread of the pandemic in the district. This district also has recorded a high density of population, at 1,318 persons per square km. Overall, the state is very dense, with a significant share of migrant population, and all these factors collectively contributed to high rates of infections all throughout. In addition, despite the government's making rules about lockdowns and the restricted openings of market spaces, people were very casual about these restrictions, and easily let down their guard, since India had done well during the first wave of the pandemic. Thus, the very casual approaches toward the dreadful effects of this virus were felt much more strongly in its second wave during March 2021 onward, which lasted until October–November 2021.

Finally, some easing of COVID-19 infections and ensuing deaths began declining in November 2021 when the vaccines started reaching the populations of India and people started getting vaccinations. This helped control the pandemic gradually, and by then people also started taking this pandemic more seriously. Overall, then, success was achieved in terms of containing this deadly pandemic. This case study of Kerala's COVID-19 infections, and the waves of its dreadful impacts across the districts of Kerala has significant lessons in terms of health-care policy and practice. It is important for people to follow the rules implemented by the government, along with taking all preventive measures as much as possible. It is also important for state governments to have funds available to meet the basic needs of the poorest migrant population who could have been contained at their original places of residence. That could have contained the disease from its viral spread and dispersal—which grabbed the nation at risk—spreading like wildfire from a few epicenters to all throughout the country. The nation needed to implement the lockdowns much sooner and much more strictly, and those not abiding by the rules needed to be punished more severely. Taking strict actions could have helped control the pandemic sooner, with far lower levels of death. Thus, lessons learned from this pandemic should be useful to Kerala and any other state or nation in terms of controlling and containing a pandemic, should such an event occur again in the future.

REFERENCES

Arashi, M., Bekker, A., Salehi, M. et al. (2021). Evaluating prediction of COVID-19 at provincial level of South Africa: A statistical perspective. *Environmental Science and Pollution Research*, 29, 21289–21302. https://doi.org/10.1007/s11356-021-17291-y

Bag, R., Ghosh, M., Biswas, B., & Chatterjee, M. (2020). Understanding the spatio-temporal pattern of COVID-19 outbreak in India using GIS and India's response in managing the pandemic. *Regional Science Policy and Practice, 12*, 1063–1103. https://doi.org/10.1111/rsp3.12359

Basu, P., & Mazumder, M. (2021). Regional disparity of covid-19 infections: An investigation using state-level Indian data. *Indian Economic Review, 56*, 215–232. https://doi.org/10.1007/s41775-021-00113-w

Bhunia, G. S., Roy, S., & Shit, P. K. (2021). Spatio-temporal analysis of COVID-19 in India: A geostatistical approach. *Spatial Information Research*, 29, 661–672. https://doi.org/10.1007/s41324-020-00376-0

Binti, H. F. A, Lau, C., Nazri, H., Ligot, D. V., Lee, G., Tan, C. L. et al. (2020). Corona tracker: Worldwide COVID-19 outbreak data analysis and prediction. *Bulletin of the World Health Organization* https://doi.org/10.2471/BLT.20.255695

Carrat, F., Figoni, J., Henny, J. et al. (2021). Evidence of early circulation of SARS-CoV-2 in France: Findings from the population-based "CONSTANCES" cohort. *European Journal of Epidemiology, 36*, 219–222. https://doi.org/10.1007/s10654-020-00716-2

Chathukulam, J., & Tharamangalam, J. (2021). The Kerala model in the time of COVID19: Rethinking state, society and democracy. *World Development, 137*, 105207.

Chen, Y., Li, Q., Karimian, H., Chen, X., & Li, X. (2021). Spatio-temporal distribution characteristics and influencing factors of COVID-19 in China. *Scientific Reports, 11*, 3717. https://doi.org/10.1038/s41598-021-83166-4

Dey, S. K., Rahman, M. M., Siddiqi, U. R., & Howlader, A. (2020). Analyzing the epidemiological outbreak of COVID-19: A visual exploratory data analysis approach. *Journal of Medical Virology, 92*, 632–638. https://doi.org/10.1002/jmv.25743

Elias, A. A. (2021). Kerala's innovations and flexibility for Covid-19 recovery: Storytelling using systems thinking. *Global Journal of Flexible Systems Management, 22*(Suppl-1), S33–S43. https://doi.org/10.1007/s40171-021-00268-8

Fang, L., Wang, D., & Pan, G. (2020). Analysis and estimation of COVID-19 spreading in Russia based on ARIMA model. *SN Comprehensive Clinical Medicine, 2*, 2521–2527. https://doi.org/10.1007/s42399-020-00555-y

Fong, S. J., Dey, N., & Chaki, J. (2021). An introduction to COVID-19. In *Artificial intelligence for coronavirus outbreak* (pp. 1-22). Springer Briefs in Applied Sciences and Technology. Singapore: Springer. https://doi.org/10.1007/978-981-15-5936-5_1

Gouda, K. C., Nikhilasuma, P., Kumari, R., Singh, P., Benke, M. et al. (2020). Analyzing spatio-temporal spread of Covid19 in India. *COVID-19 Pandemic: Case Studies & Opinions, 1*(4), 57–67.

Haleem, A., Javaid, M., Vaishya, R., Deshmukh, S. G. Areas of academic research with the impact of COVID-19. (2020). *American Journal of Emergency Medicine, 38*, 1524–1526. https://doi.org/10.1016/j.ajem.2020.04.022.HTTP1: https://dashboard.kerala.gov.in/covid/index.php

India, C. o. (2011). *Provisional population totals: Kerala*. Directorate of Census Operations. https://dhs.kerala.gov.in/

Jayesh, S., & Sreedharan, S. (2020). Analyzing the Covid-19 cases in Kerala: A visual exploratory data analysis approach. *SN Comprehensive Clinical Medicine, 2*, 1337–1348. https://doi.org/10.1007/s42399-020-00451-5

Khan, A., & Arokkiaraj, H. (2021). Challenges of reverse migration in India: A comparative study of internal and international migrant workers in the post-COVID economy. *Comparative Migration Studies, 9*(49), 1–19. https://doi.org/10.1186/s40878-021-00260-2

Liu, K. (2021). COVID-19 and the Chinese economy: Impacts, policy responses and implications. *International Review of Applied Economics, 35*(2), 308–330. https://doi.org/10.1080/02692171.2021.1876641

Mandi, J., Chakrabarty, M., & Mukherjee, S. (2020). How to ease Covid-19 lockdown? Forward guidance using a multi-dimensional vulnerability index. Ideas for India. *Macroeconomics*. www.ideasforindia.in/topics/macroeconomics/how-to-ease-covid-19-lockdown-forward-guidance-using-a-multidimensional-vulnerability-index.html

Mishra, M. K. (2020). The world after COVID-19 and its impact on global economy. Kiel: Leibniz Information Centre for Economics. www.econstor.eu/bitstream/10419/215931/1/MKM%20PAPER%20FOR%20COVID.pdf

Mondal, S., & Ghosh, S. (2020, April). Fear of exponential growth in Covid-19 data of India and future sketching. medRxiv. https://doi.org /10.13140/RG.2.2.28834.17607

Morgan, A. K., Awafob, B. A., & Quarteyb, T. (2021) The effects of COVID-19 on global economic output and sustainability: Evidence from around the world and lessons for redress. *Sustainability: Science, Practice and Policy, 17*(1), 76–80. https://doi.org/10.1080/15487733.2020.1860345

Palamim, C. V. C., Ortega, M. M., & Marson, F. A. L. (2020). COVID-19 in the indigenous population of Brazil. *Journal of Racial and Ethnic Health Disparities, 7*, 1053–1058. https://doi.org/10.1007/s40615-020-00885-6

Ray, D., & Subramanian, S. (2020). India's lockdown: An interim report. *Indian Economic Review, 55*(1), 31–79.

Rupani, P. F., Nilashi, M., Abumalloh, R. A. et al. (2020). Coronavirus pandemic (COVID-19) and its natural environmental impacts. *International Journal of Environmental Science and Technology, 17*, 4655–4666. https://doi.org/10.1007/s13762-020-02910-x

Sartorius, B., Lawson, A. B., & Pullan, R. L. (2021). Modelling and predicting the spatio-temporal spread of COVID-19, associated deaths and impact of key risk factors in England. *Scientific Reports, 11*, 5378. https://doi.org/10.1038/s41598-021-83780-2

State Urbanization Report. (2012). Department of Town and Country Planning, Government of Kerala.

Silva, C. C. da, Lima, C. L. de, Silva, A. C. G. da et al. (2021). Covid-19 dynamic monitoring and real-time spatio-temporal forecasting. *Frontiers in Public Health*. https://doi.org/10.3389/fpubh.2021.641253

Zinchenko, Y. P., Shaigerova, L. A., Almazova, O. V. et al. (2021). The spread of COVID-19 in Russia: Immediate impact on mental health of university students. *Psychological Studies, 66*, 291–302. https://doi.org/10.1007/s12646-021-00610-1

6 The 2020 Hurricanes, Internal Displacements, and COVID-19 in Latin America and the Caribbean Countries

Lessons Learned for Disaster Risk Reduction

Roberto Ariel Abeldaño Zuñiga and Gabriela Narcizo de Lima

CONTENTS

6.1 INTRODUCTION

During the 2020 hurricane season, there were 31 events in the Atlantic and 21 events in the Eastern North Pacific region (National Oceanic and Atmospheric Administration-NOAA, 2020).

These events had devastating population impacts on the most vulnerable communities in the Latin American and Caribbean (LAC) region. According to statistics from the United Nations Office for the Coordination of Humanitarian Affairs (OCHA), hurricanes Iota and Eta (the largest hurricanes in 2020) affected more than seven million people in Central America and the Caribbean (United Nations Office for Disaster Risk Reduction, 2021).

The Internal Displacement Monitoring Centre reported that 646 disaster events of various origins were reported throughout the Americas during 2020. These events

DOI: 10.1201/9781003227106-6

affected 28 countries and territories, where they caused 4.5 million new internal displacements (Internal Displacements Monitoring Centre, 2021). Of that total, the same agency has estimated that 2,255,049 new internal displacements corresponded to people affected by the 2020 hurricane season (Internal Displacements Monitoring Centre, 2021).[1]

The inhabitants of LAC countries have recurrently faced the impacts of hurricanes (Abeldaño Zúñiga, 2021; Abeldaño Zuñiga & Fanta Garrido, 2020), but 2020 was a unique year. The COVID-19 pandemic imposed new and greater challenges on those that already existed for disaster risk management in this region. The 2020 cyclone season was one of the most active in recent years. It created a crisis in the preparedness and response of risk and disaster management systems, which until then had been focused on managing the multiple challenges that the COVID-19 pandemic imposed on the region (International Organization for Migration, 2020).

Some authors described the increased vulnerability of internally displaced persons due to disasters during the COVID-19 pandemic in the year 2020 (Abeldaño Zuñiga & González Villoria, 2021; Orendain & Djalante, 2020). The impacts faced by these people are varied but can be summarized in three aspects.

The first is related to the displacement conditions they face. The shelters are often overcrowded, making it difficult to maintain physical distance. Consequently, the risk of contracting COVID-19 increased among these people. During emergency evacuation operations, it is complex to ensure physical space is maintained between people. Here, the measures taken to effect evacuations clashed with some indications to counteract the spread of COVID-19 (Pei et al. 2020; The Conversation, 2020).

The second relates to service interruptions that usually occur during the crisis phase of a disaster. Essential services such as water, gas, and electric power are often interrupted. Consequently, access to drinking water is often interrupted. This also clashes with another of the basic recommendations during a pandemic: frequent hand washing (UNICEF, 2020; International Organization for Migration, 2020).

The third aspect is related to the pressure that the COVID-19 pandemic imposed on health systems. This pressure also posed major challenges to the capacity of health services to respond to disasters (Naciones Unidas, 2020; International Organization for Migration, 2020). This means that some displaced persons may have had their access to health services disrupted or delayed.

In this context, this chapter aims to describe the impacts of hurricanes in the Central American and Caribbean region during 2020 from a "lessons learned" perspective for disaster risk management in health crises.

6.2 THE MAIN HURRICANES IN LAC IN 2020

In 2020, 31 tropical cyclones were reported in the Atlantic region (Table 6.1). The countries affected by the events that occurred in this region were the Bahamas, Belize, Colombia, Costa Rica, Cuba, the Dominican Republic, El Salvador, Guatemala, Honduras, Jamaica, Mexico, Nicaragua, Panama, Puerto Rico, Trinidad and Tobago, and Venezuela. In terms of damage caused, the events of the greatest magnitude were Tropical Storms Cristobal and Gonzalo, and Hurricanes Hanna, Isaias, Laura, Nana,

TABLE 6.1
2020 Atlantic Hurricane Season

Designation	Date	Affected Countries or Territories*	Direct Deaths
Tropical Storm Arthur	16–19 May	None	0
Tropical Storm Bertha	27–28 May	None	0
Tropical Storm Cristobal	1–9 June	Mexico	3
Tropical Storm Dolly	22–24 June	None	0
Tropical Storm Edouard	4–6 July	None	0
Tropical Storm Fay	9–11 July	None	0
Tropical Storm Gonzalo	21–25 July	Trinidad and Tobago	0
Hurricane Hanna	23–26 July	Mexico	4
Hurricane Isaias	30 July–4 August	Bahamas, Dominican Republic, Puerto Rico	2
Tropical Depression Ten	31 July–1 August	None	0
Tropical Storm Josephine	11–16 August	None	0
Tropical Storm Kyle	14–15 August	None	0
Hurricane Laura	20–29 August	Cuba, Dominican Republic, Puerto Rico	40
Hurricane Marco	21–25 August	None	0
Tropical Storm Omar	31 August–5 September	None	0
Hurricane Nana	1–3 September	Belize, Guatemala	0
Hurricane Paulette	7–22 September	None	0
Tropical Storm Rene	7–14 September	None	0
Hurricane Sally	11–17 September	None	0
Hurricane Teddy	12–23 September	Puerto Rico	2
Tropical Storm Vicky	14–17 September	None	0
Tropical Storm Beta	17–22 September	None	0
Tropical Storm Wilfred	17–21 September	None	0
Subtropical Storm Alpha	17–19 September	None	0
Hurricane Gamma	2–6 October	Mexico	5
Hurricane Delta	4–10 October	Mexico	0
Hurricane Epsilon	19–26 October	None	0
Hurricane Zeta	24–29 October	Mexico	0
Hurricane Eta	31 October–13 November	Honduras, Jamaica, Nicaragua, Mexico, Costa Rica, Belize, Guatemala, El Salvador	165
Tropical Storm Theta	10–15 November	None	0
Hurricane Iota	13–18 November	Venezuela, Colombia, Nicaragua, Honduras, El Salvador, Costa Rica, Panama, Guatemala, Mexico	67

Note: * In the LAC region.

Teddy, Gamma, Delta, Eta, and Iota. These events caused 288 direct deaths (National Hurricane Center, 2020a).

In the Pacific region, 21 tropical cyclones were reported (Table 6.2), including tropical depressions, of which four reached hurricane strength, 13 were characterized as tropical storms, and four as tropical depressions. Of the hurricanes, three were intense, given that they reached category 4 on the Saffir-Simpson scale, in order of appearance, "Douglas" in July, "Genevieve" in August, and "Marie" in September–October, all with maximum sustained winds of 215 km/h (Comisión Nacional del Agua and Servicio Meteorológico Nacional, 2021).

Of all the events that occurred in the Pacific Ocean in 2020, only Tropical Storm Amanda affected countries in the LAC region, specifically, El Salvador, Guatemala, and Honduras, where this storm caused 40 direct deaths (National Hurricane Center, 2020b).

TABLE 6.2
2020 Eastern Pacific Hurricane Season

Designation	Date	Affected Countries*	Direct Deaths
Tropical Depression One-E	20–25 April	None	0
Tropical Storm Amanda	30–31 May	El Salvador, Guatemala and Honduras	40
Tropical Storm Boris	24–27 June	None	0
Tropical Depression Four-E	29–30 June	None	0
Tropical Storm Cristina	6–12 July	None	0
Tropical Depression Six-E	13–14 July	None	0
Tropical Depression Seven-E**	20–21 July	None	0
Hurricane Douglas	20–29 July	None	0
Hurricane Elida	8–12 August	None	0
Tropical Depression Ten-E	13–16 August	None	0
Tropical Storm Fausto	16–17 August	None	0
Hurricane Genevieve	16–21 August	Mexico	6
Tropical Storm Hernan	26–28 August	Mexico	0
Tropical Storm Iselle	26–30 August	Mexico	0
Tropical Storm Julio	5–7 September	None	0
Tropical Storm Karina	12–16 September	None	0
Tropical Storm Lowell	20–25 September	None	0
Hurricane Marie	29 September-6 October	None	0
Tropical Storm Norbert	5-14 October	None	0
Tropical Storm Odalys	3-5 November	None	0
Tropical Storm Polo	17-19 November	None	0

Notes: * In the LAC region.
** Tropical Depression Seven-E was redesignated as unnamed tropical storm in the post-season re-analysis.

Considering the events that caused the most significant damage in LAC countries, three hurricanes that occurred in August, October, and November in the Atlantic region are worth mentioning: Laura, Eta, and Iota.

"Laura" (Table 6.3 and Figure 6.1) was a powerful category 4 hurricane (on the Saffir-Simpson scale) that made landfall near Louisiana, United States, accompanied by a catastrophic storm surge of up to 5.5 meters above ground level. This hurricane was responsible for 40 direct deaths in Cuba, the Dominican Republic, and Puerto Rico, and seven in the United States (Aon Benfield Analytics, 2020a).

Hurricane Eta (Table 6.3 and Figure 6.2) originated on October 31 as Tropical Depression No. 29 of the 2020 tropical cyclone season in the Atlantic Ocean, 510 km southeast of Kingston, Jamaica, and encountered favorable conditions for rapid strengthening with a sea surface temperature above 29°C and in a very light shear environment,[2] which allowed it to gain strength as it advanced toward the eastern coast of Central America.

On November 3, "Eta" made landfall on the coast of Nicaragua as a category 4 hurricane (on the Saffir-Simpson scale), with maximum sustained winds of 240 km/h and with its cloud bands covering Central America and southeastern Mexico. Hurricane Eta was responsible for at least 165 direct deaths and more than 100 missing persons in Central America and south-southeastern Mexico (Aon Benfield Analytics, 2020b; Pasch, Reinhart et al., 2021; National Hurricane Center, 2020a).

Unfortunately, this devastation was soon followed by the disastrous impacts of "Iota (Table 6.3 and Figure 6.3)," an intense category 4 hurricane that made landfall along the coast of Nicaragua just two weeks after the devastation created by "Eta."

Under the influence of Hurricane Iota, widespread freshwater flooding, exacerbated by preexisting flooding conditions caused by "Eta," resulted in 67 deaths and 41 missing persons in parts of Central America, totaling damages of up to $1.4 billion (Aon Benfield Analytics, 2020b).

Some country-specific damages and fatalities created by Hurricane Iota are listed below (Aon Benfield Analytics, 2020b; Stewart, 2021):

- Mexico – The states of Chiapas, Tabasco, and Veracruz, located in the southeast of the country, were the hardest hit by the heavy rains of "Iota." Floods and landslides damaged nearly 59,000 homes, affecting 297,000 people, and also blocked roads and cut off access to 135 communities for a few days. No deaths were reported.
- Colombia – the hurricane caused ten direct deaths, eight on the mainland and two in the Colombian archipelago of San Andres-Providencia-Santa Catalina, in addition to 8 missing persons. On the mainland, floods created by heavy rains accompanied by landslides and rockfalls were reported in many regions of the country, affecting hundreds of thousands of people in different cities, with estimated economic losses of at least US$ 100 million.
- Honduras – Hurricane Iota caused at least 13 direct deaths, and mudslides were the leading cause of death, resulting in eight deaths in San Manuel Colohete and five in Los Trapiches. More than 360,000 people were directly affected by the hurricane in this country.

TABLE 6.3
Details of Hurricanes Laura, Eta and Iota

Designation	Date	Track	Duration of the Event	Maximum Intensity of Winds	Minimum Core Pressure	Estimated Total Damage Costs*	Affected Countries or Territories*	Direct Deaths
Hurricane Laura	20–29 August	8,820 km	198 hours	240 km/h with gusts of 295 km/h	952 hPa	$100 million	Cuba, Dominican Republic, Puerto Rico	40
Hurricane Eta	31 October–13 November	5,775 km	300 hours	240 km/h with gusts of 295 km/h	923 hPa	$6.8 billion	Honduras, Jamaica, Nicaragua, Mexico, Costa Rica, Belize, Guatemala, El Salvador	165
Hurricane Iota	13–18 November	1,810 km	120 hours	260 km/h with gusts of 315 km/h	917 hPa	$1.4 billion	Venezuela, Colombia, Nicaragua, Honduras, El Salvador, Costa Rica, Panama, Guatemala, Mexico	67

Note: * In the LAC region.

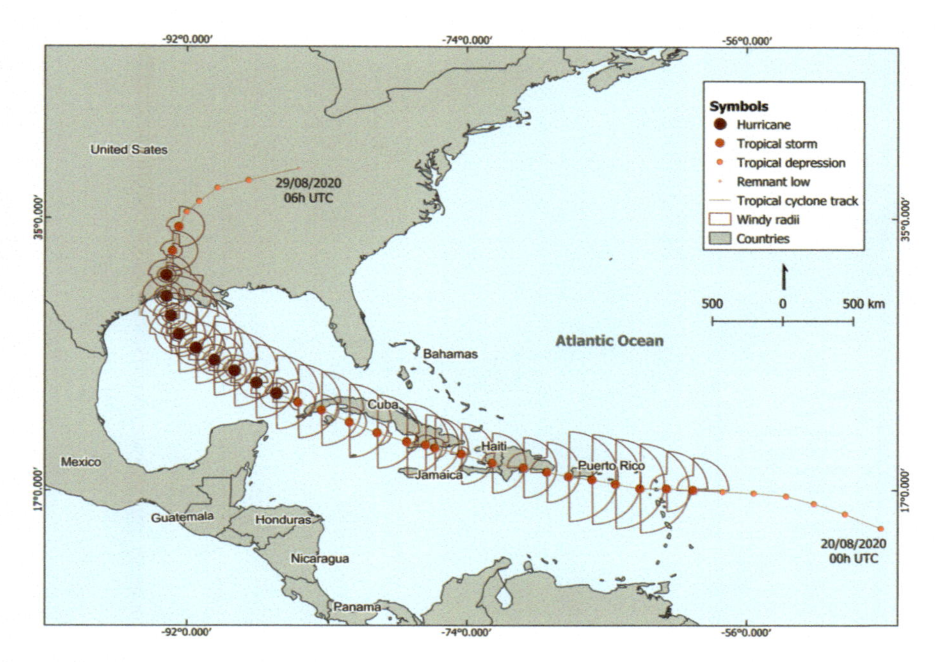

FIGURE 6.1 Hurricane Laura track, 2020.

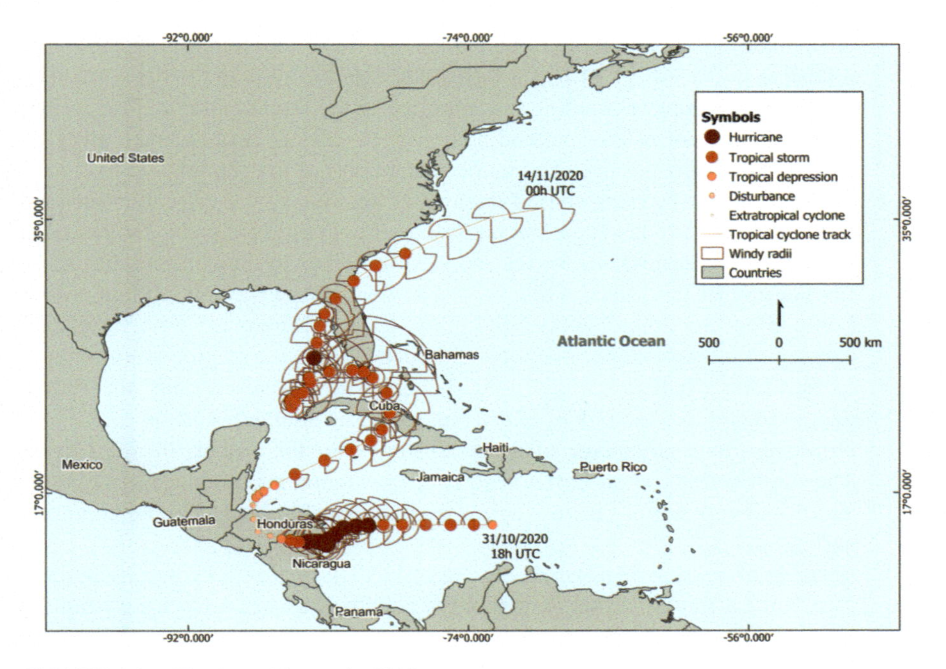

FIGURE 6.2 Hurricane Eta track, 2020.

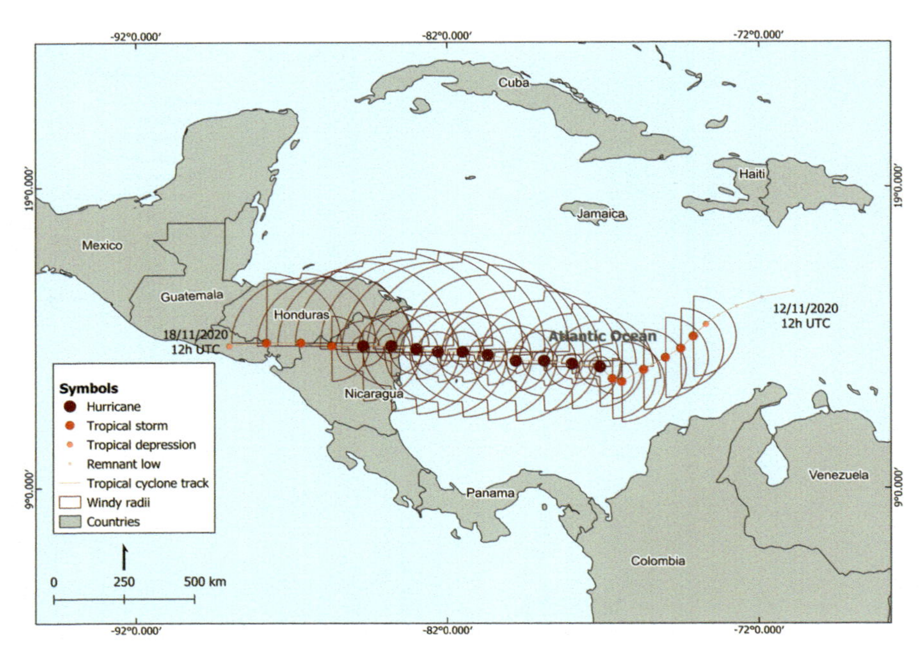

FIGURE 6.3 Hurricane Iota track, 2020.

- Guatemala – two people died, and two more were reported missing due to a mudslide in the village of El Carmen Jalauté de Purulhá, in the department of Baja Verapaz. Severe flooding was reported in the municipality of Río Hondo in the department of Zacapa, and numerous rescues of people swept away by intense water currents occurred in the eastern part of the country.
- Venezuela – heavy rains caused freshwater flooding in the extreme northwest of the country that damaged 288 houses, mainly in the Paraguaná Peninsula, in the state of Falcón. Damage was reported in homes in the communities of El Cayude and El Tranquero, while the town of Santa Ana was left without power. Some minor flooding also occurred in the state of Miranda, but no significant damage to infrastructure or deaths was reported. No loss of life was reported.
- El Salvador – had to set up 1,000 shelters with a total capacity for 30,000 people to care for victims of Hurricane Iota. The southern portion of "Iota's" circulation took advantage of deep moisture over the eastern Pacific Ocean. The southwesterly wind flow forced over the mountains produced heavy rains and flooding, resulting in two deaths and significant damage to the country's agriculture.
- Nicaragua – at least 50 direct deaths created by Hurricane "Iota" were recorded. Before the arrival of "Iota," much of the country was still recovering from the damage caused by Hurricane Eta. In particular, the areas around and south of Puerto Cabezas were devastated by "Eta," and clean-up operations had to be suspended to restart evacuations before "Iota" made landfall there.

- Costa Rica – widespread flooding and landslides occurred throughout the country, creating the evacuation of 26 people from the municipality of Corredores, located in the province of Puntarenas, near the Panama–Costa Rica border, and in the town of Parrita, which is located southwest of the capital San José. No deaths were reported in this country.
- Panama – government officials reported that flooding killed one person in Nole Duima in the county (*comarca*) of Ngäbe-Buglé. There is no specific damage estimate for this country.

6.3 INTERNAL DISPLACEMENTS AND THE EPIDEMIOLOGY OF THE PANDEMIC IN LAC IN 2020

The Internal Displacements Monitoring Centre (IDMC) is a Geneva, Switzerland-based agency that acts as an observatory, producing estimates of internally displaced persons by disasters and conflict globally. In its publicly accessible database, the IDMC records as "new IDPs"[3] the estimated number of internal displacements over a given period of time (reporting year). Figures may include individuals who have been displaced more than once (Internal Displacements Monitoring Centre, 2021). They analyze infections, deaths, and COVID-19 tests in the 30 days following the impact of the three hurricanes. This analysis is done as part of the prevailing epidemiological context in those countries at that time and not as a cause–consequence analysis. With the available data, it is impossible to determine how many internally displaced persons (IDPs) have been infected by COVID-19 or vice versa; thus, only a temporal association of events can be shown to understand the high vulnerability of IDPs in the context of COVID-19.

Hurricane Laura impacted between August 20 and 29, 2020, and produced 434,061 new IDPs in Cuba, the Dominican Republic, Haiti, and Puerto Rico. In the 30 days following the impact of the hurricane, there were 1,522 COVID-19 infections in Cuba, 18,605 in the Dominican Republic, and 635 in Haiti. At the same time, there were 23 deaths from COVID-19 in Cuba, 533 in the Dominican Republic, and 24 in Haiti.

However, it is important to note that this situation must be interpreted in the context of the testing capacities of each country. The number of COVID-19 tests per 1,000 population in the region was, in general, very low (16.4 in Cuba and 9.1 in the Dominican Republic) (Our World in Data–University of Oxford, 2021) (Figure 6.4A and 6.4B). The impact of Hurricane Laura occurred coincidentally with the peak of the first wave of COVID-19 in LAC.

The stringency index is a composite measure based on nine response indicators, including school closures, workplace closures, and travel bans, rescaled to a value from 0 to 100, where 100 is understood as the strictest level (Our World in Data–University of Oxford, 2022).

In those 30 days, the Average Stringency Index was highest in Cuba (84.6), while it was lower in the Dominican Republic (78.7) and substantially lower in Haiti (21.3). Thus, restrictions on the movement of people were greater in Cuba and the Dominican Republic and much laxer in Haiti (Our World in Data–University of Oxford, 2022).

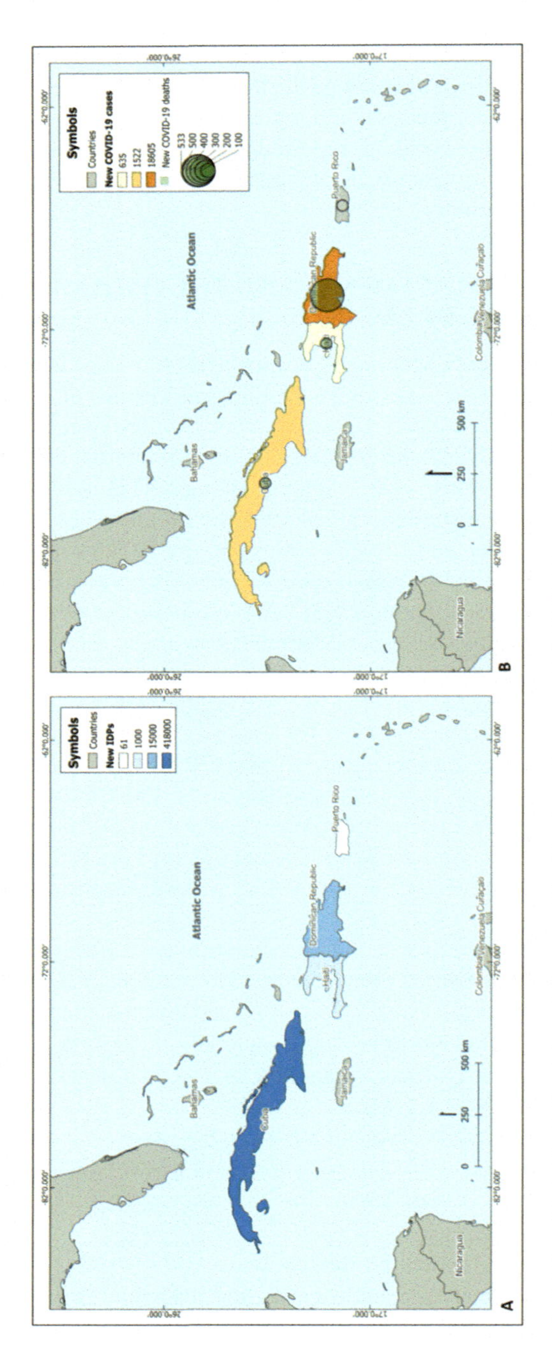

FIGURE 6.4A AND 6.4B New IDPs caused by Hurricane Laura, and COVID-19 cases and deaths.

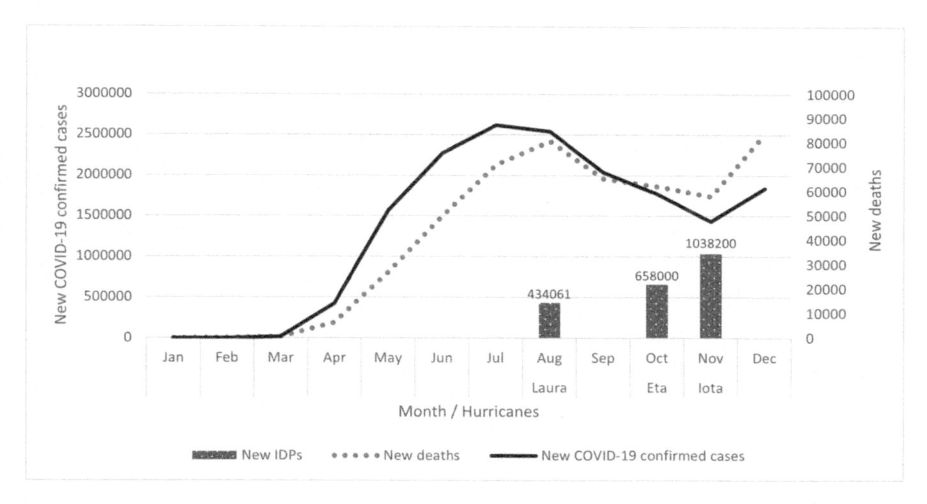

FIGURE 6.5 New IDPs due to hurricanes Laura, Eta and Iota, and COVID-19 epidemiological indicators in Latin America and the Caribbean. Year 2020.

Another important element of the epidemiological context in these countries is access to handwashing facilities, precisely because of one of the most important recommendations of the health authorities in the face of the pandemic: "frequent handwashing." In 2000, 85.2% of the total population of Cuba had this access, while only 55.2% of the people of the Dominican Republic had it, and only 22.9% of the people of Haiti had access to handwashing facilities (United Nations Department of Economic and Social Affairs, 2021).

This epidemiological context speaks of displaced persons' high exposure and vulnerability in those days in the countries mentioned. Here it is necessary to clarify that Puerto Rico does not present data on many variables since the data for this territory are attributed to the United States, except for the data on new internally displaced persons (Figure 6.5 and Figure 6.6).

With impacts between October 31 and November 13, Hurricane Eta resulted 658,000 new IDPs in Belize, Colombia, Costa Rica, Cuba, the Dominican Republic, El Salvador, Guatemala, Honduras, Mexico, Nicaragua, and Panama. Of these, the most significant impacts in terms of the displaced population were observed in Cuba (188,000), Guatemala (184,000), Honduras (175,000), Nicaragua (71,000), and Mexico (15,000); while in countries such as Colombia, the inhabitants of the island of San Andres were affected, with 8,000 new IDPs (Internal Displacements Monitoring Centre, 2021) (Figure 6.7A and 6.7B).

The number of COVID-19 infections occurring in these countries during the 30 days following the impact was alarming. There were 245,225 infections and 5,449 deaths in Colombia, while the testing rate was 27.4 per 1,000 population. Mexico, on the other hand, had 188,581 infections and 14,187 deaths, while testing was only 3.6 per 1,000 inhabitants (one of the lowest testing rates in the region due to the type of epidemiological surveillance adopted by this country, called "sentinel surveillance") (Our World in Data–University of Oxford, 2022).

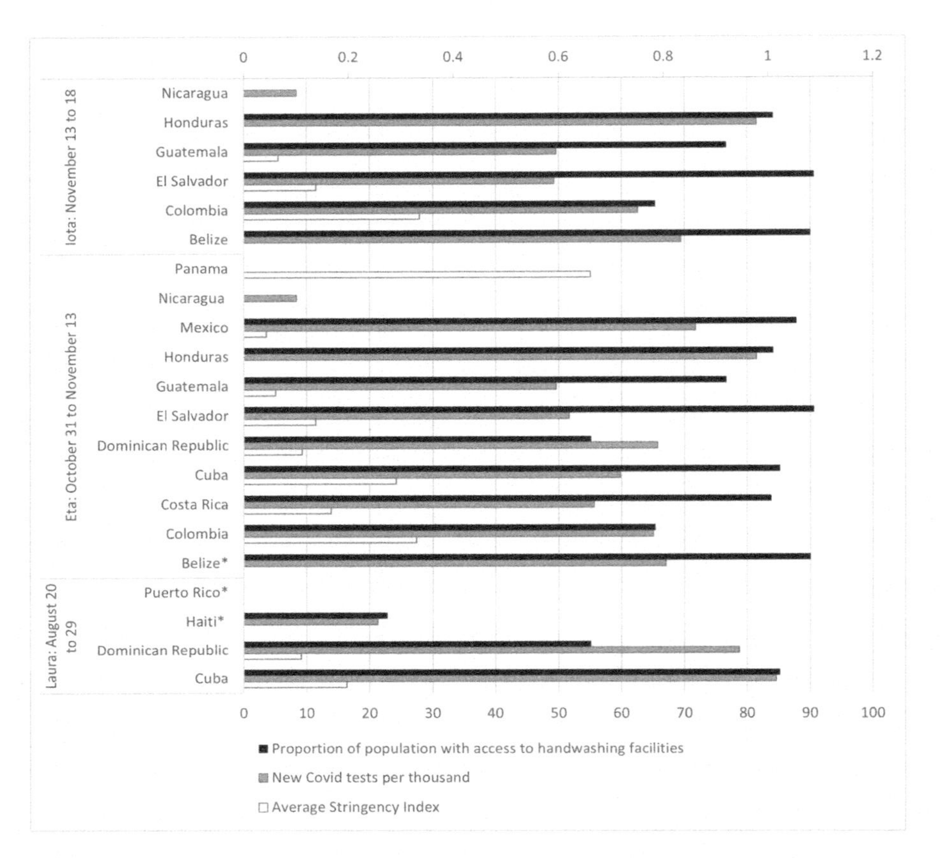

FIGURE 6.6 Proportion of population with access to handwashing facilities, new covid-19 cases per thousand, and Average Stringency Index, by country. Year 2020.

The average Stringency Index of the countries affected by the hurricanes during those days decreased in some of those countries due to the reestablishment of some essential activities in the region. Thus, the Stringency Index presented by the 11 affected countries ranged from 8.3 (the lowest, in Nicaragua) to 81.5 (the highest, in Honduras) (Our World in Data–University of Oxford, 2022). Many countries were beginning to reopen some activities to reactivate the economy. At the same time, there was a rise to the peak of the pandemic's second wave in this region.

Regarding handwashing facilities access, countries such as the Dominican Republic had 55.2% of its population with access to these facilities; Colombia had 65.4% with access; and Guatemala had 76.7% having access, while Costa Rica, Cuba, Honduras, and Mexico had approximately 90% of their populations with access to these facilities (United Nations Department of Economic and Social Affairs, 2021).

The low access to handwashing facilities is another sign of the high exposure that IDPs had to COVID-19, but following Hurricane Eta, Hurricane Iota formed, which had an even greater impact in terms of new IDPs (Figure 6.5 and Figure 6.6).

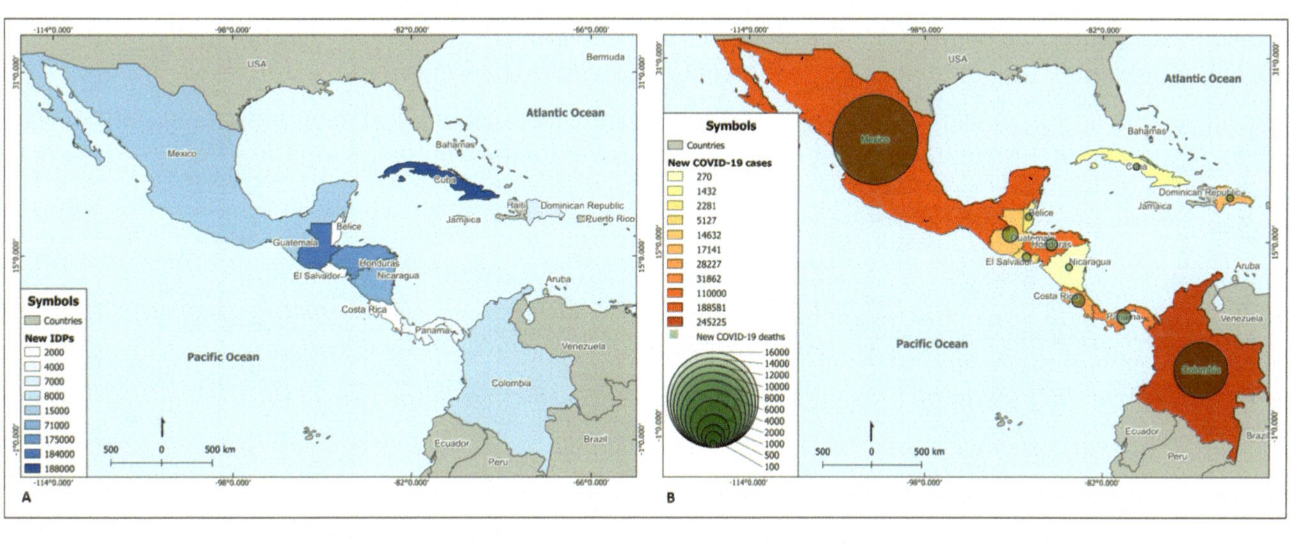

FIGURE 6.7A AND 6.7B New IDPs caused by Hurricane Eta, and COVID-19 cases and deaths.

"Iota," with impacts between November 13 and 18, caused 1,038,200 new IDPs. The countries affected were Belize, Colombia, El Salvador, Guatemala, Honduras, and Nicaragua. Honduras recorded 743,000 new IDPs, while Nicaragua recorded 160,000, Guatemala 126,000, and Colombia 8,000 (Internal Displacements Monitoring Centre 2021) (Figure 6.8A and 6.8B). The epidemiological context was the same as mentioned in the previous paragraph, as two hurricanes hit consecutively (Figure 6.5 and Figure 6.6).

Regarding the differences in infections and deaths between countries, Banik et al. (2020) suggest that factors such as public health system, population age structure, poverty level, and BCG vaccination are powerful contributory factors in determining fatality rates, and can explain those differences between countries.

Regarding other health impacts experienced by IDPs in the COVID-19 pandemic, several studies explored the wide range of health impacts of the COVID-19 pandemic on displaced populations, migrants, and refugees (Aragona et al., 2020; Abeldaño Zuñiga and González Villoria, 2021). Compared to several non-migrant reference groups, incidence (infection) risks among migrants and displaced persons tend to be higher (Hintermeier et al., 2021). At the same time, ICU admissions and hospitalization rates appear to be lower among migrants. However, the last data comes from mathematical modeling, as data that consider migration or displacement variables are unavailable in most countries (Hintermeier et al., 2021). For pregnant women, COVID-19 morbidity and mortality rates have been disproportionately higher than for the rest of the population, as has the burden of severe and fatal respiratory illness, as observed in previous outbreaks (SARS in 2003 and MERS in 2012) and influenza (Hantoushzadeh et al., 2020).

However, some authors suggest that case fatality rates are higher in populations with a higher proportion of migrants (Hayward et al., 2021). Migrants and displaced persons in these studies were often exposed to poor living and working conditions (Schenker, 2010; Hayward et al., 2021). Therefore, they were at increased risk of becoming infected or developing severe symptoms. In addition, evidence consistently suggests that mental health has been negatively affected by the pandemic in all migrant groups (Hargreaves et al., 2019; Aragona et al., 2020; Hintermeier et al., 2021).

Thousands of migrants in high-income countries are excluded or have restricted access to conventional health systems due to their immigration status. This is probably a significant barrier to accessing tests and treatment and the eventual implementation of vaccines in low- and middle-income countries (Hayward et al., 2021).

In countries with disaggregated data that consider immigration status as a variable, the incidence of COVID-19 has been higher for migrants, despite the low rates of tests among them. They also have had the highest positivity rates and a higher risk of death from COVID-19 (Chew et al., 2020; Jaqueti Arocaet al., 2020; Kim et al., 2020). This indicates that when disaggregated data exist, it is possible to visualize the differential impacts on migrant and displaced populations.

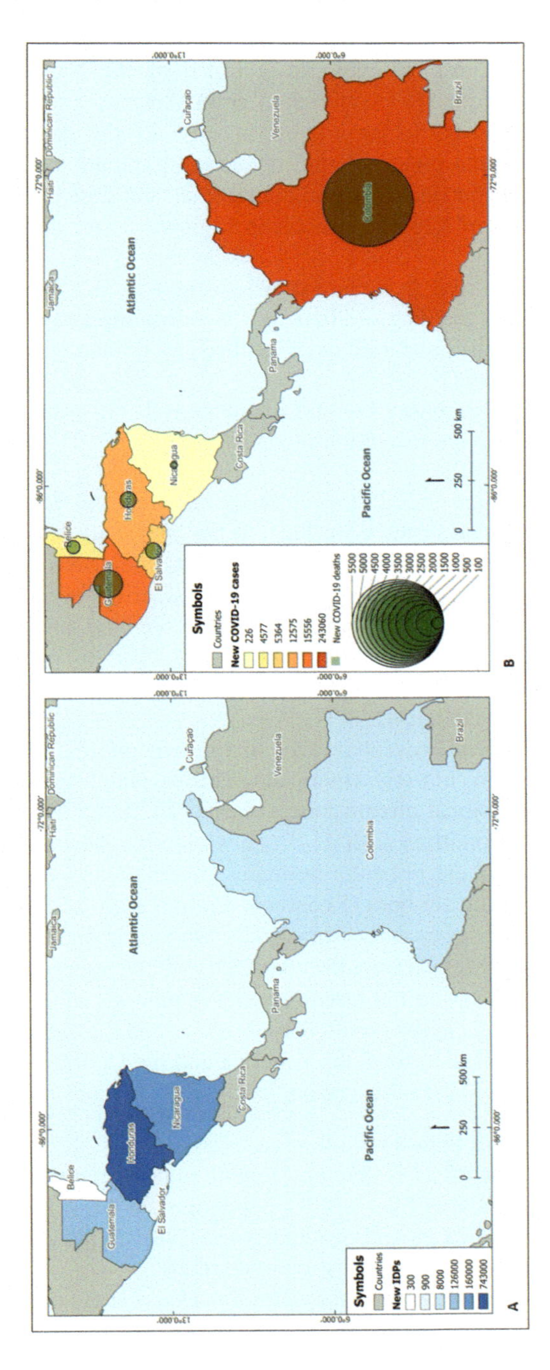

FIGURE 6.8A AND 6.8B New IDPs caused by Hurricane Iota, and COVID-19 cases and deaths.

6.4 CONCLUSIONS AND LESSONS LEARNED FOR DISASTER RISK REDUCTION IN LAC

In 2020, there were 52 tropical cyclones recorded by the National Oceanic and Atmospheric Administration (NOAA) in the Central America and Caribbean region. Of these, three were of particular interest because of the damage they caused and mainly because of the large volume of internal population displacement they provoked. Those three hurricanes were Laura, Eta, and Iota, which collectively caused the displacement of 2,130,261 people in 13 countries (or in the case of Puerto Rico, territory).

Hurricane Laura occurred in a context where the first wave of the COVID-19 pandemic (between June and August 2020) was in full swing and where the pressure on health services from the pandemic itself has been more than many health systems have been able to withstand.

The impact of both Hurricanes Eta and Iota occurred consecutively and also in a context overshadowed by the rise of the second wave of COVID-19 in several countries of the region.

Although the hurricanes of 2020 have already happened, there were no disasters of the same magnitude during the year 2021. What happened in 2020 should leave some lessons learned for disaster risk reduction in LAC countries in the face of future complex scenarios, such as the convergence of two concurrent events: a global health crisis such as a pandemic) with the impact of one or more disasters as a humanitarian crisis.

In this regard, some lessons learned for disaster risk reduction in a global health emergency are summarized below.

The onset of the pandemic revealed in the first months of 2020 the deficiencies of many of the health systems in Latin American countries, which have been recurrently overtaken by epidemiological circumstances during 2020 and 2021. In addition to high infection rates in countries such as Brazil, Colombia, Mexico, Argentina, and Chile, the lack of investment in health by many governments has placed excessive pressure on existing health services (Economic Commission for Latin America and the Caribbean, 2021; Enríquez and Sáenz, 2021; Organización Panamericana de la Salud, 2021). Brazil and Mexico were the countries with the highest number of deaths in the LAC region, and during the last two years, situations of collapse have been observed in some hospitals in the region that have alarmed the authorities.

The Economic Commission for Latin America and the Caribbean has found that, during this period of crisis, the LAC region

> has accumulated more than 44 million cases and almost 1.5 million deaths from COVID-19. This effect is disproportionate to other regions of the world, given that, with only 8.4% of the world's population, in Latin America and the Caribbean, COVID-19 cases and deaths from the virus represent almost 20% and 30%, respectively, of the world's recorded totals.
>
> *Economic Commission for Latin America and the Caribbean, 2021*

These figures put pressure on the infrastructure of health services and the very health of the countries' health system workers (Abeldaño Zuñiga et al., 2021).

In this context of an overwhelmed health system, the people displaced by the hurricanes studied in this chapter probably had limited access to health services.

Some authors assert that national planning systems did not show the necessary capacities to adapt to a global crisis scenario. In this sense, there was no capacity to create scenarios or anticipate and manage risks (Enríquez & Sáenz, 2021). The pandemic already showed the effects of the first wave in mid-2020, while the impacts of Hurricane Laura occurred in August. These experiences have not been used to anticipate the possible scenarios that would come later when Eta and Iota impacted.

Hurricanes often hit several countries in the same region. Therefore, integrated actions are needed. In the LAC context, in 2017, Belize, Costa Rica, Guatemala, Honduras, Mexico, and Panama adopted the San Pedro Sula Declaration. These countries agreed to work together to implement the Regional Comprehensive Framework for Protection and Solutions (MIRPS) for refugees (Marco Integral Regional para la Protección y Soluciones, 2020a; 2020b). In 2019, El Salvador joined this initiative. In 2020, in the context of this framework and in the face of the spread of the coronavirus in Central America and Mexico, these countries adopted the following priorities for dealing with refugees, which could also be applied to IDPs (Marco Integral Regional para la Protección y Soluciones, 2020a). First, ensuring medical assistance to the populations of interest, including primary health care, providing affordable and appropriate access to sexual and reproductive health services, mental health, nutrition, and vaccination for children, among others (World Health Organization, 2019). Second, enabling call centers to attend consultations and manage appointments. Third, including populations of interest in social support initiatives. Fourth, allowing access to vaccination and continuity of care, which is more difficult for people on the move, so coordination with shelters to ensure continuity of services and care is crucial (Zard et al., 2021).

In Colombia, which hosts millions of migrants from Venezuela, the situation of displaced people is worrisome (Perez-Brumer et al., 2021). The population is mostly undocumented, and national authorities have publicly declared the exclusion of migrants from the national vaccination strategy (The Guardian, 2020).

Regarding vaccination of IDPs, identity and address validation processes for accessing COVID-19 vaccines often require personal documents. In that sense, it is necessary to contemplate the obstacles faced by IDPs who have lost their personal documents during disasters (Zard et al., 2021).

Only three countries were manufacturing COVID-19 vaccines in LAC as of December 2021: Argentina, Brazil, and Mexico (Duke Global Health Innovation Center, 2021). This means that the rate of vaccinations in the poorest countries in the region continues to be slow. In countries like Nicaragua, Guatemala, or Honduras, however, less than 50% of the population is fully vaccinated (Our World in Data–University of Oxford, 2021). Increased efforts should be made to donate vaccines to Central American countries that cannot manufacture or package them locally (González Franco & Ojeda Chamba, 2021).

Economic vulnerability is increased when a disaster destroys a community due to the financial losses associated with disasters, a situation that has been exacerbated by the pandemic and ensuing economic recession (Platform on Disaster Displacement, 2020). There are also other dimensions of social impacts, as both disaster-related displacements and the pandemic can impact all life's facets, such as access and continuity in education, and food insecurity, among others (Platform on Disaster Displacement, 2020).

Economic vulnerability is caused by the interruption of daily activities due to a disaster situation. The International Monetary Fund highlighted that these difficulties have deepened during the pandemic, generating a crisis in LAC countries. For individuals employed in the formal sector, workplace closures, loss of working hours, and decreased labor income have disproportionately affected IDPs because they generally lack alternative sources of income (International Labour Organization-ILO, 2020). The most vulnerable families often cope with economic hardship through various mechanisms of reduced spending, for example, on food, interruption of housing rent, indebtedness, sale of assets, and interruption of children's school attendance (United Nations High Commissioner for Refugees, 2021).

Concerning education, it is recommended for schools not to require proof of "lessons learned" or "years completed" in order not to interrupt children's education (Caarls et al., 2021). This is a measure that UNICEF (Caarls et al., 2021) recommended applying in countries which are host to Venezuelan migrant children. The inclusion of displaced girls with disabilities is also important. Full educational inclusion is needed that makes available programs and qualified teachers in the provision of inclusive education.

NOTES

1 The Migration Glossary of the International Organization for Migration (IOM) (2006) defined internally displaced persons as "people or groups of people who have been forced or obliged to flee or leave their homes or their habitual residence, particularly as a result of, or to avoid, the effects of armed conflict, a situation of generalized violence, violation of human rights, or natural or human disasters and have not crossed a border of an internationally recognized State."

2 Wind shear is the difference in wind speed or direction between two points in the earth's atmosphere.

3 The total number of movements (one single person can have one or more movements) for a given event or that has occurred over a period of time is considered a "flow," and the most common type of this metric is called a "new displacement." The total number of IDPs at a given time is referred to as a "stock" metric. It represents a static snapshot of the number of IDPs in a given location at a specific point in time (Internal Displacements Monitoring Centre, 2021).

REFERENCES

Abeldaño Zúñiga, & Ariel, R. 2021. Cambio Climático y Desastres En América Latina, El Caribe y Europa: Un Análisis Comparado de La Incidencia de Desplazamientos Internos de Población. In M. Palacios Sanabria, M. Torres Villarreal, & F. Navas Camargo (Eds.), *Desafíos Migratorios: Realidades Desde Diversas Orillas* (pp. 263–290). Bogotá: Editorial Universidad del Rosario. https://doi.org/https://doi.org/10.12804/urosario9789587845068

Abeldaño Z., Ariel, R., & Garrido, J. F. (2020). Internal displacement due to disasters in Latin America and the Caribbean. In W. L. Filho, G. J. Nagy, M. Borga, D. Chávez Muñoz, & A. Magnuszewski (Eds.), *Climate change, hazards and adaptation options: Handling the impacts of a changing climate* (pp. 389–409). Cham: Springer. https://doi.org/10.1007/978-3-030-37425-9_21

Abeldaño Zuñiga, R. A., & González Villoria, A. M. 2021. Still ignored and still invisible: The situation of displaced people and people affected by disasters in the COVID-19 pandemic. *Sustainability Science*, 2019(February), 1–4. https://doi.org/10.1007/s11625-021-00949-4

Abeldaño Zuñiga, R. A., Juanillo-Maluenda, H., Sánchez-Bandala, M. A., Burgos, G. V., Andrea Müller, S. A., & Rodríguez López, J. R. (2021). Mental health burden of the COVID-19 pandemic in healthcare workers in four Latin American countries. *INQUIRY: The Journal of Health Care Organization, Provision, and Financing*, *58*(January), 004695802110610. https://doi.org/10.1177/00469580211061059

Aon Benfield Analytics. (2020a). Global Catastrophe Recap: September 2020.

Aon Benfield Analytics . (2020b). Global Catastrophe Recap: November 2020.

Aragona, M., Barbato, A., Cavani, A., Costanzo, G., & Mirisola, C. (2020). Negative impacts of COVID-19 lockdown on mental health service access and follow-up adherence for immigrants and individuals in socio-economic difficulties. *Public Health, 186*(September), 52–56. https://doi.org/10.1016/j.puhe.2020.06.055

Banik, A., Nag, T., Chowdhury, S. R., & Chatterjee, R. (2020). Why do COVID-19 fatality rates differ across countries? An explorative cross-country study based on select indicators. *Global Business Review, 21*(3), 607–625. https://doi.org/10.1177/0972150920929897

Caarls, K., Cebotari, V., Karamperidou, D., Alban Conto, M., Zapata, J., & Yang Zhou, R. Y. (2021). *Lifting barriers to education during and after COVID-19: Improving education outcomes for migrant and refugee children in Latin America and the Caribbean.* Florence, Italy: UNICEF. https://orbilu.uni.lu/bitstream/10993/46627/1/Caarls%2C Cebotari%2C et all. 2021.pdf

Chew, M. H., Koh, F. H., Wu, J. T., Ngaserin, S., Ng, A., Ong, B. C., & Lee, V. J. (2020). Clinical assessment of COVID-19 outbreak among migrant workers residing in a large dormitory in Singapore. *Journal of Hospital* Infection, 106(1), 202–203. https://doi.org/10.1016/j.jhin.2020.05.034

Comisión Nacional del Agua, and Servicio Meteorológico Nacional. (2020a). Huracán “LAURA” Del Océano Atlántico: Del 19 Al 28 de Agosto de 2020.

Comisión Nacional del Agua, and Servicio Meteorológico Nacional. (2020b). Reseña Del Huracán “Eta” Del Océano Atlántico: 31 de Octubre–13 de Noviembre de 2020. Mexico City. https://smn.conagua.gob.mx/tools/DATA/Ciclones Tropicales/Ciclones/2020-Eta.pdf

Comisión Nacional del Agua, and Servicio Meteorológico Nacional. (2020c). Reseña Del Huracán “Iota”- Categoría V Del Océano Atlántico: 13 Al 18 de Noviembre de 2020.

Comisión Nacional del Agua, and Servicio Meteorológico Nacional. (2021). Resumen de La Temporada de Ciclones Tropicales Del Año 2020. Mexico City. https://smn.conagua.gob.mx/tools/DATA/Ciclones Tropicales/Resumenes/2020.pdf

The Conversation. (2020). Hurricanes and wildfires are colliding with the COVID-19 pandemic – and compounding the risks. https://theconversation.com/hurricanes-and-wildfires-are-colliding-with-the-covid-19-pandemic-and-compounding-the-risks-145003

Duke Global Health Innovation Center. (2021). Vaccine Manufacturing. https://launchandscalefaster.org/covid-19/vaccinemanufacturing

Economic Commission for Latin America and the Caribbean. (2021). COVID-19 Sanitaria y Su Impacto En La Salud, La Economía. Santiago, Chile. https://repositorio.cepal.org/bitstream/handle/11362/47301/1/S2100594_es.pdf

Enríquez, A., & Sáenz, C. (2021). Primeras Lecciones y Desafíos de La Pandemia de COVID-19 Para Los Países Del SICA. *Estudios y Perspectivas*. Santiago, Chile. Cepal. www.cepal.org/apps%0Ahttps://www.cepal.org/es/publicaciones/46802-primeras-lecciones-desafios-la-pandemia-covid-19-paises-sica

González Franco, J., & Chamba, J. 2021. COVID-19 pandemic: A promissory opportunity for a stronger European Union – Latin America partnership. *Revista de Estudios Sociales, Política y Cultura, 1*(10), 146–167. https://uspt.edu.ar/uspt-revistadigital/index.php/iespyc/article/view/35

The Guardian. (2020, December 22). Alarm at Colombia plan to exclude migrants from coronavirus vaccine. www.theguardian.com/global-development/2020/dec/22/colombia-coronavirus-vaccine-migrants-venezuela-ivan-duque

Hantoushzadeh, S., Shamshirsaz, A. A., Aleyasin, A., Seferovic, M. D., Soudabeh K. A., Arian, S. E., Pooransari, P. et al. 2020. Maternal death due to COVID-19. *American Journal of Obstetrics and Gynecology, 223*(1), 109.e1–109.e16. https://doi.org/10.1016/j.ajog.2020.04.030

Hargreaves, S., Rustagek, Nellums, L. B., McAlpine, A., Pocock, N., Devakumar, D., Aldridge, R. W. et al. (2019). Occupational health outcomes among international migrant workers: A systematic review and meta-analysis. *The Lancet Global Health, 7*(7), e872–882. https://doi.org/10.1016/S2214-109X(19)30204-9

Hayward, S. E., Deal, A., Cheng, C., Crawshaw, A., Orcutt, M., Vandrevala, T. F., Norredam, M. et al. (2021). Clinical outcomes and risk factors for COVID-19 among migrant populations in high-income countries: A systematic review. *Journal of Migration and Health, 3*, 100041. https://doi.org/10.1016/j.jmh.2021.100041

Hintermeier, M., Gencer, H., Kajikhina, K., Rohleder, S., Hövener, C., Tallarek, M., Spallek, J., & Bozorgmehr, K. (2021). SARS-CoV-2 among migrants and forcibly displaced populations: A rapid systematic review. *Journal of Migration and Health, 4*, 100056. https://doi.org/10.1016/j.jmh.2021.100056

Internal Displacements Monitoring Centre. (2021). Displacements Data. www.internal-displacement.org/

International Labour Organization (ILO). (2020). ILO Monitor: COVID-19 and the world of work. Sixth Edition. *Part I: Latest labour market developments: continuing workplace closures, working-hour losses and decreases in labour income*. Geneva, Switzerland.

International Organization for Migration. (2020.) Una Temporada de Huracanes Activa: Desafíos Para Las Personas Desplazadas Durante La Pandemia. https://rosanjose.iom.int/es/blogs/una-temporada-de-huracanes-activa-desafios-para-las-personas-desplazadas-durante-la-pandemia

Jaqueti Aroca, J., Molina Esteban, L. M., García-Arata, I., & García-Martínez, J. (2020). COVID-19 in Spanish and immigrant patients in a sanitary district of Madrid. *Revista Española de Quimioterapia, 33*(4), 289–291. https://doi.org/10.37201/req/041.2020

Kim, H. N., Lan, K. F., Nkyekyer, E., Neme, S., Pierre-Louis, M., Chew, L., & Duber, H. C. (2020). Assessment of disparities in COVID-19 testing and infection across language groups in Seattle, Washington. *JAMA Network Open, 3*(9), e2021213. https://doi.org/10.1001/jamanetworkopen.2020.21213

Marco Integral Regional para la Protección y Soluciones. (2020a). Aplicando El Marco de Respuesta Integral Para Los Refugiados En Centroamérica y México. Respuesta Al COVID-19. Vol. 1. New York. www.acnur.org/es-mx/op/op_fs/5eb5c2454/aplicando-el-marco-de-respuesta-integral-para-los-refugiados-en-centroamerica.html

Marco Integral Regional para la Protección y Soluciones. (2020b). II Informe Anual Del Marco Integral Regional Para La Protección y Soluciones (MIRPS). Mexico City. www.oas.org/es/sadye/inclusion-social/docs/II_Informe_Anual_MIRPS.pdf

Naciones Unidas. (2020). COVID y Huracanes: La Doble Amenaza Que Enfrentan Los Niños de América Central y El Caribe. https://news.un.org/es/story/2020/08/1478292

National Centers for Environmental Information. (2022). *The global anomalies and index data – global surface temperature anomalies* | National Centers for Environmental Information (NCEI). National Oceanic and Atmospheric Administration. 2022.

National Hurricane Center. (2020a). 2020 Atlantic Hurricane Season. NOAA. www.nhc.noaa.gov/data/tcr/index.php?season=2020&basin=atl

National Hurricane Center. (2020b). 2020 Eastern Pacific Hurricane Season. NOAA. www.nhc.noaa.gov/data/tcr/index.php?season=2020&basin=epac

National Oceanic and Atmospheric Administration. (2020). 2020 Hurricane Season. NOAA. www.nhc.noaa.gov/data/tcr/index.php?season=2020&basin=epac

Orendain, D. J. A., & Djalante, R. 2020. Ignored and invisible: Internally displaced persons (IDPs) in the face of COVID-19 pandemic. *Sustainability Science, 16*(1), 337–340. https://doi.org/10.1007/s11625-020-00848-0

Organización Internacional para las Migraciones (OIM). (2006). Glosario Sobre Migración. *Derecho Internacional Sobre Migración. Glosario Sobre Migración.* Vol. 7. Ginebra. https://doi.org/ISSSN: 1816-1014

Organización Panamericana de la Salud. (2021). OPS: América Latina y El Caribe Podrían Enfrentar Una "Avalancha de Problemas de Salud" Si Continúa Interrupción de Servicios de Salud. www.paho.org/es/noticias/28-7-2021-ops-america-latina-caribe-podrian-enfrentar-avalancha-problemas-salud-si

Our World in Data–University of Oxford. (2021). Coronavirus Pandemic (COVID-19) – Statistics and Research. https://ourworldindata.org/coronavirus

Our World in Data–University of Oxford. (2022). COVID-19 Stringency Index. https://ourworldindata.org/grapher/covid-stringency-index

Pasch, R. J., Berg, R., Roberts, D. P., & Papin, P. P. (2021). *Tropical cyclone report: Hurricane Laura.*

Pasch, R. J., Reinhart, B. J., Berg, R., & Roberts, D. P. (2021). *Tropical cyclone report: Hurricane Eta.*

Pei, S., Dahl, K. A., Yamana, T. K., Licker, R., & Shaman, J. (2020). Compound risks of hurricane evacuation amid the COVID-19 pandemic in the United States. *MedRxiv*. https://doi.org/10.1101/2020.08.07.20170555

Perez-Brumer, A., Hill, D., Andrade-Romo, Z., Solari, K., Adams, E., Logie, C., & Silva-Santisteban, A. (2021). Vaccines for all? A rapid scoping review of COVID-19 vaccine access for Venezuelan migrants in Latin America. *Journal of Migration and Health, 4*, 100072. https://doi.org/10.1016/j.jmh.2021.100072

Platform on Disaster Displacement. (2020). *Internal displacement in the context of disasters and the adverse effects of climate change*. Geneva, Switzerland. www.un.org/internal-displacement-panel/sites/www.un.org.internal-displacement-panel/files/27052020_hlp_submission_screen_compressed.pdf

Schenker, M. B. (2010). A global perspective of migration and occupational health. *American Journal of Industrial Medicine, 53*(4), 329–337. https://doi.org/10.1002/ajim.20834

Servicio Meteorológico Nacional. (2022). Información Histórica.

Stewart, S. R. (2021). *Tropical cyclone report: Hurricane Iota.*

UNICEF. (2020). Niños y Niñas de América Central y El Caribe Enfrentan La Doble Amenaza de Una Temporada de Huracanes Más Fuerte y La COVID-19. 2020. www.unicef.org/es/comunicados-prensa/ninos-america-central-caribe-enfrentan-temporada-huracanes-mas-fuerte-covid19

United Nations Department of Economic and Social Affairs. (2021). SDG Indicators Database. https://unstats.un.org/sdgs/dataportal

United Nations High Commissioner for Refugees. (2021). COVID-19: Las Personas Desplazadas y Sus Medios de Vida. https://storymaps.arcgis.com/stories/cd0e2f535c994d79b085a4dee02dd79f

United Nations Office for Disaster Risk Reduction. (2021). El Impacto de Los Huracanes Obliga a Miles de Desplazados a Volver a Empezar. www.undrr.org/es/news/los-efectos-de-iota-y-eta-aun-reverberan-el-impacto-de-los-huracanes-obliga-miles-de

World Health Organization (WHO). (2019). Promoting the health of refugees and migrants: Draft global action plan, 2019–2023. Vol. 1. Geneva, Switzerland. www.unhcr.org/uk/figures-at-a-glance.html,%0Ahttp://www.unhcr.org/uk/figures-at-a-glance.html,%0Ahttp://www.unhcr.org/uk/figures-at-a-glance.html,%0Ahttp://www.who.int/migrants/news-events/A70_24-en.pdf?ua=1

Zard, M., Lau, L. S., Bowser, D. M., Fouad, F. M., Lucumí, D. I., Samari, G., Harker A. et al. (2021). Leave no one behind: Ensuring access to COVID-19 vaccines for refugee and displaced populations. *Nature Medicine, 27*(5), 747–749. https://doi.org/10.1038/s41591-021-01328-3

7 Hot Spot Tracker

Detecting and Visualizing the Types of Spatiotemporal Hot Spots of COVID-19 in the United States

Xiaolu Zhou, Guize Luan, Fei Zhao, and Dongying Li

CONTENTS

7.1 BACKGROUND

The COVID-19 (coronavirus disease of 2019) was declared a pandemic by the World Health Organization on March 11, 2020 (1). By the end of April, there were 3.09 million total confirmed cases and 0.22 million total deaths globally (2). In the United States, more than one million people were confirmed as having contracted the virus and more than 60,000 passed away, with the number continuing to grow.

Faced with this severe situation, a large number of epidemiological and public health studies have been carried out to investigate the at-risk populations, spread means, exposure, and prevention. From geospatial and public health perspectives, mapping the distribution and the spread of the disease in a timely and accurate way is critical to guide policy and resource allocations and raise public awareness. Most public data portals provide the absolute number of confirmed, recovered, and death cases. Although the counts are helpful, it is difficult for the general public and local

DOI: 10.1201/9781003227106-7

health officials to understand the spatial and temporal trends of the development of the disease.

Another neglected aspect of case surveillance is related to spatial dependence. Outbreaks of infectious diseases tend to occur in clustered patters spatially and temporally (3, 4), which has critical implications for examining mechanisms of spread, surveillance, and control. The pattern of outbreak in Asia and Europe clearly showed that spatial proximity to areas where cases cluster is a high-risk factor in adjacent areas. However, most current reports and information portals convey information by individual counties and neglect the clustering effects beyond a single county.

Clustering methods are especially well situated to detecting hot spots of diseases. For instance, Zulu et al. (2014) applied local Moran's I and a Getis-Ord statistic to delineate the distribution of core spatial clusters and hot spots of HIV prevalence (5). Sherman et al. applied scan statistics to find geographical targets for colorectal cancer screening interventions (6). Desjardins et al. applied a prospective space–time scan statistic to detect emerging clusters of COVID-19 based on daily case data at the county level. These approaches are very helpful in detecting the spatial and temporal clusters of diseases, which provide critical information to officials and support the decision-making process. Compared to studies of clusters identified for specific periods, less discussed in the literature is how the intensity and spatial extent of the clusters change over time. For instance, while two areas may present the same case count, the cluster with an intensifying trend is at higher risk than the one with diminishing intensities. Areas that have recently turned into a cluster, or an emerging cluster warrant public attention for disease control. A few studies have considered the changing trends in clusters based on a space-time cube (7, 8). However, limited discussion has been documented regarding the definition of neighbors and temporal changes in hot spots.

In addition, visualization of epidemic information plays an important role in quick and effective communication of disease-related information to the public. Zhou et al. (2020) have summarized and demonstrated how the combination of geographic information systems (GIS) and big data technologies can contribute to the efforts to combat the COVID-19 pandemic. One important advantage of GIS and big data is that they provide dynamic visualization and analysis that integrate data at multiple scales (9). Since the outbreak of COVID-19, some informative geospatial dashboards have been created to visualize the virus's spatial patterns and human mobility patterns. For instance, an interactive web-based dashboard that tracks reported COVID-19 cases in real time has been developed at Johns Hopkins University (2). They also publish the raw data that were collected from a variety of sources (e.g., US Centers for Disease Control and Prevention, European Center for Disease Prevention and Control) to the Github (https://github.com/CSSEGISandData/COVID-19), which have been used in a range of studies. Gao et al. (2020) developed a data dashboard to show the county-level mobility patterns in the United States (10). Many local authorities also display information related to case updates through web maps to inform the public. For instance, the Harris County Public Health Department, Texas, provides a web portal with total cases, counts by region, and charts showing case distribution by demographics (11). Thus far, most available dashboards have focused on reporting the development of case numbers and geographical distribution by administrative boundary. To our best knowledge, no dashboard reveals the trends in hot spot changes over space and time.

To fill this gap, this study aims to contribute to the surveillance of COVID-19 cases spread through detecting geospatial hot spots and their spatiotemporal dynamics on a publicly available dashboard. This dashboard uses mapping and charting to display the information, and supports fast and dynamic queries of the hot spots. We expect our work can provide a platform for the tracking of outbreaks and inform public health decision-making. More importantly, we aim to break the knowledge barrier and equip the general public with statistically proven information to better understand the dynamics, evaluate local risks, and practice proper preparedness and personal protection.

7.2 DATA AND METHODS

This study uses data published by the Center for Systems Science and Engineering at Johns Hopkins University, Baltimore, Maryland, USA. Data are available for the United States and other countries, and dates back to January 22, 2020, for counties in the United States. We used the county-level data to analyze the hot spots.

We used the reported cumulative case counts in each county. To analyze the new cases by day, we subtracted the case count of the previous date from the current date. Occasionally data for some counties and dates show negative values. To correct the negative values, and also identify the critical trends, we applied a weighted moving average to the previous two days to dampen the fluctuation. All the remaining negative case counts were set to zeros. Since the distributions of case counts are heavily skewed, we applied the natural logarithm of one plus (natural logarithm of $1 + x$) to input values to emphasize the hot spots in the entire United States.

7.2.1 G_i^* Statistic

Clusters were identified using a Getis-Ord $\mathbf{G}_i^*$ statistic, which is an inferential statistic to identify a concentration of high values. G statistics were developed by Getis and Ord (12, 13), which can be used to analyze spatial patterns of events or phenomena. As with Moran's *I*, the general G statistic is a global index to reflect the degree of spatial autocorrelation for the whole study area. In contrast, the G_i^* statistic is a local index, which is able to discern clusters of high or low values. The G_i^* statistic considers neighbor interactions instead of just looking at individual counties. If one county has a high number of confirmed cases, attention is needed, but this county alone may not be necessarily qualified as a statistically significant hot spot. To be a statistically significant hot spot, such a county should have a high case number and be surrounded by other high-case counties as well.

A simple form of the G_i^* statistic can be expressed as:

$$G_i^* = \frac{\sum_{j=1}^{n} w_{ij} x_j}{\sum_{j=1}^{n} x_j},$$

where G_i^* is the statistic reflecting the spatial dependency of incident *i to all events;* x_j is the value of incident at location *j*; and w_{ij} is the weight between event *i* and *j*, indicating the mutual spatial relationship, expressed as an n x n elements of weight matrix; n is the number of observations (14).

How to define w_{ij} is an important step in modeling the spatial relationship. In many cases, the spatial relationship can be defined as a function of distance *d* between two points. In this study, since our data is area-based (i.e., county polygons), we used the first-order queen's neighborhood to define the basic weight matrix. Polygons were considered as neighbors if they shared any common boards or vertices. If two polygons were neighbors, the weights were adjusted by the interaction intensity between the two polygons. The interaction was modeled based on the gravity model:

$$I_{ij} = G_{ij}\frac{m_i m_j}{d_{ij}^2}$$

where I_{ij} is the interaction between adjacent polygons of *i* and *j*; *m* is the population of the county; d_{ij} is the distance between the centroid of i and j; and G_{ij} is a variable to normalize the interaction.

The statistical significance of local autocorrelation between neighbors is assessed by the z-score test. A standardized G_i^* z-score based on its sample mean and variance can be expressed as:

$$Z\left(G_i^*\right) = \frac{\sum_{j=1}^{n} w_{ij} w_j - \bar{X}\sum_{j=1}^{n} w_{ij}}{S\sqrt{\frac{n\sum_{j=1}^{n} w_{ij}^2 - \left(\sum_{j=1}^{n} w_{ij}\right)^2}{n-1}}}.$$

The results of the z-scores indicated the locations of spatial clusters in the study area. Significant positive z-scores represent hot spots, while significant negative z-scores represent cold spots.

7.2.2 Mann-Kendall test

The G_i^* statistic identified the local clusters in geographical space, but it did not account for the time dimension. We followed the idea of the space–time cube introduced in ArcGIS Pro (15), stacking the case number for each county in a z-dimension. We choose one day as the temporal step since the data are updated daily. As a result, each county's daily confirmed case count can be sliced into bins. We adapted this method to calculate the G_i^*statistic for each bin based on our polygon-based neighborhood definition.

Once we obtained the G_i^*statistic at each county every day, we used the Mann-Kendall test to detect the statistical increasing or decreasing patterns in the time dimension (16, 17). The null hypothesis of the Mann-Kendall test is no monotonic trend in a series of data. It is a non-parametric test and it first replaces time series values X_i to their relative ranks R_i. Then the statistic S is given as:

$$S = \sum_{i=1}^{n-1} \sum_{j=i+1}^{n} sgn\left(R_j - R_i\right),$$

where sgn is a sign function, returning +1, 0, −1 if the difference between R_j, R_i is positive, 0, or negative respectively.

Based on the statistic S, a z-score can be calculated. A positive z-score shows an upward trend and vice versa. We applied the Mann-Kendall test on the temporal G_i^* scores. The significant trends were recorded.

7.2.3 Trend Classification

Based on the G_i^* and z-score of the Mann-Kendall test, we classified the clusters into seven categories: Emerging Hot Spot, Intensifying Hot Spot, Persistent Hot Spot, Diminishing Hot Spot, Historical Hot Spot, Sporadic/Other Hot Spot, and Never Detected. We adapted the definitions of the hot spots listed on the ESRI website (15) and we used the previous ten days as the time window in this analysis:

- Emerging hot spot refers to hot spots that are statistically significant for any of the recent 30% of time window (in our case three days), but have never been significant during the rest of time window (in our case seven days).
- Intensifying hot spots are hot spots that have been significant for at least 70% of the time window, and the increasing trend in G_i^* scores or case numbers needs to be statistically significant for the last seven days.
- Persistent hot spots are places that have been statistically significant hot spots for at least 70% of the time window, and no significant temporal trends were observed (neither increasing or decreasing).
- Diminishing hot spots are places that have been statistically significant for at least 70% of the time window, and the decreasing trend in G_i^* scores or case numbers needs to be statistically significant for the last seven days.
- Historical hot spots are places where the most recent 30% time period is not hot, but that were ever persistent hot spots in history.
- Sporadic/Other hot spots are places where hot spots were ever detected, but that do not fall into the above five categories.

7.2.4 System Implementation

We cleaned, preprocessed, and computed the statistics using Python 3.7. ArcGIS Pro 2.4 were used to set the map symbology and publish the maps. The interactive web dashboard was designed and developed using the ArcGIS Operational Dashboard. To make the visualization faster, we displayed the symbols representing the different types of clusters on the centroid of each county on the map.

7.3 RESULTS

7.3.1 The Dashboard Interface

Figure 7.1 shows the dashboard interface. The top drop-down allows the users to select the state and date of interest. By default, the initial spatial display is set to the entire United States and the time is set to the date when most recent data are available. The left panel displays the top counties by case count. The map in the middle shows the six different types of clusters (i.e., emerging, intensifying, diminishing, persistent, historical, sporadic/other hot spots) denoted by different colors. As the user selects a state, the corresponding indicators and map elements are updated for the selected state. Users can also use the selection tool to select individual or multiple counties to display more information pertaining to them. The line charts below show the historical data, including case counts, hot spot intensity, and number of counties, representing each type of hot spot. Key statistics about the hot spots are displayed on the right panel.

7.3.2 Hot Spots Identified across the United States

The platform can be used to derive information at multiple scales. For the entire United States, we extracted maps showing analyses for six days between March and May in the United States and examine the dynamics (Figure 7.2). On March 1, there were a few emerging clusters in Rhode Island, Washington, Oregon, California, and Illinois. No other types of hot spots were detected. A week later, on March 8, emerging hot spots spread quickly from coastal areas to many counties in the central United States. Hot spots in the states of Washington and California turned into a persistent category. Another week later, on March 15, the emerging hot spots kept growing, and a few intensifying hot spots were witnessed in counties in New York, Connecticut, and Southern California. On April 1, the biggest hot spots were concentrated in the New England area, with a number of intensifying hot spots. Many emerging hot spots either disappeared or turned into persistent hot spots. On May 1, although a lot of areas still showed persistent hot spots, a number of diminishing hot spots hot spots had appeared in New York, New Jersey, and Massachusetts. The number of historical hot spots started growing. However, places in Illinois, Texas, and Minnesota started showing some intensifying hot spots. On May 10, although the overall confirmed cases gradually leveled off or even dropped, a few intensifying hot spots were observed. Counties in Minnesota and Texas still showed intensifying hot spots and these trends warranted attention. The temporal distribution at the bottom of Figure 7.2 shows the change patterns of different hot spots. Emerging hot spots grew very fast at the beginning and quickly turned into persistent or intensifying hot spots. The number of historical hot spots kept increasing in recent months, but the majority of hot spots were still persistent around May 10.

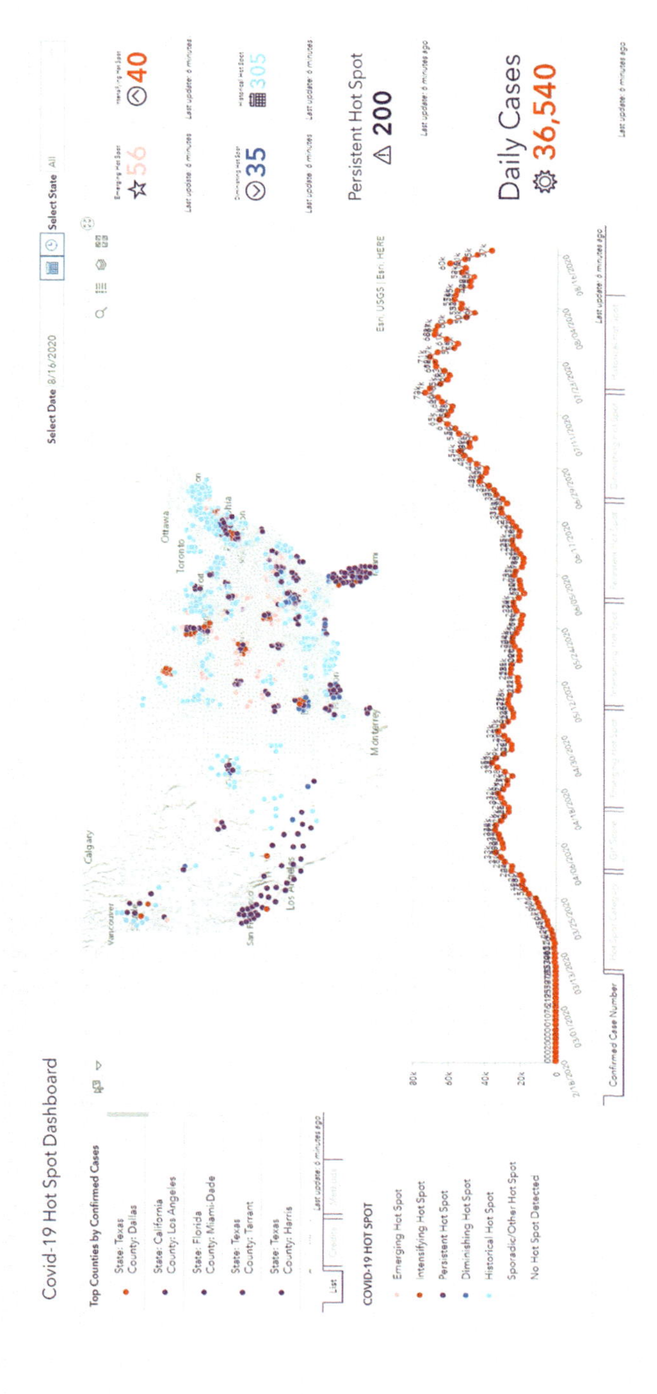

FIGURE 7.1 The overall interface for the dashboard.

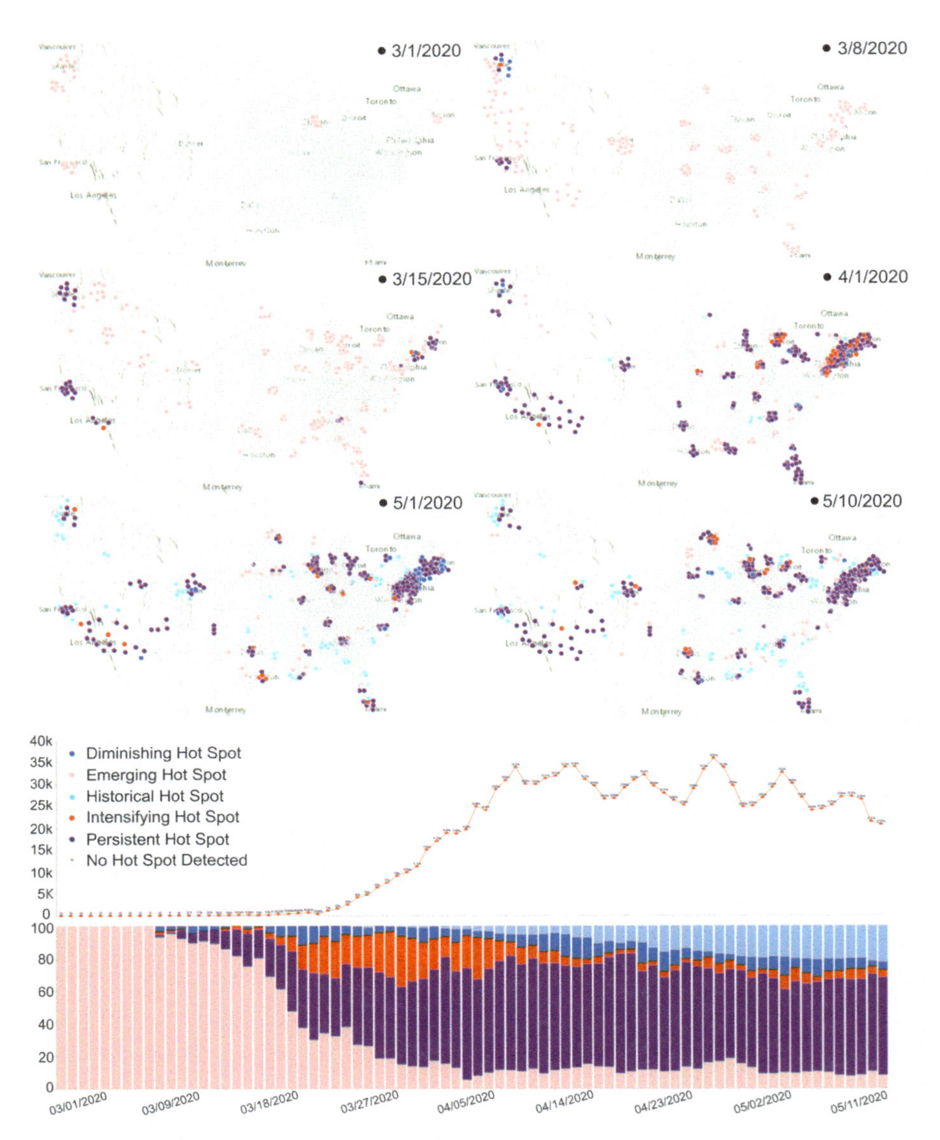

FIGURE 7.2 The spatiotemporal changes of hot spots in the United States since March 1, 2020.

7.3.3 Harris County versus Dallas County, Texas, as a Test Case

The platform also allows us to look into local trends. We compared the patterns of Harris County and Dallas County in Texas as a demonstration (see Figure 7.3). Harris and Dallas Counties are two major counties with the largest populations in Texas. The dashboard shows the emerging hot spots appeared on March 5 and March 9 in Harris and Dallas County respectively. In the following days, hot spots in Dallas

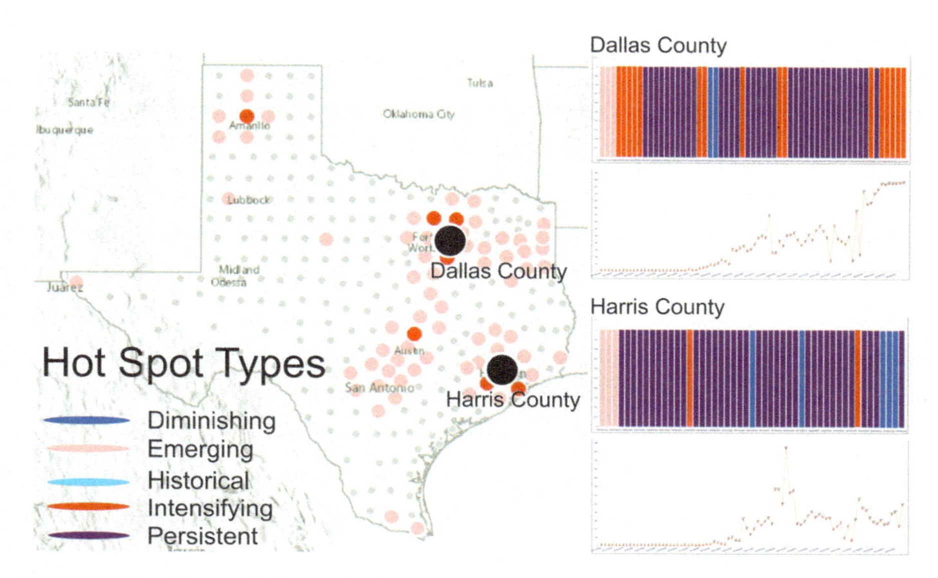

FIGURE 7.3 Temporal change in the confirmed cases in Dallas and Harris Counties, Texas.

County turned into intensifying hot spots, while hot spots in Harris County turned into persistent hot spots. In April, these two counties mostly stayed as persistent hot spots. Between May 1 and May 10, the hot spots in Harris County showed an overall diminishing pattern, while Dallas county revealed an intensifying trend. Even though these two counties are similarly populated and they are within the same state, we can clearly find different trends in hot spots. This warrants close attention to local variations in disease transmission.

7.4 DISCUSSION AND CONCLUSION

The aim of this research was to utilize clustering approaches to detect the spatio-temporal trends of a COVID-19 breakout and develop a dashboard that visualizes the results. We applied spatial statistics and a space–time cube to detect COVID-19 hot spots and their developing trends in the United States. We also visualized our results through a data dashboard to convey the findings in real time using infographics that are easy to understand. Our methods identify the spatiotemporal statistically significant hot spots, and classify them into six different trends: emerging, intensifying, persistent, diminishing, sporadic/other, and historical. We believe this tool is helpful in informing policy makers to monitor situations from adjacent counties, develop and implement social distancing and at-home rules, increase testing, and provide timely warnings and information to raise awareness. During a time period in which different states have cautiously and strategically reopened businesses, this dashboard can be an effective tool to help monitor the dynamics of cases in different locations.

The dashboard not only demonstrates the overall pattern in the United States, but also provides detailed spatiotemporal information for smaller regions or individual counties. As showed in the Harris County example, this tool can demonstrate the temporal profile of the hot spot for any county, state, or region of interest. Our findings provide statistically supported evidence for deciding whether there has been a decreasing trend in cluster intensity. Although we do not recommend relying on a single source for policy decisions, these findings, along with other considerations such as case counts, preparedness and the response of health-care systems, and other capacities and vulnerabilities, can inform public policies. We also expect this dashboard to provide easily understandable information for the general public, as they may not know how to interpret the random daily variations in case counts. This will help individuals, especially those who suffer from immune deficiencies and whose life routines expose them to higher risks, better prepare and adjust to the situations. In addition, the data and the dashboard are maintained separately, and information can be easily updated once data become available. We plan to update the website every day once John Hopkin University publishes the new data.

We suggest our approach can be applied to COVID-19 and the surveillance of other infectious diseases. In infectious disease epidemiology, detecting hot spots and the developing trends in clusters are equally important. We agree with Desjardins et al. (2020) that data sharing enables a variety of researchers from different disciplines to collaboratively combat COVID-19 (18). Geographers and urban scholars can contribute knowledge on spatiotemporal techniques to monitor, trace, and potentially predict disease transmission.

We also acknowledge a few limitations in this study. First, although we used a population-adjusted adjacent weight matrix to define the neighborhood matrix, this is just one way to model the interaction or connection between counites. Future studies will explore other ways to quantify the spatial interactions. Second, as of now, we have only visualized the confirmed cases. Future work will incorporate both death and recovery cases as both of these indicators are also important in public health. Third, understanding how public policies and local measures affect the developing patterns of the hot spots are important knowledge that is relevant. We plan to explore more policy information and incorporate it into the dashboard.

REFERENCES

1. WHO. WHO Timeline–COVID-19 2020. [Internet]. Available from: www.who.int/news-room/detail/27-04-2020-who-timeline---covid-19
2. Dong E, Du H, Gardner L. Dong, E., Du, H., & Gardner, L. (2020). An interactive web-based dashboard to track COVID-19 in real time. The Lancet. Infectious Diseases. 2020;20(5):533–534.
3. Famulare M, Hu H. Extracting transmission networks from phylogeographic data for epidemic and endemic diseases: Ebola virus in Sierra Leone, 2009 H1N1 pandemic influenza and polio in Nigeria. International Health. 2015;7(2):130–138.
4. Lee SS, Wong NS. The clustering and transmission dynamics of pandemic influenza A (H1N1) 2009 cases in Hong Kong. Journal of Infection. 2011;63(4):274–280.
5. Zulu, L. C., Kalipeni, E., & Johannes, E. (2014). Analyzing spatial clustering and the spatiotemporal nature and trends of HIV/AIDS prevalence using GIS: the case of Malawi, 1994-2010. *BMC infectious diseases*, *14*(1), 1–21.

6. Sherman RL. Applying spatial analysis tools in public health: an example using SaTScan to detect geographical targets for colorectal cancer screening interventions. Prev Chronic Dis. 2014 Mar;11:E41.
7. Mo C, Tan D, Mai T, Bei C, Qin J, Pang W et al. An analysis of spatiotemporal pattern for COIVD-19 in China based on space-time cube. Journal of Medical Virology. 2020;92(9):1587–1596.
8. Nakaya T, Yano K. Visualising crime clusters in a space-time cube: An exploratory data-analysis approach using space-time kernel density estimation and scan statistics. *Transactions in GIS*. 2010;14(3):223–239.
9. Zhou C, Su F, Pei T, Zhang A, Du Y, Luo B et al. COVID-19: challenges to GIS with big data. Geography and Sustainability. 2020;1(1):77–87.
10. Gao S, Rao J, Kang Y, Liang Y, Kruse J. Mapping county-level mobility pattern changes in the United States in response to COVID-19. Available at SSRN 3570145. 2020.
11. Harris County Public Health. Harris County/Houston COVID-19 Cases 2020. [Internet]. Available from: https://harriscounty.maps.arcgis.com/apps/opsdashboard/index.html#/c0de71f8ea484b85bb5efcb7c07c6914
12. Getis A, Ord JK. The analysis of spatial association by use of distance statistics. Perspectives on Spatial Data Analysis Geographical Analysis. 1992; 24:189–206.
13. Ord JK, Getis A. Local spatial autocorrelation statistics: distributional issues and an application. Geographical Analysis. 1995;27(4):286–306.
14. Songchitruksa P, Zeng X. Getis–Ord spatial statistics to identify hot spots by using incident management data. Transportation Research Record. 2010;2165(1):42–51.
15. ESRI. Emerging Hot Spot Analysis (Space Time Pattern Mining). [Internet]. Available from: https://pro.arcgis.com/en/pro-app/tool-reference/space-time-pattern-mining/emerginghotspots.htm
16. Mann HB. Nonparametric tests against trend. Econometrica: Journal of the Econometric Society. 1945:245–259.
17. Kendall MG. Rank correlation methods. Oxford(GB): Griffin, 1948.
18. Desjardins M, Hohl A, Delmelle E. Rapid surveillance of COVID-19 in the United States using a prospective space-time scan statistic: Detecting and evaluating emerging clusters. Applied Geography. 2020:102202.

8 Lifestyle Effects on the Risk of Transmission of COVID-19 in the United States

Evaluation of Market Segmentation Systems

Esra Ozdenerol and Jacob Daniel Seboly

CONTENTS

8.1 INTRODUCTION

As the US health-care system moves closer to a value-based approach, there is a constantly growing need for market intelligence that provides health-care providers, health plan providers, major employers, and policy makers with insights into their constituents and positions them to anticipate their needs and behaviors. This study demonstrates the use of market intelligence tools (i.e., lifestyle segments, market segmentation, geodemographic segmentation) to identify a population, determine its

DOI: 10.1201/9781003227106-8

lifestyle clusters, and estimate their propensity for various diseases. Throughout this chapter, we l use the terms lifestyle segments, lifestyle segmentation, market segmentation, and geodemographic segmentation interchangeably, but they convey the same meaning. Owing to the pandemic, we chose to do our analysis with COVID-19 as a health outcome. This chapter offers a unique spatial and temporal approach, proving indispensable for timely and effective ways of analyzing the impact of COVID-19 on American households by their lifestyle characteristics, which are summarized as LifeMode groups based on lifestyle and life stage. We specifically tested two hypotheses: "Is there a difference in average COVID-19 rates among different LifeModes?" and "Which LifeModes have COVID-19 rates that are higher/lower than average?" We focused on comparing each LifeMode's mean to the nation's mean to see spatial and temporal patterns of high risk and the lifestyle effects on the risk of transmission of COVID-19 in the United States.

Lifestyle is an important factor, along with genes, behavior, and environment, that influences humans' health and their risk for diseases. Lifestyle data are collected at the household level through geodemographic segmentation typically used for marketing purposes to identify consumers' lifestyles and preferences by private sector marketing (Harris et al., 2005). Since this information is used by firms to identify new customers and potential business locations, geodemographic segmentation is a common marketing strategy that involves grouping potential customers into lifestyle segments by state, region, city, or neighborhood. Reflecting the diversity among American neighborhoods, lifestyle segments reflect demographic shifts over the last decade to establish consumer markets, as well as the emergence of new markets due to population growth and geographic, demographic, and socioeconomic change. When composing lifestyle segments, neighborhoods are classified into unique segments nested under LifeMode groups based not only on demographics but also on socioeconomic and behavioral characteristics. For example, geographical data represent where the focal groups are located and where they are buying and using products. Behavioral data focus on when the groups are more likely to buy, under what circumstances the groups are more likely to buy, and how the groups choose to consume or use the product. Demographics represent the race, gender, age group, and marital status of customer/consumers. Psychographic data concentrate on their uniqueness, personal preferences, and lifestyle choices; what they do in their spare time and what products they choose to free up more spare time; how they see themselves and their communities; and identify careers, opinions, and income parameters (People2 People & Places, 2020).

Considering the geographical heterogeneity of the US population or the ways in which to uniquely characterize households and their lifestyle, such as retirement communities or diverse urban immigrant enclaves, our approach may provide actionable information for key stakeholders with respect to the focus of interventions and reveal the underlying factors involved in differential health outcomes. Furthermore, such an understanding of the COVID-19 crisis may be instrumental in the implementation of prevention and control policies to those specific households exhibiting spatial and temporal patterns of high risk, and may prepare us for future pandemics affecting populations with different lifestyles.

8.2 MATERIALS AND METHODS

8.2.1 Data

We combined data from multiple sources and merged them in a geographic information system (GIS) to create a visual representation through maps. We used the ESRI Tapestry segmentation system to associate lifestyle clusters to COVID-19 (ESRI-Tapestry, 2020). The COVID-19 data set is twofold: infection and mortality rates. We explicitly described both ESRI Tapestry segmentation and COVID-19 data sets under separate headings below.

With the advance of GIS technology and cloud computing, progressively better data sets and tools have become available to improve the identification of market segments and the operationalization of a lifestyle segment scheme in the marketplace. Lifestyle segmentation (i.e., market segmentation) describes the division of a market into homogeneous groups, which will respond differently to promotions, communications, advertising, product, pricing, and other marketing mix variables. Examples of the most common commercial or market segmentation systems available today include MOSAIC, ACORN, ESRI Tapestry, and many others (Esri-Tapestry, 2020; Acorn, 2020; Claritas, 2020; Mosaic, 2020). These segmentation systems utilize consumer surveys (e.g., Experian's Consumer View database) and apply traditional customer profiling techniques such as relationships between purchased products and consumers' beliefs and life patterns (Experian Segmentation, 2020). We used the ESRI Tapestry segmentation system that is available on an annual basis as population and household counts by Tapestry segment are updated each year (Esri-Tapestry, 2020).

8.2.1.1 ESRI Tapestry Segmentation System

The ESRI Tapestry segmentation system uses Experian's Consumer View database, the Survey of the American Consumer from GfK MRI (Survey of the American Consumer, 2019), and the US Census American Community Survey (ACS, 2020). The GIS that supports the ESRI Tapestry Segmentation platform illustrates relationships, connections, and visual patterns that are not necessarily obvious in any one data set and enables different demographic data sets to be brought together to create a complete picture of local communities and neighborhoods across the United States (Esri-Tapestry, 2020; Esri Data, 2020).

ESRI Tapestry segmentation classifies US neighborhoods into 67 unique market segments, based on socioeconomic and demographic factors, then consolidates these 67 segments into 14 LifeModes with names such as "High Society," "Senior Styles," and "Factories and Farms" that have commonalities based on lifestyle and life stage (Esri-Tapestry, 2020). ESRI Tapestry segmentation data were downloaded from ESRI (Esri-Tapestry, 2020). Our data set contains a variable denoting the dominant tapestry segment within each US county. Table 8.1 shows the number of counties within each LifeMode.

8.2.1.2 COVID-19 Infection and Mortality Rates

To associate lifestyle clusters to COVID-19, we downloaded two data sets: COVID-19 infection and mortality rates from USAFacts.com (USA Facts, 2020). One data set contained the number of confirmed positive COVID-19 cases in each US county by

TABLE 8.1
Dominant Life Mode within Each US County

LifeMode Name	Code	Counties
Affluent Estates	1	71
Upscale Avenues	2	41
Uptown Individuals	3	13
Family Landscape	4	159
GenXUrban	5	163
Cozy Country Living	6	1261
Ethnic Enclaves	7	106
Middle Ground	8	80
Senior Styles	9	69
Rustic Outposts	10	965
Midtown Singles	11	21
Hometown	12	141
Next Wave	13	5
Scholars & Patriots	14	45
Total		**3140**

date. The other data set contained the number of COVID-19 deaths in each county by date. These data sets covered the period from January 22, 2020, when the first case in the United States was discovered, to June 30, 2020. Using population data obtained from the US Census Bureau, we calculated the rate of infections and deaths per 100,000 residents for each county and date and further mapped the COVID-19 cases by county (US Census Bureau, 2020). This facilitated a fairer and more accurate analysis of the impact of COVID-19 on different lifestyles.

8.2.2 Statistical Methodology

We used analysis of variance (ANOVA) statistical test to determine whether there is any association between COVID-19 infection and mortality rates and lifestyles (Kaufmann & Schering, 2014). We further used analysis of means (ANOM) to determine which lifestyles have higher risk (Rao, 2005). Since there are many similarities and overlaps between lifestyle segments within the same LifeModes, and testing at the segment level would drastically reduce sample sizes—thus curtailing the power of the statistical tests for the statistical analysis—we chose to use the broader tapestry LifeModes rather than lifestyle segments (Esri-Tapestry, 2020).Table 8.2 contains a summary of the statistical tests performed with the corresponding hypothesis tested for the analysis.

Our nationwide analysis included all the counties in the United States. A one-way ANOVA was performed with LifeModes as the factor variable and infection rates as the response variable to determine whether there is a difference in average infection rate among different LifeModes. The one-way ANOVA compares the means between the LifeModes and determines whether any of those means are statistically significantly different from each other.

TABLE 8.2
Summary of Statistical Tests Performed

Test	Purpose
ANOVA	Is there a difference in average COVID-19 rate among different LifeModes?
ANOM	Which LifeModes have COVID-19 rates which are higher/lower than average?

Specifically, it tests the null hypothesis: "All means are equal":

$$Ho: \mu 1 = \mu 2 = \mu 3 = \cdots = \mu k,$$

where μ = group mean and k = number of groups. If the one-way ANOVA returns a statistically significant result, we accept the alternative hypothesis (HA), "Not all means are equal," which means that there are at least two group means that are statistically significantly different from each other. For infection rates, we ran the ANOVA analysis once for each date during the data time frame. For every date between March 1 and June 30, 2020, the ANOVA analysis revealed a statistically significant association between COVID-19 infections and LifeMode classification. We then performed the same analysis with mortality rates for each day in the period. According to the ANOVA analysis, a significant association between COVID-19 deaths and LifeModes emerged on April 1, 2020, and was sustained through to June 30, 2020. As some of the assumptions for the one-way ANOVA test were violated, specifically the normality of residuals and homogeneity of variances, we also performed the Welch ANOVA to increase confidence in the results. The Welch ANOVA test confirmed the statistical significance of the association between COVID-19 infections/deaths and LifeModes, with strong p-values ranging from 10–4 to 10–40 depending on the date.

ANOM was also performed to determine which LifeModes have incidence rates that are significantly above/below the overall mean incidence rate. For each LifeMode and each date, we calculated a confidence interval for the mean COVID-19 infection rate. For each LifeMode with a confidence interval on a date that was entirely above the overall mean infection rate on that date, we can conclude that this LifeMode had an above-average risk of COVID-19 infection on that date. Conversely, for each LifeMode with a confidence interval on a date that was entirely below the overall mean for that date, we can conclude that this LifeMode had a below-average risk of COVID-19 infection on that date. The same process was used to determine which LifeModes had above- and below-average risks of COVID-19 death on which dates.

8.3 RESULTS

8.3.1 COVID-19 Spread in the United States

In this visual COVID-19 time line (Figure 8.1), we delve into some significant milestones that occurred in the United States in 2020 and further map COVID-19

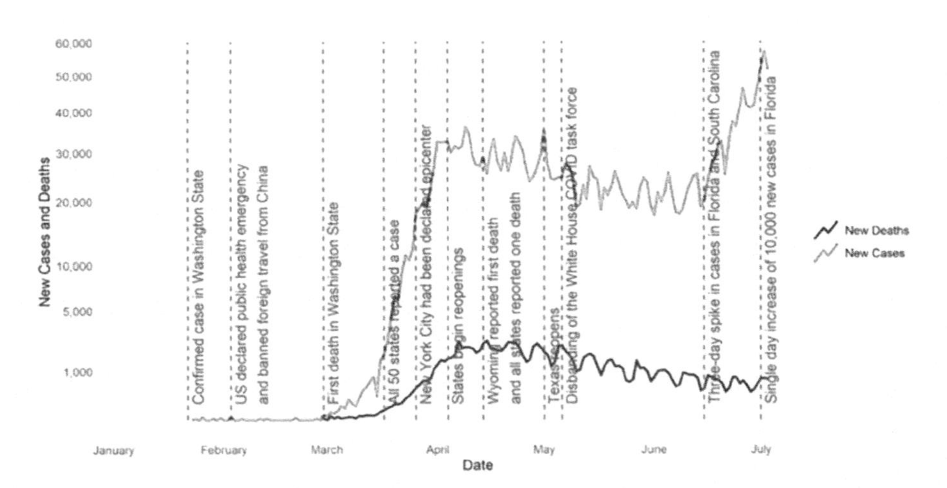

FIGURE 8.1 New COVID-19 cases and deaths in conjunction with the timeline of COVID-19 developments.

cases and deaths (Figure 8.2). The first cases of a severe respiratory illness that would come to be known as SARS-CoV-2 or COVID-19 were reported in Wuhan, China, in December of 2019 (Stawicki et al., 2020). Due to the interconnectedness of the modern world, it took only 21 days for the first case of COVID-19 in the United States to be reported. The first case of COVID-19 was confirmed in Washington State on January 21, 2020 (AJMC, 2020). The United States declared a public health emergency on February 3, 2020, and banned foreign individuals who had recently traveled to Wuhan from entering the United States. The first death in the United States was recorded in the state of Washington on February 29, 2020. By March 17, 2020, all 50 states had reported a case (Council on Foreign Relations, 2020). Figure 8.2 shows the number of cumulative COVID-19 cases per county on March 15, 2020.

Upon reaching the milestone of having COVID-19 present in every state, it was estimated that more than 6,300 cases had been diagnosed in the United States and that the global death toll had surpassed 7,900. (CBS News, 2020). By March 26, New York City had been declared the epicenter of the outbreak with more than 20,000 cases, and by March 28, 2020, there were more than 115,000 cases and 1,891 deaths in the United States (Dong et al., 2020). By April 7, 395,926 COVID-19 cases had been reported, with two-thirds of cases coming from just eight geographical jurisdictions: New York City (NY), New York, New Jersey, Michigan, Louisiana, California, Massachusetts, and Pennsylvania (CDC, 2020). By April 14, Wyoming reported its first COVID-19-related death, which meant that every state had reported at least one death (Council on Foreign Relations, 2020). Figure 8.2 shows the number of cumulative COVID-19 cases per county on April 15, 2020.

Throughout the summer, there was a consistent rise in cases throughout the United States. This increase coincided with the disbanding of the White House COVID-19 task force, and the phased reopening of states. In early April, the US president

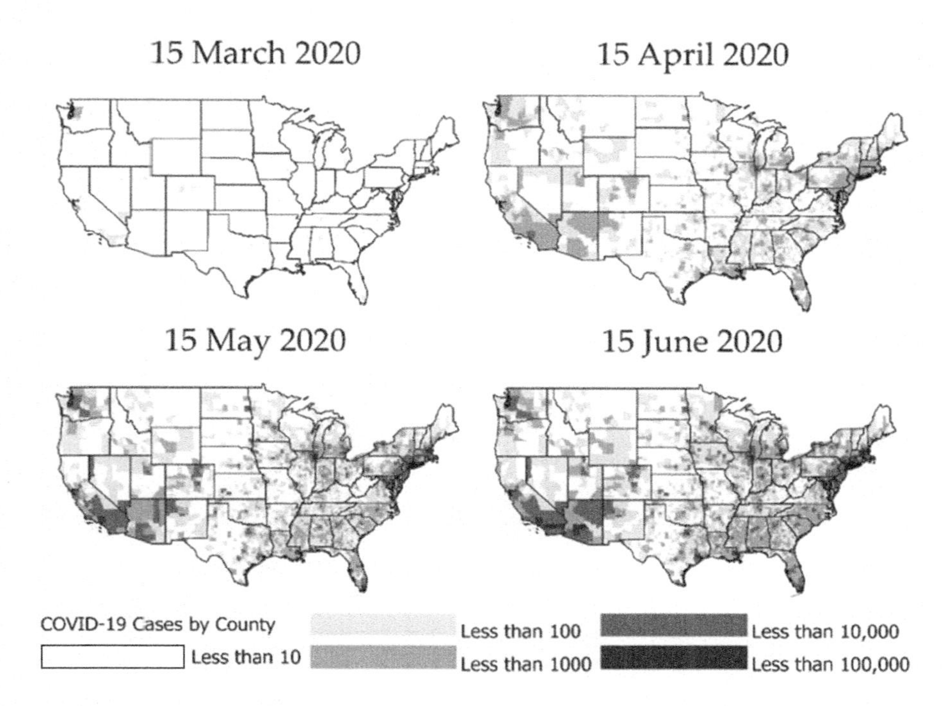

FIGURE 8.2 The number of cumulative COVID-19 cases per county from March 15 to June 15, 2020.

gave permission to state governors to decide how reopening would function in their particular states. Texas was one of the first states to decide to reopen in early May (Council on Foreign Relations, 2020). Figure 8.2 shows the number of cumulative COVID-19 cases per county on May 15, 2020. By mid-June, Florida and South Carolina were both recording a three-day-long spike in cases. Later, on July 2 , Florida reported a single-day increase of 10,000 new cases, and by July 8, hospitals in both Florida and Arizona began to reach capacity due to their COVID-19 caseload (Council on Foreign Relations, 2020). Figure 8.2 shows the number of cumulative COVID-19 cases per county on June 15, 2020.

8.3.2 High-Risk LifeModes

The results of Welch ANOVA analysis show statistically significant associations between LifeModes and COVID-19 infection rate and COVID-19 mortality rate for all dates in the study period. P-values ranged from 10−4 to 10−40 depending on the date. Table 8.3 shows the high-risk LifeModes for infection and death.

Figures 8.3 and 8.4 graphically illustrate when each LifeMode had above-/below-average risk for COVID-19 infection/death on what dates. Figure 8.3 represents infection rates. LifeModes are displayed on the vertical axis of these charts, while dates are displayed on the horizontal axis. Where each LifeMode intersects with a

TABLE 8.3
High-risk LifeModes for COVID-19 Infection and Death

Affluent Estates (1)
Upscale Avenues (2)
Uptown Individuals (3)
Rustic Outposts (10)
Hometown (12)
Next Wave (13)

*Senior Styles (9) is low risk for infection only. *Ethnic Enclaves (7) is high risk for infection only. *GenXUrban (5) is high risk for death only. *Scholars and Patriots (14) is low risk for death only.

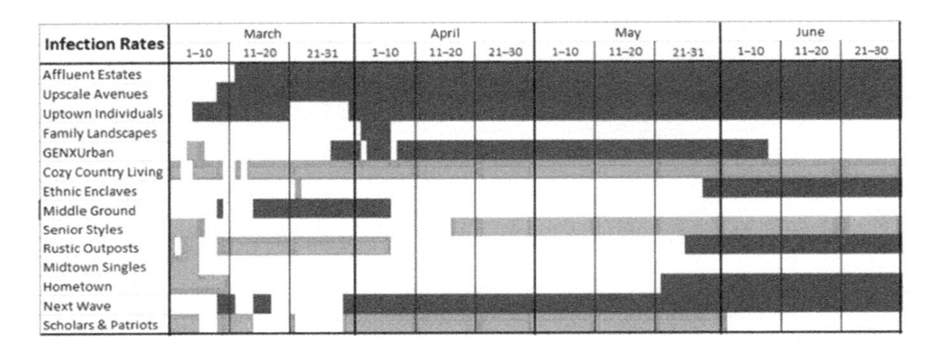

FIGURE 8.3 COVID-19 infection risk by LifeMode and date. Red color indicates high risk for that LifeMode and date; green color indicates low risk for that LifeMode and date; white color indicates inconclusive risk.

date, the color of the appropriate grid point represents the risk for that LifeMode on that date. Red represents above-average risk, green represents below-average risk, and gray represents inconclusive results.

In the cases where a particular LifeMode's status as high/average/low risk changes over time, we emphasize that the change is not caused by any intrinsic changes in the makeup of the groups, but instead by changes in the relative behavior, mobility, and exposure of the groups. The following paragraphs provide examples and speculation as to how this might be the case with explanations of Figures 8.3 and 8.4.

According to Figure 8.3, Affluent Estates, Upscale Avenues, and Uptown Individuals exhibited a consistently high risk of COVID-19 infection from early March through to June. These are more affluent people who travel much more than the average American and are active in fitness pursuits such as bicycling, jogging, yoga, and hiking. Next Wave also exhibited high risk from late March through to June. These are mostly people of international origin and immigrants who frequently travel overseas. Thus, travel may explain why these lifestyles were more likely to be exposed to COVID-19 in the early stages of the pandemic. GenXUrban became

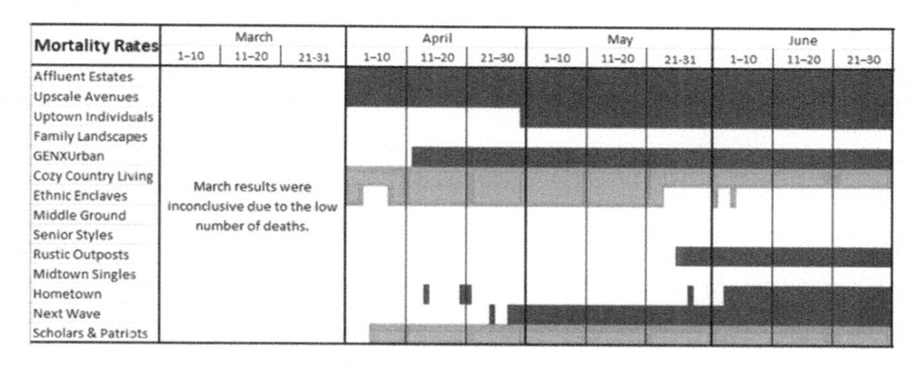

FIGURE 8.4 COVID-19 mortality risk by LifeMode and date. Red color indicates high risk for that LifeMode and date; green color indicates low risk for that LifeMode and date; white color indicates inconclusive risk.

a high-risk LifeMode in late March but went back to inconclusive in early June. The households in this LifeMode are generally 38–47 years old, married, living in single-family housing, are predominantly white, and make USD 47,000–68,000 annually. They own older single-family homes in urban areas, live and work in the same county, and enjoy going to museums and rock concerts, dining out, and walking for exercise.

LifeModes representing poorer households, including Ethnic Enclaves, Rustic Outposts, and Hometown, became high risk in late May and stayed in that category through the rest of our study period. While members of these LifeMode segments do not travel as often, they are more likely to work at essential (blue-collar) jobs, less likely to have the opportunity to work from home, and probably have less access to quality health care. Members of these LifeModes, such as Latino communities of Ethnic Enclaves, may have also been more skeptical of government and public health recommendations regarding COVID-19 safety and afraid to seek out public services, including medical care (Rodriguez-Diaz, 2020). All these factors could have combined to ensure that these lifestyles became high risk once the pandemic was more established in the United States.

LifeModes which have remained consistently at low risk throughout the pandemic include Cozy Country Living, Senior Styles, and Scholars and Patriots. Cozy Country Living represents the most rural LifeMode. It is found to be low risk for both infection and death. These are the last communities that the pandemic reached, as these are the most disconnected from urban society and international travel. Senior Styles represents mostly retirement communities, nursing homes, and neighborhoods where the population is predominantly made up of those over 65. Given the elevated risk of COVID-19 to the elderly, these communities generally took extra precautions to prevent the spread of COVID-19, which explains their low risk of infection. Senior Styles (9) is low risk for infection only, but not for death. Scholars and Patriots communities consist mainly of university towns and military bases. Both universities and military bases took significant steps to prevent outbreaks of COVID-19, with university towns even sending many of their students home. This would have limited opportunities for COVID-19 to affect these LifeModes.

Figure 8.4 is drawn according to the same scheme but represents mortality rates instead. According to Figure 8.4, the analysis of mortality rates yields many of the same results as the infection rate analysis. However, Ethnic Enclaves was high risk for infection but low risk for death. This might be observed because Ethnic Enclaves mostly consists of younger families, and COVID-19 is much less likely to cause death in young and middle-aged people. Meanwhile, Senior Styles is low risk for infection but not for death. This occurs because the mortality rate for the elderly is much higher, so even if these populations see fewer than average cases, they can still reach an average rate of deaths.

8.3.3 Spatial Variation of High-Risk Life Modes

We mapped the high-risk life modes associated with COVID-19 incidence and mortality. These maps demonstrate that the lifestyle characteristics of these LifeModes account for some of the spatial variation and significantly contribute to viral transmission and incidence/mortality rates. According to Figure 8.5, Affluent Estates, Upscale Avenues, and Uptown Individuals exhibited consistently high risk of COVID-19 infection during the pandemic's early days in March and throughout April in the West Coast cities. Upscale Avenues were the first and the most impacted by COVID-19 in Seattle, Washington, and Los Angeles, California. Upscale Avenues was also the first group impacted by COVID-19 in the states of Wyoming and Colorado, which are both vacation destinations and retirement communities of these prosperous LifeModes.

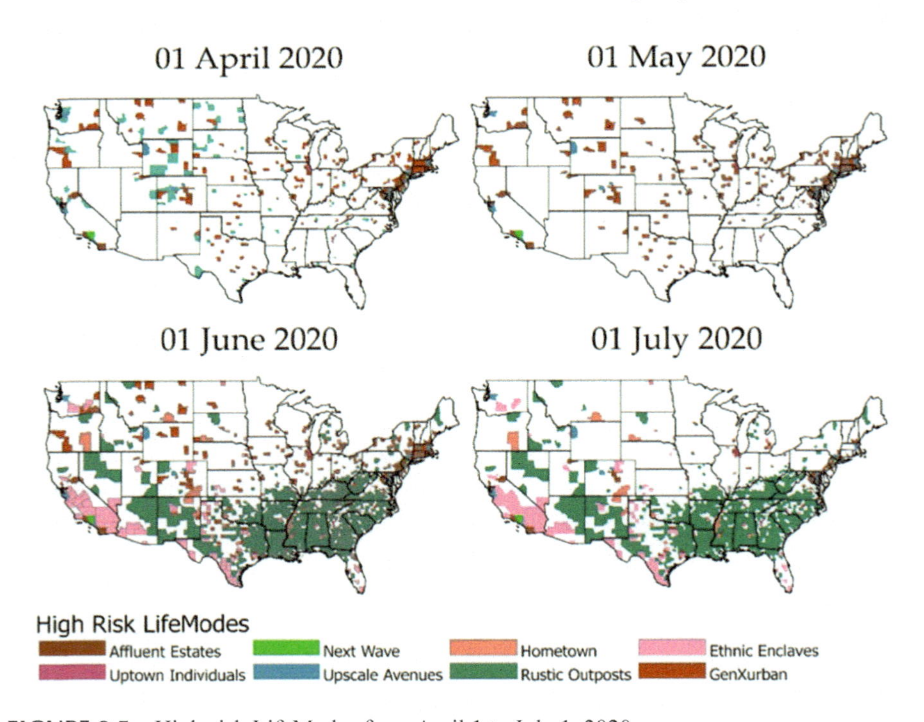

FIGURE 8.5 High-risk LifeModes from April 1 to July 1, 2020.

GenXUrban was only impacted in Washington and Oregon States early in the pandemic in inland areas of these states. It is possible that they may have been exposed to a returned traveler who was infected. GenXUrban became the most dominant LifeMode in less dense areas of inland states in March and April. Our results indicate that GenXUrban (5) is a high-risk LifeMode for death only. Therefore, this merits attention as to what makes this LifeMode high risk for death. They could face health challenges such as lack of health care and social services. Affluent Estates were the most impacted by COVID-19 in San Francisco, Los Angeles, and San Diego in California, and major East Coast cities such as New York, Boston, and Washington DC, where the virus arrived early and spread quickly. These dynamics of the outbreak swept through American cities in early March. We see a hierarchical spread of the pandemic from the coastal gateway cities to major cities of inland areas by plane and by car travel. For example, Lake Tahoe is a weekend getaway for vacationers, and this explains the hot spot of the members of Affluent Estates in this area and the transmission to the GenXUrban members of the local community.

The New Wave LifeMode, consisting of frequent overseas travelers, was also a dominant life mode in Los Angeles, in Southern California, that has been the most impacted and exposed to COVID-19 in the early stages of the pandemic. The New Wave LifeMode is the most racially and ethnically diverse among LifeMode groups with a Hispanic majority. A large share is foreign born and speak only their native language. Most are renters in older multiunit structures and live-in crowded homes. Uptown Individuals only exhibited in Chicago and Atlanta, as they are the gateway cities of international travel and represent young successful singles who are also frequent travelers and reside in highly dense cities. This LifeMode represents a younger population in comparison to the prosperous married couple members of Affluent Estates and Upscale Avenues. This pattern continued through to June.

As states moved to ease lockdown restrictions and reopened in May, new LifeModes emerged as being impacted by COVID-19 and the growth subsumed rural areas of California, Texas, and Southern states. This time, LifeModes represented poorer households, including Ethnic Enclaves, Rustic Outposts, and Hometown. The members of these LifeModes are more likely to work at essential (blue-collar) jobs, less likely to have the opportunity to work from home, and have less access to quality health care. Young, Hispanic families, and multilingual and multigenerational households of Ethnic Enclaves in Texas and southern Florida became targets of COVID-19. The Hometown LifeMode was impacted by COVID-19 in Native American communities in New Mexico, Colorado, and Oregon. COVID-19 hit the poor communities of the Mississippi River delta and impacted members of the Hometown LifeMode. Rustic Outposts emerged as a dominant Life Mode in the entire southern United States impacted by COVID-19, and rural areas became high risk.

Figure 8.6 shows the shift from no outbreak to low-risk to high-risk LifeModes quickly spreading and covering the coastal and southern counties of the nation from the beginning of the outbreak in March to July. Table 8.4 shows that the mean case rate by LifeMode is highest among Next Wave (1,965), Uptown Individuals (1,865), Upscale Avenues (1,479), Midtown Singles (1,429), Hometown (1,387), and lowest among Scholars and Patriots (555) and Cozy Country Living (477). Contrary to the

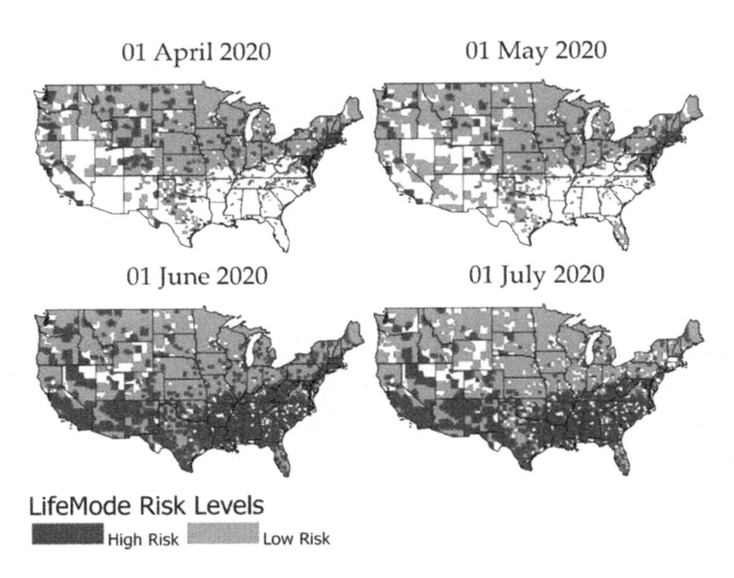

FIGURE 8.6 High- and low-risk LifeModes from April 1 to July1, 2020.

TABLE 8.4
Mean Case Rate by LifeMode

LifeMode	Code	Counties	Mean Infection Rate* (cases per 100,000)	Mean Mortality Rate* (deaths per 100,000)
Affluent Estates	1	87	(766, 1134)	(39.4, 62.7)
Upscale Avenues	2	33	(1029, 1966)	(48.5, 102.6)
Uptown Individuals	3	9	(1178, 2084)	(55.2, 131.3)
Family Landscapes	4	111	(584, 742)	(14.4, 22.4)
GenXUrban	5	195	(585, 750)	(26.8, 38.6)
Cozy Country Living	6	1383	(359, 438)	(10.1, 12.5)
Ethnic Enclaves	7	89	(1004, 1729)	(14.0, 23.5)
Middle Ground	8	82	(510, 784)	(11.6, 37.4)
Senior Styles	9	66	(369, 592)	(10.6, 25.0)
Rustic Outposts	10	971	(732, 870)	(23.6, 29.6)
Midtown Singles	11	11	(594, 1645)	(12.2, 107.6)
Hometown	12	65	(812, 1423)	(28.2, 59.3)
Next Wave	13	5	(2308, 4198)	(146.5, 364.1)
Scholars and Patriots	14	35	(367, 974)	(4.6, 13.6)

* 95% confidence interval bounds generated via 1000-member bootstrap ensembles
* Cases per 100,000 people

high mortality rates of Senior Styles, this LifeMode was low risk for infection only in comparison to Ethnic Enclaves, which was high risk for infection only. GenXUrban was high risk for death only, while Scholars and Patriots was low risk for death only.

8.3.4 Demographic Profile of High-Risk LifeModes

We analyzed the resultant high-risk LifeModes by their demographic profile, specifically by their diversity index, population density, population by age groups, college education, median age, and household size (see Table 8.5).

The diversity index summarizes racial and ethnic diversity. The index shows the likelihood that two persons, chosen at random for the same area, belong to different racial or ethnic groups. The index ranges from 0 (no diversity) to 100 (complete diversity). For example, the diversity core for the whole of the United States is 64.0, which means there is a 64.0% probability that two people randomly chosen from the US population would belong to a different race or ethnic group. Looking at the demographic profile of high-risk LifeModes for COVID-infection, the diversity index is found to be high among high-risk LifeModes. For example, the New Wave LifeMode has a diversity index of 89.5, followed by Ethnic Enclaves (82.4) and Midtown Singles (78.4). These LifeModes are urban neighborhoods in which immigrant groups or ethnic minorities are residentially concentrated. In the COVID-19 literature, people residing in immigrant neighborhoods were less likely to be tested, but the likelihood that a test was positive, was larger in those neighborhoods, as well as in neighborhoods with larger households or predominantly Black populations. COVID-19 diagnoses were associated with counties with higher numbers of monolingual Spanish speakers (Rodriguez-Diaz, 2020; Chen, 2020; Bai, 2020)

We see more cases of COVID-19 and high mortality rates in high-risk LifeModes with high population densities. The Uptown Individuals (7,614.5), Next Wave (4,252.3), Midtown Singles (2,398.5), and Upscale Avenues (1,105.5) LifeModes have high COVID-19 mortality rates and they reside in highly dense urban areas and live in higher-density housing. The GenXurban LifeMode, which is found to be high risk for death only, has the highest percent population of 65 and above (20.4%), followed by 45–64 at 27.9%. This LifeMode has a growing population of retirees. In contrary, Ethnic Enclaves, which is high risk for infection only, has a high percentage of the population below 18 years of age at 29.1%, followed by 18–44 at 40%. Ethnic Enclaves also stands out by average household size being 3.35, followed by Next Wave (3.31). Scholars and Patriots is low risk for death only and has the lowest population of 65 and above at 4.8%. The population of this LifeMode has the highest proportion of individuals between 18 and 44 at 76.9%, including college campuses and military populations. College enrollment is highest among members of Scholars and Patriots (49.5), followed by Midtown Singles (7.9) and Uptown Individuals (6.8). Senior Styles are low risk for infection only, with a high percentage of 65 and above population (37.9%). Senior Style households are commonly married empty nesters or singles living alone; homes are single family (including seasonal getaways), retirement communities, or high-rise apartments.

TABLE 8.5
Demographic Profile of LifeModes in Comparison to Overall US Averages

	COVID-19 Cases per 100,000*	COVID-19 Deaths per 100,000*	Diversity Index	Median Age	Ave. HH size	Pop Density (Persons per mi²)	Pop <18	Pop 18-44	Pop 45-64	Pop 65+	College enrollment
United States	848	41.00	64	38.2	2.59	92.7	22.3	36.1	25.9	15.6	6.3
Affluent Estates	1114	59.87	43.8	43	2.89	503.2	24.7	28.0	31.2	16.1	4.9
Upscale Avenues	1479	83.56	66.5	40.6	2.70	1105.5	21.1	34.8	28.3	15.9	6.1
Uptown Individuals	1865	109.8	65	34.7	1.86	7614.5	11.7	55.8	21.4	11.2	6.8
Family Landscapes	716	18.12	54	36.8	2.86	294.4	25.4	36.0	26.9	11.7	5.6
GenXUrban	893	52.04	41.2	43.5	2.43	409.8	19.7	32.0	27.9	20.4	5.5
Cozy Country Living	477	17.74	26.6	45	2.53	21.9	20.5	29.5	30.6	19.4	3.9
Ethnic Enclaves	1087	19.05	82.4	31.8	3.35	134.2	29.1	40.0	21.6	9.3	6.1
Middle Ground	1209	88.44	68.9	36	2.41	396.8	21.7	40.2	24.2	13.8	6.7
Senior Styles	842	31.26	47.6	57.4	1.94	89.3	12.3	23.9	25.9	37.9	4.7
Rustic Outposts	818	25.33	49.1	40.7	2.60	29.2	22.0	33.3	28.0	16.7	3.8
Midtown Singles	1429	94.15	78.4	30.9	2.38	2398.5	25.0	45.2	20.4	9.4	7.9
Hometown	1387	89.11	65.5	38	2.48	200.5	23.3	34.8	25.4	16.5	5.4
Next Wave	1965	116.1	89.5	29.8	3.31	4252.3	29.3	43.0	19.8	7.9	5.9
Scholars and Patriots	555	9.04	58.2	22.8	2.28	464.2	10.4	76.9	7.9	4.8	49.5

* COVID-19 data obtained June 30, 2020

Source: ESRI, Inc. [23, 28]

8.3.5 Economic Profile of the High-Risk LifeModes

When we review the economic profile of the high-risk LifeModes in Table 8.6, members of Affluent Estates (31.7), Upscale Avenues (29.5), and Uptown Individuals (36.9) that exhibited consistently high risk of COVID-19 infection during the pandemic's early days were professionals by occupation, in comparison to Next Wave (30.4), Hometown (25.1), and Midtown Singles (26.5), who engage primarily in service occupations. Table 8.7 shows that the unemployment rate is highest among Hometown (9.5), Midtown Singles (8.0), and Next Wave (7.8). The service industry is the industry with the highest percentage of members of high-risk LifeModes employed, followed by the finance, real estate, and retail trade industries. Next Wave (9.8), Rustic Outposts (8.9), Ethnic Enclaves (8.3), and Cozy Country Living (8.0) are the high-risk life modes that have the highest percentage of employment in the construction industry.

8.4 DISCUSSION

During a time of increased attention on social determinants of health (SDOH), our finding that human social behaviors such as lifestyle preferences affect the risk of contracting COVID-19 proved very useful in comparison to the index-based approaches that quantify SDOH. With this increased understanding of SDOH and with market intelligence tools, we can target vulnerable populations that could be impacted by COVID-19 for prevention and control strategies. Lifestyle segmentation classifies neighborhoods into unique segments (i.e., 67 segments) and LifeMode groups (i.e., 14 summary groups) based not only on demographics but also socioeconomic and behavioral characteristics. With a greater depth of understanding of these at-risk households based on lifestyle, we can explore the localized households and predict expansion of the geographical spread of the pandemic. These households can be targeted for better social services for COVID-19 at clinical settings. Public health messages and clinical information could be issued to the public and medical practitioners for these at-risk households to provide better assistance in clinical diagnoses and to implement preventive measures such as social distancing and reducing unnecessary travel. This chapter makes a unique contribution to the public health literature by associating lifestyle characteristics to COVID-19 infection and mortality rates at the US county level and sequentially mapping the impact of COVID-19 on different lifestyles.

Even though SDOH are broadly defined by the World Health Organization as the conditions in which people are born, grow, live, work, and age, the phenomena are often represented solely by socioeconomic indicators, such as income and education (Office of Disease Prevention and Health Promotion, 2020). SDOH indicators, such as income, are associated with greater life expectancy in the United States; however, these associations may change based on the underlying area characteristics and health behaviors (Chetty et al., 2016). Index-based approaches are used as proxy methods to quantify SDOH (Singh & Siahpush, 2002; Singh, 2003, p. 93; Singh et al., 2013; Kind et al., 2014). The area deprivation index (ADI) by Singh et al. (Singh & Siahpush, 2002; Singh, 2003, p. 93; Singh et al., 2013), which was extended by Kind

TABLE 8.6
Employment by occupation (%)

	COVID-19 Cases per 100,000	COVID-19 Deaths per 100,000	Mgmt. Business Financial	Professional	Sales	Admin Supp.	Services	Farming Forestry Fishing	Construction Extraction	Installation Main. Repair	Production	Transp. Material moving
United States	848	41.00	14.6	22.2	10.5	13.2	18.5	0.8	4.9	3.2	5.8	6.2
Affluent Estates	1114	59.87	25.1	31.7	12.2	10.6	10.5	0.2	2.6	1.9	2.4	2.9
Upscale Avenues	1479	83.56	19.6	29.5	10.6	12.8	14.5	0.2	3.5	2.4	3.1	3.9
Uptown Individuals	1865	109.8	24.5	36.9	10.0	9.4	12.8	0.1	1.6	1.0	1.6	2.2
Family Landscapes	716	18.12	15.9	23.0	11.0	14.3	15.8	0.4	4.7	3.8	5.4	5.8
GenXUrban	893	52.04	14.5	24.1	10.6	14.4	17.0	0.4	4.4	3.3	5.5	5.6
Cozy Country Living	477	17.74	13.8	19.1	9.5	13.1	16.7	1.6	6.3	4.4	8.2	7.4
Ethnic Enclaves	1087	19.05	10.5	15.7	10.3	14.2	21.2	2.4	6.7	3.9	6.9	8.2
Middle Ground	1209	88.44	12.9	21.6	10.6	13.9	21.2	0.4	4.6	2.9	5.6	6.3
Senior Styles	842	31.26	16.0	23.9	11.8	13.3	19.1	0.6	3.8	2.5	3.9	5.0
Rustic Outposts	818	25.33	9.6	15.3	9.3	13.0	19.0	1.9	8.0	5.2	9.9	8.9
Midtown Singles	1429	94.15	9.7	17.9	10.7	14.7	26.5	0.4	4.6	2.7	5.5	7.2
Hometown	1387	89.11	8.1	15.2	9.8	14.6	25.1	0.6	4.9	3.4	9.1	9.1
Next Wave	1965	116.1	6.3	10.5	9.4	11.9	30.4	1.3	8.7	3.0	8.8	9.7
Scholars and Patriots	555	9.04	9.1	28.0	11.5	14.0	26.1	0.6	2.3	1.8	2.9	3.9

TABLE 8.7
Employment by Industry (%)

	COVID-19 Cases per 100,000	COVID-19 Deaths per 100,000	Unemp. Rate	Agri Mining	Construction	Manuf.	Wholesale Trade	Retail Trade	Trans. Utilities	Information	Finance Real estate	Service	Public Admin
United States	848	41.00	5.5	1.9	6.4	10.1	2.6	11.0	5.1	1.8	6.7	50.0	4.5
Affluent Estates	1114	59.87	3.3	0.9	4.8	9.3	3.2	9.0	3.8	2.3	10.1	52.0	4.6
Upscale Avenues	1479	83.56	4.0	0.5	5.3	7.8	2.8	9.6	4.6	2.7	8.3	53.3	5.1
Uptown Individuals	1865	109.8	3.4	0.4	2.7	5.1	2.0	7.4	2.7	4.4	10.4	61.0	4.0
Family Landscapes	716	18.12	4.4	1.3	6.6	10.5	3.0	11.5	5.7	1.8	7.1	47.2	5.4
GenXUrban	893	52.04	4.5	1.1	5.9	10.3	2.6	11.4	4.9	1.6	6.9	50.4	4.9
Cozy Country Living	477	17.74	4.4	4.6	8.0	14.1	2.6	10.9	5.7	1.2	5.2	43.2	4.6
Ethnic Enclaves	1087	19.05	6.6	3.4	8.3	9.4	3.0	11.7	6.3	1.5	5.6	46.4	4.4
Middle Ground	1209	88.44	5.7	0.9	5.8	8.7	2.5	11.8	4.9	1.8	6.6	52.8	4.2
Senior Styles	842	31.26	5.7	1.2	5.4	7.2	2.4	11.3	4.3	1.7	6.7	53.9	4.4
Rustic Outposts	818	25.33	6.7	4.9	8.9	14.3	2.4	11.7	6.3	1.0	8.1	41.7	4.9
Midtown Singles	1429	94.15	8.0	0.7	5.5	7.3	2.1	12.5	5.4	1.7	4.0	54.8	3.9
Hometown	1387	89.11	9.5	1.2	5.5	12.4	2.1	12.2	6.0	1.3	6.1	50.0	4.6
Next Wave	1965	116.1	7.8	1.5	9.8	10.1	2.8	11.2	5.7	1.3	4.7	51.1	2.2
Scholars and Patriots	555	9.04	7.0	0.9	2.7	4.4	1.2	12.3	2.0	1.6	4.3	67.1	3.9

et al. (Kind et al., 2014), focused on socioeconomic disadvantage and the differing dimensions of poverty. The SDOH index by the Carolinas Health Care system includes additional dimensions such as food accessibility, though these have not yet been validated against actualized health outcomes and cannot reveal the underlying factors involved in differential health outcomes (Fuchs, 2017.) Kolak et al. modeled SDOH as multivariate indices rather than as a singular deprivation index (Kolak et al., 2020). Using variables of advantage, isolation, opportunity, mixed immigrant cohesion, and accessibility, they imported their findings into seven distinct multidimensional neighborhood typologies. Even though they attempted to address spatial heterogeneity, their data only allowed for cross-sectional analyses, which introduced the risk of missing changes in socioeconomic patterns at the census tract level and associated health outcomes over time.

These index-based approaches that quantify SDOH overlook the geographical heterogeneity of the US population or the ways in which uniquely characterized households and their lifestyle, such as retirement communities or diverse urban immigrant enclaves, behave and contribute to spread of the virus. We addressed this challenge with market intelligence tools (i.e., lifestyle segments, market segmentation) by identifying a population, determining its lifestyle clusters, and estimating their propensity for COVID-19 infection and death.

SDOH indicators have different associations with differing health outcomes. SDOH interacts with health outcomes differently in different places. Through a scoping analysis of the literature on COVID-19 and SDOH, we identified studies that have thus far found SDOH associated with COVID-19 transmission and mortality (Millett et al., 2020; Li et al., 2020; Wiemers, 2020; Cyrus, 2020; Fielding-Miller, 2020; Scarpone et al., 2020). COVID- 19 is disproportionately impacting certain populations according to race-ethnicity and socioeconomic status (Rodriguez-Diaz et al., 2020; Millett et al., 2020; Li et al., 2020). Large disparities across race-ethnicity and socioeconomic status exist in the prevalence of conditions that are associated with the risk of severe complications from COVID-19 (Millett et al., 2020; Li et al., 2020). Counties with higher proportions of Black people have a higher prevalence of comorbidities and had more COVID-19 diagnoses and deaths, after adjusting for county-level characteristics such as age, poverty, comorbidities, and epidemic duration (Millett et al., 2020). COVID-19 deaths were higher in disproportionately Black rural and small metro counties(Millett et al., 2020). Li et al. also found that US counties with a higher proportion of Black residents are associated with increased COVID-19 cases and deaths; however, the various suggested mechanisms, such as socioeconomic and health-care predispositions, did not appear to drive the effect of race in their model (Li et al., 2020). Counties with higher average daily temperatures are also associated with decreased COVID-19 cases but not deaths.

Several theories are posited to explain these findings, including the prevalence of vitamin D deficiency (Li et al., 2020). While vulnerability is highest among older adults regardless of their race-ethnicity or socioeconomic status, the findings suggest particular attention should also be given to the risk of adverse outcomes in midlife for non-Hispanic Blacks, adults with a high school degree or less, and low-income Americans (Wiemers, 2020; Cyrus, 2020; Fielding-Miller, 2020). COVID-19 diagnoses rates were greater in Latino counties nationally. In multivariable analysis, COVID-19 cases were greater in Northeastern and Midwestern Latino counties. COVID-19 deaths were greater in Midwestern Latino counties. COVID-19 diagnoses were associated with counties with greater numbers of monolingual Spanish speakers, higher employment rates, more heart disease deaths, less social distancing, and days since the first reported case. COVID-19 deaths were associated with household occupancy density, air pollution, employment, days since the first reported case, and age (fewer < 35 y.o.) (Rodriguez-Diaz, 2020). Studies modeling and mapping intercounty transmission risk of COVID-19 in New York City merged information on the number of tests and the number of infections at the zip code level with demographic and socioeconomic information from the decennial census and the American Community Surveys. The rate of infection in the population depends on both the frequency of tests and on the fraction of positive tests among those tested. People residing in poor or immigrant neighborhoods were less likely to be tested, but the likelihood that a test was positive was larger in those neighborhoods, as well as in neighborhoods with larger households or predominantly Black populations (Chen, 2020; Bai et al., 2020).

In addition to race-ethnicity and socioeconomic status, some studies focused on other SDOH indicators. A study in Germany found the strongest predictors of COVID-19 incidence at the county scale were related to community interconnectedness, geographical location, transportation infrastructure, and labor market structure (Scarpone et al., 2020). Location, densities of the built environment, and socioeconomic variables are important predictors of COVID-19 incidence rates; however, these SDOH associations are complex and may change based on underlying area characteristics and lifestyle behaviors.

This chapter contributes to the SDOH and COVID-19 literature by offering a unique spatial and temporal approach that proves indispensable for timely and effective way of analyzing impact of COVID-19 on American households by their lifestyle characteristics. We demonstrated that COVID-19 was first introduced to West Coast Metropolitan cities by frequently traveling LifeModes such as Affluent Estates and Uptown Individuals, and then was introduced through community spread to GenXUrban and successively poorer American households, including Ethnic Enclaves, Rustic Outposts, and Hometown segments. Our findings reveal that the affluent and mobile lifestyles exhibited the highest risk of infection during the early stages of the outbreak, and then the risk shifted to the poor, isolated, and vulnerable lifestyles during the mature stages.

The most important limitation is that this is a population study. The dominant LifeMode in a county does not represent every single household in that county; each county contains a unique mixture of all 14 LifeModes. Additionally, while Tapestry descriptions provide accurate representations of today's consumers, they are generalizations about consumers with no guarantee that a given household will fit perfectly into a specific segment or LifeMode. Rather, Tapestry segmentation is best used as a set of common characteristics among the typical consumer households. LifeMode groups represent markets that share a common experience—born in the same generation or immigration from another country—or a significant demographic trait, such as affluence. All the consumers in the proposed county will not fall into a specific LifeMode summary group but a representative one, composed of a population—a complete set of people with a specialized set of consumer characteristics. Our ultimate decision on using the dominant LifeMode so that the results of our study can be generalized to a larger population such as a county depends on this understanding.

8.5 CONCLUSIONS

We can only build a culture of health by engaging health and related non-health sectors. The geographic information science coupled with health information technology can help develop links between the two sectors that match lifestyle segments from private sector marketing data and those of geographically identified patients in health-care delivery systems. By correlating COVID-19 data with specific lifestyle segments, it becomes possible to identify patterns of transmission and predict the demand and utilization of health services.

We conclude that there needs to be more research done to translate scientific data into real-world solutions. Given that virtually every household in the United States has been assigned a lifestyle segment, linking segments to geographically identified patients (e.g., incidence, morbidity) in health-care delivery systems could support the ability to estimate morbidity levels for COVID-19 and subsequently predict the demand for health services. Such an understanding of the COVID-19 crisis could be instrumental in better preparing us for a future pandemic or future wave, as it could be used to predict how future epidemics might be introduced to American households, particularly in the early stages of an outbreak, with the affluent and mobile lifestyles, or during late stages with the poor, isolated, and vulnerable lifestyles.

Our approach may provide actionable information for key stakeholders with respect to the focus of interventions and reveal the underlying factors involved in differential health outcomes. Sequentially mapping and geographically illustrating when and where each LifeMode had above-/below-average risk for COVID-19 infection/death provided clues regarding at-risk households and the timing of their infection and possible intervention strategies for future scenarios.

As the US health-care system moves closer to a value-based approach, there will be a constantly growing need for market intelligence that provides health-care providers, health plans, major employers, and policy makers with insights into their constituents and positions them to anticipate their needs and behaviors. This could be accomplished by identifying a population, determining its lifestyle clusters, and estimating the propensity for various diseases. Targeting key lifestyle modes/segments and their locations enables the health-care system to focus on communities that might have been missed when straight demographic criteria were used.

REFERENCES

Acorn. http://acorn.caci.co.uk/
American Community Survey. www.census.gov/programs-surveys/acs
American Journal of Managed Care (AJMC). A Timeline of COVID-19 Developments in 2020. www.ajmc.com/view/a-timeline-of-covid19-developments-in-2020

Bai, S., Junfeng, J., & Chen, Y. (2020). Mapping the intercounty transmission risk of COVID-19 in New York State. https://ssrn.com/abstract=3582774

CBS News. Corona Virus Updates from March 17, 2020. www.cbsnews.com/live-updates/coronavirus-disease- covid-19-latest-news-2020-03-17/

Centers for Disease Control and Prevention (CDC). *Morbidity and Mortality Weekly Report.* www.cdc.gov/mmwr

Chen, Y., Junfeng, J., Bai, S., & Lindquist, J. (2020, May 19). Modeling the spatial factors of COVID-19 in New York City. https://ssrn.com/abs

Chetty, R., Stepner, M., Abraham, S., Lin, S., Scuderi, B., Turner, N. Bergeron, A., & Cutler, D. (2016). The association between income and life expectancy in the United States, 2001–2014. *JAMA, 315*, 1750–1766.

Claritas MyBestSegments. https://segmentationsolutions.nielsen.com/mybestsegments/Default.jsp?ID=7020&menuOption=learnmore&pageName=PRIZM%2BSocial%2BGroups&segSystem=CLA.PNE

Council on Foreign Relations. (2020). Updated: Timeline of the Coronavirus. Think Global Health. www.thinkglobalhealth.org/article/updated-timeline-coronavirus

Cyrus, E., Clarke, R., Hadley, D., Bursac, Z., Trepka, M. J., Devieux, J. G., Bagci, U., Furr-Holden, D., Coudray, M., Mariano, Y. et al. (2020). The impact of COVID-19 on African American communities in the United States. *Health Equity, 4*, 476–483.

Dong, E., Du, H., & Gardner, L. An interactive web-based dashboard to track COVID-19 in real time. (2020). *Lancet. Infectious Diseases, 20*, 533–534.

Esri-Tapestry. 2020. www.esri.com/landing-pages/tapestry

Esri Data–Current Year Demographic & Business Data–Estimates & Projections. www.esri.com

Experian Segmentation. www.segmentationportal.com

Fielding-Miller, R. K., Sundaram, M. E., & Brouwer, K. (2020). Social determinants of COVID-19 mortality at the county level. *PLoS One, 15*, e0240151.

Fuchs, V. R. (2017). Social determinants of health: Caveats and nuances. *JAMA , 317*, 25–26.

Harris, R., Sleight, P., & Webber, R. (2005). Introducing geodemographics. In *Geodemographics, GIS and neighborhood targeting* (pp. 1–25). London: Wiley.

Kaufmann, J., & Schering, A. G. (2014). Analysis of variance ANOVA. Wiley StatsRef: Statistics Reference Online. https://onlinelibrary.wiley.com/doi/abs/10.1002/9781118445112.stat06938

Kind, A. J., Jencks, S., Brock, J., Yu, M., Bartels, C., Ehlenbach, W., Greenberg, C., & Smith, M. Neighborhood socioeconomic disadvantage and 30-day rehospitalization: A retrospective cohort study. (2014). *Annals of Internal Medicine, 161*, 765–774.

Kolak, M., Bhatt, J., Park, Y. H., Padrón, N. A., & Molefe, A. (2020). Quantification of neighborhood-level social determinants of health in the continental United States. *JAMA Network Open, 3*, e1919928.

Li, A. Y., Hannah, T. C., Durbin, J. R., Dreher, N., McAuley, F. M., Marayati, N. F., Spiera, Z., Ali, M., Gometz, A., Kostman, J. T., & Choudhri, T. F. Multivariate analysis of black race and environmental temperature on COVID-19 in the US (2020). *American Journal of the Medical Sciences, 360*(4), 348–356. https://doi/10.1016/j.amjms.2020.06.015

Millett, G. A., Jones, A. T., Benkeser, D., Baral, S., Mercer, L., Beyrer, C., Honermann, B., Lankiewicz, E., Mena, L., Crowley, J. S. et al. (2020). Assessing differential impacts of COVID-19 on black communities. *Annals of Epidemiology, 47*, 37–44.

Mosaic. (2020). USA Consumer Lifestyle Segmentation by Experian [Internet]. www.experian.com/marketing- services/consumer-segmentation.html

Office of Disease Prevention and Health Promotion. Healthy People 2020. Updated December 3, 2019. www.healthypeople.gov/

P2 People & Places. Geodemographic Classification. www.p2peopleandplaces.co.uk/

Rao, C. V. (2005). Analysis of means—a review. *Journal of Quality Technology, 37*, 308–315.

Rodriguez-Diaz, C. E., Guilamo-Ramos, V., Mena, L., Hall, E., & Millett, G. A. (2020). Risk for COVID-19 infection and death among Latinos in the United States: Examining heterogeneity in transmission dynamics. *Annals of Epidemiology, 52*, 46–53.

Scarpone, C., Brinkmann, S. T., Große, T., Sonnenwald, D., Fuchs, M., & Walker, B. B. (2020). A multimethod approach for county-scale geospatial analysis of emerging infectious diseases: A cross-sectional case study of COVID-19 incidence in Germany. *International Journal of Health Geographics, 19*, 32. https://doi.org/10.1186/s12942-020-00225-1

Singh, G. K. (2003). Area deprivation and widening inequalities in US mortality, 1969–1998. *American Journal of Public Health, 93*, 1137–1143.

Singh, G. K., Azuine, R. E., Siahpush, M., & Kogan, M. D. (2013). All-cause and cause-specific mortality among US youth: Socioeconomic and rural-urban disparities and international patterns. *Journal of Urban Health, 90*, 388–405.

Singh, G. K.;, & Siahpush, M. (2002). Increasing inequalities in all-cause and cardiovascular mortality among US adults aged 25–64 years by area socioeconomic status, 1969–1998. *International Journal of Epidemiology, 31*, 600–613.

Stawicki, S. P., Jeanmonod, R., Miller, A. C., Paladino, L., Gaieski, D. F., Yaffee, A. Q., Wulf, A. D., Grover, J., Papadimos, T. J., Bloem, C. et al. (2020). The 2019–2020 novel coronavirus (severe acute respiratory syndrome coronavirus 2) pandemic: A joint American College of Academic International Medicine-World Academic Council of Emergency Medicine Multidisciplinary COVID-19 Working Group Consensus Paper. *Journal of Global Infectious Diseases, 12*, 47–93.

Survey of the American Consumer®. www.mrisimmons.com/solutions/national-studies/survey-american-consumer/

US Census Bureau. (2020). Population Indicators. www.census.gov/

USA Facts. (2020). Government Data to Drive Fact-Based Discussion. https://usafacts.org/

Wiemers, E. E., Abrahams, S., AlFakhri, M., Hotz, V. J., Schoeni, R. F., & Seltzer, J. E. (2020). Disparities in vulnerability to severe complications from COVID-19 in the United States. *Research in Social Stratification and Mobility, 69*, 1–38. Version 3. medRxiv.

9 A Bibliometric Review of Research on the Role of GIS in COVID-19 Pandemic Control and Management

Science Mapping the Literature, 2020–2022

Emily A. Fogarty

CONTENTS

9.1 INTRODUCTION

As people worldwide welcomed the new 2020 year, an outbreak risk software company, BlueDot, leveraging a global infectious disease surveillance system, detected an unidentified respiratory illness emerging in Central China (Stieg, 2020). On January 9, 2020, the World Health Organization released official guidance on the National Capacities Review Tool for a novel coronavirus (World Health Organization, 2020).

DOI: 10.1201/9781003227106-9

Shortly after that, Dr. Lauren Gardner (associate professor) and Ensheng Dong (then a PhD candidate) created a map dashboard for tracking global coronavirus cases at Johns Hopkins University. This map (COVID19 Dashboard) quickly became a resource to guide the understanding of the SARS-CoV-2 virus, prepare policy makers to direct a response, and inform the public (Dong et al., 2020). This innovative map dashboard inspired state and local public health departments worldwide to create their own disease surveillance tracking systems.

Geospatial research and application development provide valuable insights that help government stakeholders monitor, manage, and make decisions at different stages while fighting the pandemic and during post-pandemic recovery. This chapter demonstrates the utilization of science mapping analysis, a quantitative way to investigate bibliographic records. It examines current research investigating the role of geographic information science in pandemic control and management by showcasing examples of scholarship demonstrating innovative uses of geospatial techniques. The objective of the study is two-fold. First, it reviews related scholarship to reveal the main themes and trends. Second, this research then summarizes the intellectual, conceptual, and social research structures involving the role of a geographic information system (GIS) in COVID-19 pandemic management and control and their evolution and dynamic aspects. Further discussion is provided, along with a description of the review's limitations and recommendations for future science mapping analysis studies.

9.2 RELEVANT LITERATURE

Geography has a long tradition of concern with integration, exploring the links that exist between disciplines and with problems whose solution requires knowledge that extends across many disciplines (Goodchild, 2022). Analyzing geographical information involves seeking patterns, relationships, and connections. The primary tool for capturing, organizing, and analyzing geographical information is a GIS. GIS is an overarching name for software systems implemented to create, manage, analyze, and map all data types. Often considered only as a data visualization tool, the value of GIS reaches much deeper into automatically harnessing the spatial relationships between various geographical data inputs.

Geographic information science (GISc) is the science behind these systems; it intersects with computer science, analytical cartography, remote sensing, and imagery science. A widely accepted definition of GISc is the scientific discipline that studies geographical information. Examples of GIS options include Esri, QGIS, and Mapbox.

There are many uses for GIS and GISc techniques and methodologies in research. Examples range from transit optimization, urban tree management, and climate change mapping to public decision-making studies (Caspari et al., 2021; Sandre et al., 2021). Brimblecombe et al. (2020) examined the fluctuating levels of visitors to museums and historic buildings in the Tokyo area, responding to the perceived threat of environmental issues, from SARS and COVID-19 to earthquakes. The authors found that GIS maps offer a way for the spatial spread of risks to be expressed for a wide range of natural and human-induced hazards. Szczepańska et al. (2021) implemented a virtual reality platform as a tool for public consultations in spatial planning and management. They argued that easy-to-understand visualizations encourage the community to participate in local matters and enable citizens to make better decisions. The inclusion of social media data as a supplement to geographical information may strengthen the researcher's ability to investigate their data. For example, Jiang et al. (2021) created a novel analytical framework to investigate the tourist's sentiment changes between different attractions based on geotagged social media data.

In recent years, GIS and related technologies like remote sensing have become increasingly popular tools for analyzing disease distributions globally (Kanga et al., 2020). The National States Geographic Information Council (NSGIC) and ten other organizations issued a joint statement on the value of GIS in pandemic management (Mann, 2022). A GIS and GISc are standard tools used in asking geographical questions about COVID-19 management and control at local, regional, and global scales. Collectively, GIS and GISc are considered visualization and analytic tools that play a central role in medical geography, public health, spatial analysis, infectious diseases, COVID-19, machine learning, predictive analysis, health disparities, and social determinants of health research. GIS tools and techniques also play a critical role in preparing spatial data in preparation for spatial statistical analysis. Several fields, such as public health, epidemiology, and the study of human diseases, employ mapping and spatial analysis (Koch, 2017). Pandemic response management is essential for decreasing the spread of a virus and its associated illness and death. Establishing new protocols and estimating the demand for resources such as physicians, hospital beds, personal protective equipment, ventilators, emergency transport vehicles, and nurses can help decision makers control the virus (Ibrahim et al., 2020). Kumar et al. (2020) argued that the advancement in clinical diagnosis techniques like real-time polymerase chain reaction, immunological, microscopy, and GIS mapping technology helped with quick diagnosis and tracking of viral infection in a short period. GIS and GISc techniques and methodologies are essential tools for informing local and regional resource mapping and subsequent planning for COVID-19. Leveraging these tools allows the user to predict confirmed case numbers and specific locations where the outbreak risk is higher and identify the areas underserved by the existing infrastructures (Ahasan & Hossain, 2021).

Social media can complement the use of community health workers for community or public engagement (Agbor et al., 2021). Qin et al. (2020) demonstrated that social media search indexes can be an effective early predictor, which would enable governments' health departments to locate potential and high-risk outbreak areas. Al-Ramahi et al. (2021) demonstrated the potential of mining social media for understanding public discourse about public health issues, such as wearing masks during the COVID-19 pandemic. They explored the influence of social media mask-wearing discourse on the impact of management and control of the pandemic. Furthermore, they recommended that policy makers pay attention to social media discourse and proactively address public perceptions to implement policy intervention toward the most prevalent topics.

9.3 METHODS AND RESULTS: THE ROLE OF GIS IN COVID-19 PANDEMIC CONTROL AND MANAGEMENT

To study the collaboration patterns across levels (individual, institution, and country), the bibliometric science mapping workflow suggested by Cobo et al. (2015) and Aria and Cuccurullo (2017) is used: study design, data retrieval, data analysis (descriptive analysis, network matrix creation, and normalization), data visualization (mapping), and interpretation. Descriptive statistics, tables, and the main information are generated using the R, Bibliometrix package. The VOSviewer software tool was employed for constructing and visualizing bibliometric networks (Van Eck & Waltman, 2014).

9.3.1 Bibliographic Collection

A bibliometric network consists of nodes and edges. The nodes can be, for instance, publications, journals, researchers, or keywords. The edges indicate relations between pairs of nodes. Bibliometric networks are usually weighted networks, meaning the edges among the nodes have weights assigned to them (van Eck & Waltman, 2014). A network analysis examines relationships between distinct entities, such as collaborations between researchers or communications between people within a company. It is possible to use network analysis for various purposes, including studying a community's structure or solving complex math and engineering problems through graph theory. In published research, science mapping (aka bibliometric science analysis) includes quantitative and qualitative data collection, results, and integration methods (Aria & Cuccurullo, 2017). A quantitative approach for the description, evaluation, and monitoring of published research, bibliometric network analysis is helpful to visualize and quantify relationships between and within specific field tags found in the collection of documents. The most commonly studied types of relations are citation relations, keyword co-occurrence relations, and coauthorship relations.

The first steps in this study were to perform a descriptive analysis of the bibliographic data frame, referred to as a collection. The documents in the collection are a culmination of two Web of Science (WoS) search queries using the following keywords: COVID, GIS, COVID CONTROL, COVID MANAGEMENT, SOCIAL MEDIA. Each document contains several elements, such as authors' names, titles, keywords, and other information. These elements constitute the bibliographic attributes of a document, also called metadata. Listed in Table 9.1 are the document collection columns, named using the standard Clarivate Analytics WoS Field Tag codify. For each document, cited references are in a single string stored in the column "CR" of the collection data frame. An automatic computer algorithm generates Keywords Plus (ID) column keywords associated with the WoS database, which are words or phrases that frequently appear in the titles of an article's references and not necessarily in the article's title or as Author Keywords (DE).

9.3.2 Descriptive Analysis

Bibliometrics is an academic science founded on statistical methods used to measure the quality and impact of scientific production. The descriptive analysis provides

TABLE 9.1
Field Tag Descriptions

Field Tag	Description
AU	Authors
TI	Document Title
SO	Publication Name (or Source)
JI	ISO Source Abbreviation
DT	Document Type
DE	Authors' Keywords
ID	Keywords associated by SCOPUS or ISI database
AB	Abstract
C1	Author Address
RP	Reprint Address
CR	Cited References
TC	Times Cited
PY	Year
SC	Subject Category
UT	Unique Article Identifier
DB	Bibliographic Database

snapshots of the annual research development, the top productive authors, papers, countries, subject categories, and most relevant keywords. Moreover, it is also helpful in displaying and analyzing the intellectual, conceptual, and social structures of research and their evolution and dynamic aspects. The network structure of a bibliographic collection is the interaction among authors, documents (e.g., papers or articles), references, and keywords. The study collection includes 357 sources that span over two years, 2020–2022. There are 619 documents in the query results; the main information found in Table 9.2 describes the collection size in terms of the type and number of documents, number of authors, number of sources, number of keywords, time span, and average number of citations.

The WoS schema comprises approximately 250 science, social sciences, and arts & humanities subject areas. Every journal and book covered by the WoS core collection is assigned to at least one subject category. Every record in the WoS core collection contains the subject category of its source publication in the WoS Source Category (SC) field. The current collection contains subjects ranging from virology and religion to environmental sciences and ecology and public, environmental, and occupational health. There are 95 unique subject categories found in the collection, and of those, only 37 subject categories were recorded once. The top ten WoS search query results Subject Categories (SC) are listed in Table 9.3. The top subject category listed in the collection is *Environmental Sciences & Ecology*, followed at a close second by *Public, Environmental & Occupational Health.* Also found in the collection are source subjects related to *Computer Science* and *Geography.*

Relevant sources in a collection represent the number of documents produced by a particular source. Table 9.4 summarizes the top ten most relevant sources. The

TABLE 9.2
Main Information

Main Information about Data	
Timespan	2020: 2022
Sources (Journals, Books, etc.)	357
Documents	619
Average years from publication	1.21
Average citations per document	7.926
Average citations per year per doc	3.377
References	25483
Document Types	
Article	497
Article; data paper	3
Article; early access	64
Article; proceedings paper	1
Correction	2
Editorial material	7
Letter	5
Meeting abstract	1
Proceedings paper	12
Review	25
Review; early access	2
Document Contents	
Keywords Plus (ID)	1059
Author's Keywords (DE)	1967
Authors	
Authors	3858
Author Appearances	4187
Authors of single-authored documents	31
Authors of multiauthored documents	3827
Authors Collaboration	
Single-authored documents	33
Documents per Author	0.16
Authors per Document	6.23
Coauthors per Document	6.76
Collaboration Index	6.53

International Journal of Environmental Research and Public Health leads the way at number one with 38 articles, followed by the *Journal of Medical Internet Research* at a distant second with 15 articles. Although the count of a document produced from one source is interesting, a better measure is local cited sources, drawn on from the references list from each document in the collection. The sources listed in the local cited sources tease out the most cited sources within the collection. For example, the *Lancet, Infectious Diseases* is not listed in the most relevant sources; however, likely due to a highly cited within the collection (Dong et al., 2020), the source is found in

TABLE 9.3
WoS Source Subject Category (SC)

Subject	Number of times used in Subject Category (SC)
Environmental Sciences & Ecology	127
Public, Environmental & Occupational Health	107
Science & Technology—Other Topics	61
Health Care Sciences & Services	50
Computer Science	47
Geography	36
Business & Economics	33
Remote Sensing	25
Engineering	24
General & Internal Medicine	21

TABLE 9.4
Top 10 Most Relevant Sources

Sources	Articles
International Journal of Environmental Research and Public Health	38
Journal of Medical Internet Research	15
Transactions in GIS	13
Science of the Total Environment	12
Sustainability	11
Scientific Reports	10
Frontiers in Public Health	9
JMIR Public Health and Surveillance	9
ISPRS International Journal of Geo-Information	8
BMC Public Health	7

the top 20 local cited sources. Table 9.5 shows the top 20 WoS search query results local cited sources from reference lists.

9.3.3 The Intellectual Structure of the Field—Co-citation Analysis

The traditional measure of evaluation of scientific research is external peer review. However, due to increased costs and the limited scope of peer review, research institutions increasingly turn to bibliometrics to supplement this process. The number of times cited in papers written by other scientists or citation statistics is used more often in the academic evaluation of research publications. Table 9.6 summarizes the collection's top ten manuscripts per citation for the entire collection of documents and the WoS subject category. Overall, there were 95 research areas for all papers

TABLE 9.5
Top 20 Most Local Cited Sources (From Reference Lists)

Sources	Articles
Science of the Total Environment	714
The Lancet	420
International Journal of Environmental Research and Public Health	358
PLoS One	303
Science	265
New England Journal of Medicine	215
Journal of the American Medical Association	212
Journal of Medical Internet Research	209
Proceedings of the National Academy of Sciences of USA	176
Sustainability	166
Nature	159
The Lancet, Infectious Diseases	155
Computers in Human Behavior	130
International Journal of Health Geographics	130
Scientific Reports-UK	125
Emerging Infectious Diseases	109
British Medical Journal	107
International Journal of Infectious Diseases	105
Landscape and Urban Planning	96
Sustainable Cities and Society	90

groups and five research areas for the highly cited manuscripts group. The most relevant research areas were *Environmental Science & Ecology*, *Public, Environmental & Occupational Health, Science & Technology*, and *Health Care Sciences & Service.*

A predominant technique in bibliometrics is citation analysis. It shows the structure of a specific field through the linkages between nodes (e.g., authors, papers, journals), while the edges can be differently interpreted depending on the network type, namely co-citation, direct citation, and bibliographic coupling (Aria & Cuccurullo, 2017). A co-citation is an indication for papers dealing with related topics. The analysis of such a network yields insights into the structure of knowledge in a given field (Yang et al., 2017). The co-citation network is a valued and undirected network in which the strength of the tie indicates the number of such co-citations. Local citations measure how many times an author (or a document) included in this collection has been cited by the documents also included in the collection.

The co-citation frequencies for the cited papers, journals, and author co-citation networks were determined based on citations in articles that were published between 2020 and 2022. A cited paper, journal, and/or author was included in the network if it had at least 20 co-citations. The author co-citation network is visualized in Figure 9.1. The sources' co-citation network is shown in Figure 9.2, and the cited papers' co-citation network is displayed in Figure 9.3. These networks help visualize the predominantly cited papers in the collection. Authors like the *World Health Organization,*

TABLE 9.6
Top Manuscripts Per Citations

	Author/Year/ Journal	Paper Title/DOI	Subject Category (SC)	TC	TC per Year
1	(Kamel Boulos & Geraghty, 2020), International Journal of Health Geographics	Geographical tracking and mapping of coronavirus disease COVID-19/ severe acute respiratory syndrome coronavirus 2 (SARS-CoV-2) epidemic and associated events around the world: how 21st century GIS technologies are supporting the global fight against outbreaks and epidemics DOI:10.1186/ s12942-020-00202-8	Public, Environmental & Occupational Health	237	79.0
2	(Kakodkar et al., 2020), Cureus Journal of Medical Science	A Comprehensive Literature Review on the Clinical Presentation, and Management of the Pandemic Coronavirus Disease 2019 (COVID-19) DOI:10.7759/cureus.7560	General & Internal Medicine	217	72.3
3	(Mollalo et al., 2020), Science of The Total Environment	GIS-based spatial modeling of COVID-19 incidence rate in the continental United States DOI:10.1016/ j.scitotenv.2020.138884	Environmental Sciences & Ecology	192	64.0
4	(Zhou et al., 2020), Geography and Sustainability	COVID-19: Challenges to GIS with Big Data DOI:10.1016/ j.geosus.2020.03.005	Science & Technology—Other topics; Physical Geography	170	56.7
5	(Kim & Bostwick, 2020), Health Education & Behavior	Social Vulnerability and Racial Inequality in COVID-19 Deaths in Chicago DOI:10.1177/ 1090198120929677	Public, Environmental & Occupational Health	144	48.0
6	(Franch-Pardo et al., 2020), Science of The Total Environment	Spatial analysis and GIS in the study of COVID-19. A review DOI:10.1016/ j.scitotenv.2020.140033	Environmental Sciences & Ecology	142	47.3

(*continued*)

TABLE 9.6 (Continued)
Top Manuscripts Per Citations

	Author/Year/ Journal	Paper Title/DOI	Subject Category (SC)	TC	TC per Year
7	(Jain & Sharma, 2020), Aerosol and Air Quality Research	Social and Travel Lockdown Impact Considering Coronavirus Disease (COVID-19) on Air Quality in Megacities of India: Present Benefits, Future Challenges, and Way Forward DOI:10.4209/ aaqr.2020.04.0171	Environmental Sciences & Ecology	103	34.3
8	(Zhang & Schwartz, 2020), Journal of Rural Health	Spatial Disparities in Coronavirus Incidence and Mortality in the United States: An Ecological Analysis as of May 2020 DOI:10.1111/jrh.12476	Health Care Sciences & Services; Public, Environmental & Occupational Health	91	30.3
9	(Zhao et al., 2020), Journal of Medical Internet Research	Chinese Public's Attention to the COVID-19 Epidemic on social media: Observational Descriptive Study DOI:10.2196/18825	Health Care Sciences & Services; Medical Informatics	86	28.7
10	(Guo et al., 2020 International Journal of Environmental Research and Public Health	Coping with COVID-19: Exposure to COVID-19 and Negative Impact on Livelihood Predict Elevated Mental Health Problems in Chinese Adults DOI:10.3390/ ijerph17113857	Environmental Sciences & Ecology; Public, Environmental & Occupational Health	78	26.0

Centers for Disease Control, Dong et al. 2020, and *Franch-Pardo et al. 2020* are featured. Top co-citing sources include *Science of the Total Environment, PLoS One,* and *Lancet*.

9.3.4 Conceptual Structure—Term Frequency and Co-word Analysis

The conceptual structure represents relationships, concepts, or words in a set of publications. This structure helps to understand the topics covered in a research field to determine the most important and most recent issues, and it could also help in the

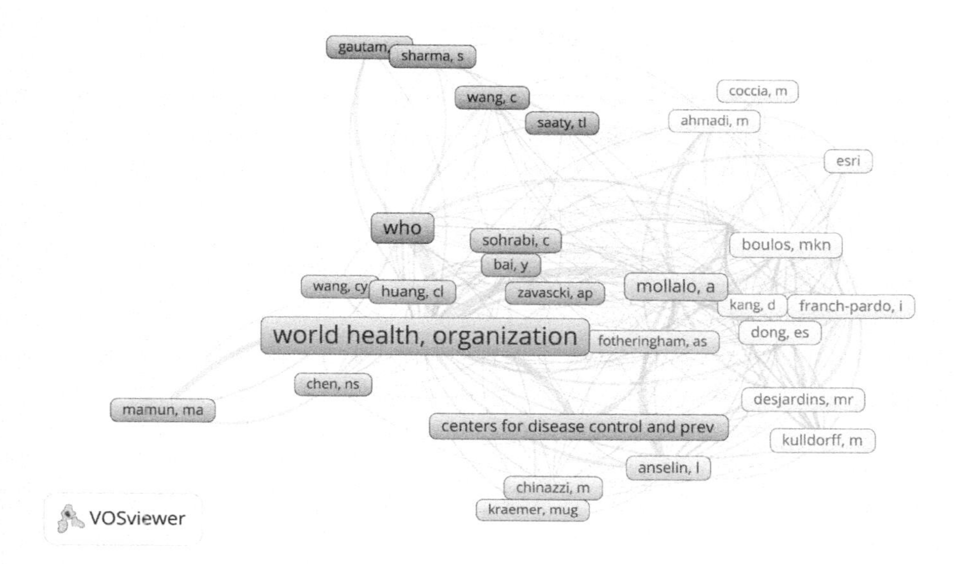

FIGURE 9.1 Co-citation network of authors with at least 20 citations.

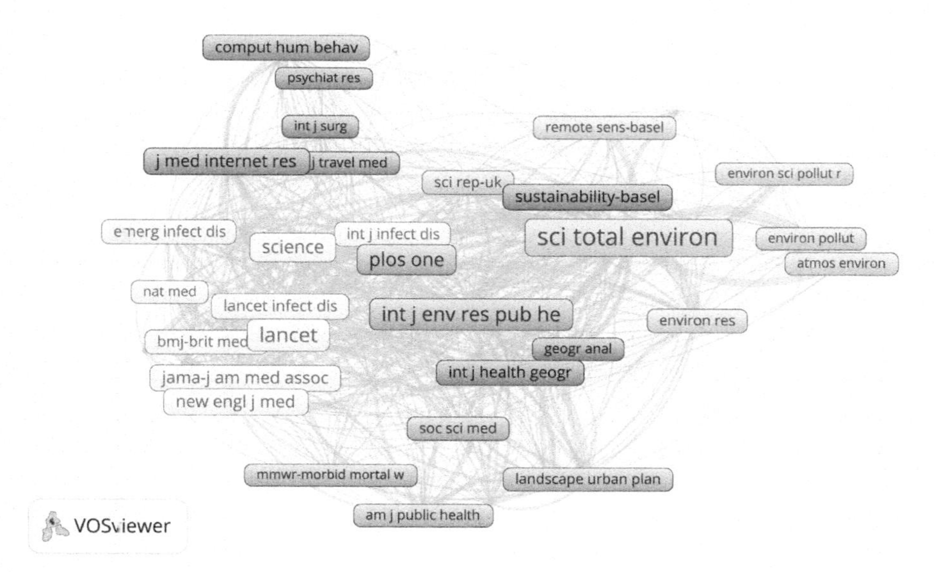

FIGURE 9.2 A co-citation network that uses cited sources as the unit of analysis with at least 20 citations.

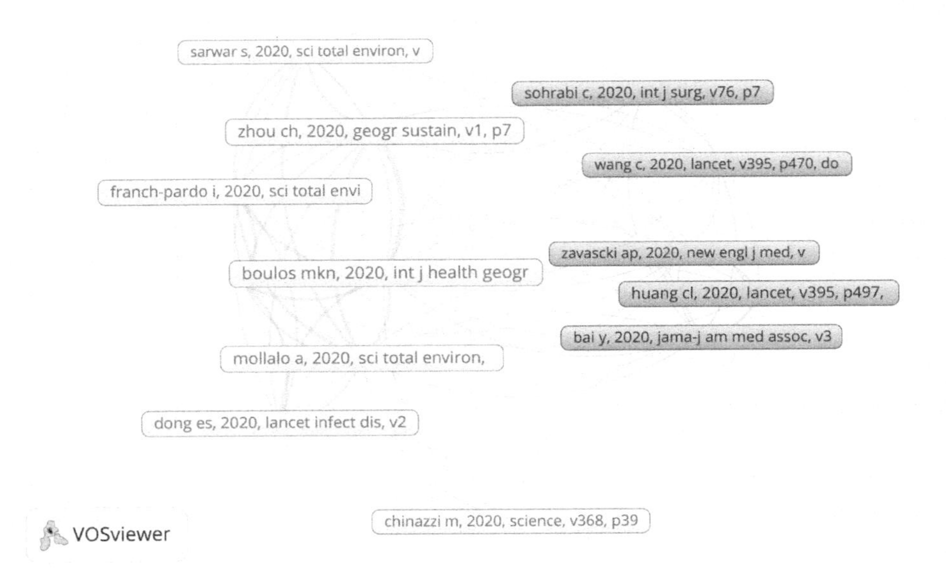

FIGURE 9.3 Co-citation network of papers that were cited at least 20 times within the collection.

study of the evolution of subjects over time. This study undertook a keyword analysis that investigates the keywords plus or author keywords found in the collection of documents. Keywords Plus (ID) are words or phrases listed in the titles of an article's references, therefore reflecting the content of the references used by the author. The Author Keywords (DE) represent what authors feel are the most important words in their research paper. Using scientific papers about how GIS is leveraged during the COVID-19 pandemic retrieved from WoS, a comparative assessment of Keywords Plus and Author Keywords was performed. In this study, 1,967 author keywords and 1,059 keywords plus were used in the analysis from 2020 to 2022, as shown in Table 9.7. This study yielded more Author Keywords terms than Keywords Plus terms, and the Keywords Plus terms were more broadly descriptive. Figure 9.4 and Figure 9.5 list the yearly top five terms found in the ID and DE fields, respectively. Keywords Plus is as effective as Author Keywords in terms of bibliometric analysis investigating the knowledge structure of scientific fields, but it is less comprehensive in representing an article's content (Zhang et al., 2016). Moreover, GIS is more prevalent in literature in 2021–2022 than in 2020, denoting increased scholarship utilizing GIS concepts and techniques.

Word clouds are valuable tools to visualize a quick synopsis of the most frequently occurring words within a particular field tag (e.g., Keyword Plus (ID), Author Keywords (DE), Document Title (TI), and Abstract (AB)) of the collection. Shown in Figure 9.6 are the most frequent words that appear in the field tag columns of Keywords Plus (ID) (left) and Author Keywords (DE) (right) within the collection; this figure helps visualize the information found in Table 9.7. The larger the word,

TABLE 9.7
Keyword Count

Author Keywords (DE)	Articles	Keywords Plus (ID)	Articles
COVID-19	354	Impact	50
GIS	105	Health	45
Pandemic	47	GIS	39
Social media	44	Management	24
Coronavirus	38	Outbreak	24
SARS-CoV-2	36	Model	23
Public health	27	China	19
Spatial analysis	26	Transmission	19
COVID-19 pandemic	18	Risk	18
Mental health	14	Social media	18

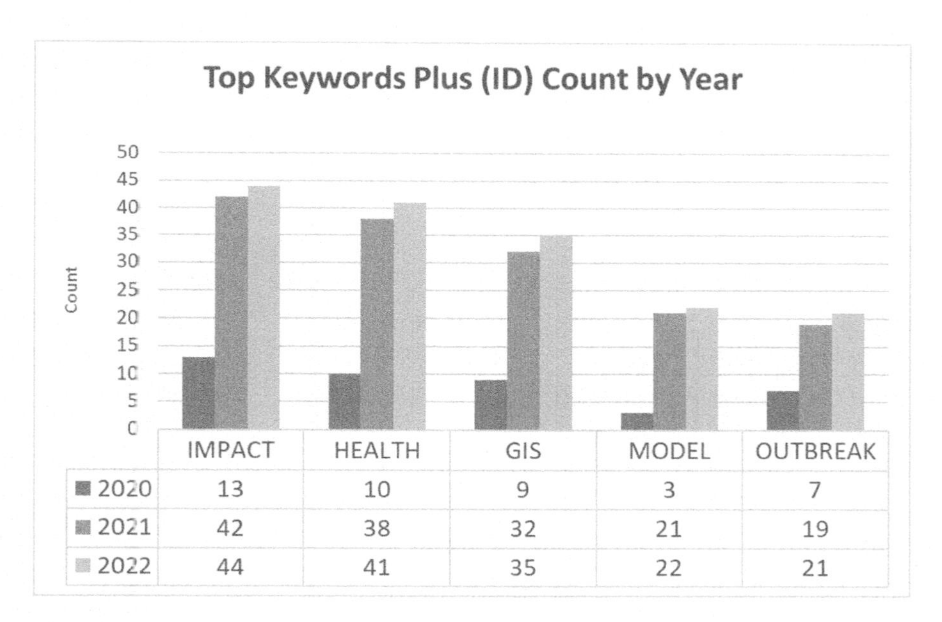

FIGURE 9.4 Top Keywords Plus (ID) count by year.

the more often it occurred within the Keywords Plus (ID) and Author Keywords (DE) collection. The keywords found in the DE collection represent the author's personal selection of keywords; in contrast, the ID collection comprises keywords selected from the reference titles used by the author.

Exploring the frequency of key terms in the collection reveals patterns investigated further by visualizing the n-gram or contiguous sequence of single, pair, or triple word combinations found within the field. The trigrams or combination of three words in a row in a sentence or title offer a quick glimpse of the commonly found terminologies

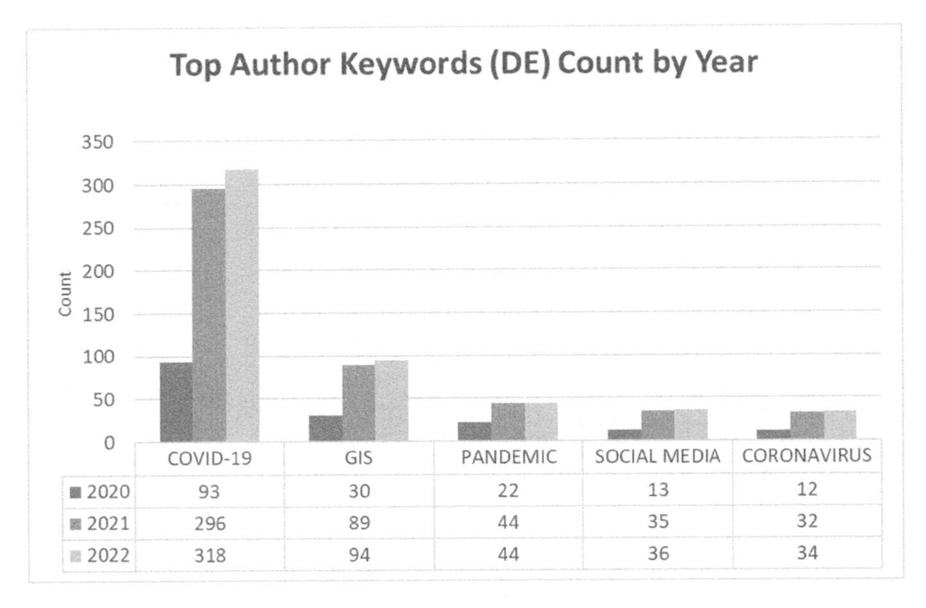

	COVID-19	GIS	PANDEMIC	SOCIAL MEDIA	CORONAVIRUS
2020	93	30	22	13	12
2021	296	89	44	35	32
2022	318	94	44	36	34

FIGURE 9.5 Top Author Keywords (DE) count by year.

FIGURE 9.6 Keywords Plus (ID) shown on the left and Author Keywords (DE) shown on the right.

within the collection. Figure 9.7 displays the top 50 three combinations of words found in Document Title (TI) and Abstract (AB) of the collection.

The larger the word combinations, the more often the three words appeared in either the titles or within the abstract of each document found in the collection. The trigrams found in the titles collection (Figure 9.7 right side) point to methodologies, locations, and the main topic COVID. In contrast, the trigrams found in the document abstracts (Figure 9.7 left side) represent what the theme or content of the document contained. Word frequencies are helpful for quick assessment of the main themes found within the collection; however, exploring how these words are connected is more informative in teasing out clusters of themes within the collection.

FIGURE 9.7 Top 50 trigram combos for Titles (TI) shown on the right and Abstract (AB) shown on the left.

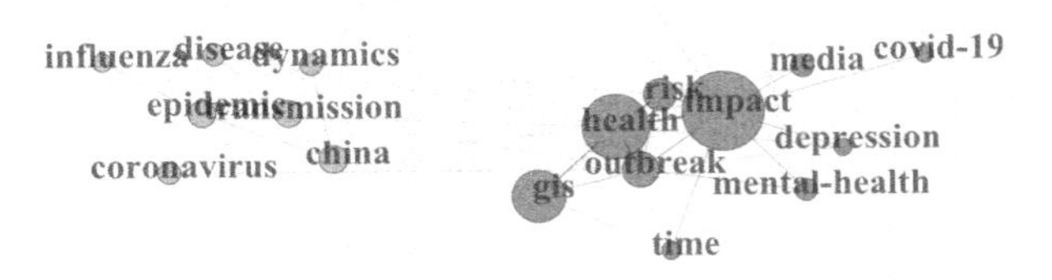

FIGURE 9.8 Co-occurrence Network—Keywords Plus (ID).

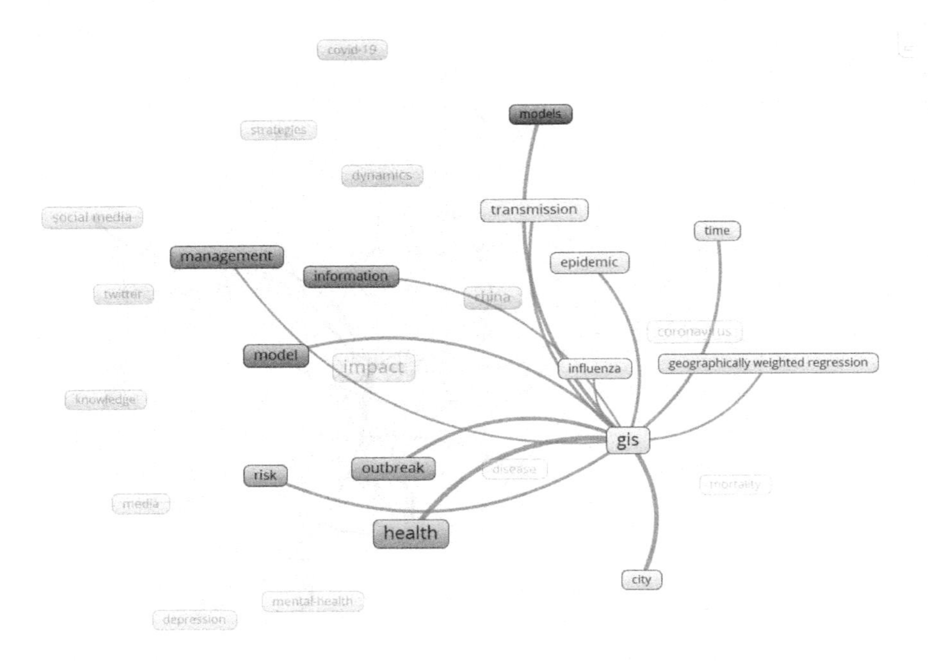

FIGURE 9.9 Co-occurrence Network—Keywords Plus (ID)—GIS.

The Keywords Plus collection of keywords contains high-frequency words like *health*, *impact*, and *GIS*, which tend to co-occur with other documents within the collection (Figure 9.8). A co-occurrence network also reveals the clusters of words within the collection. For example, *influenza*, *disease*, and *transmission* are in a cluster, whereas *GIS*, *impact*, *outbreak*, and *media* are in a different cluster. These clusters represent different themes or research ideas within the document collection. The term "GIS" (Figure 9.9) in the co-occurrence network reveals the connections to other document keywords that help us begin to understand the underlying subgroups or clusters within the collection.

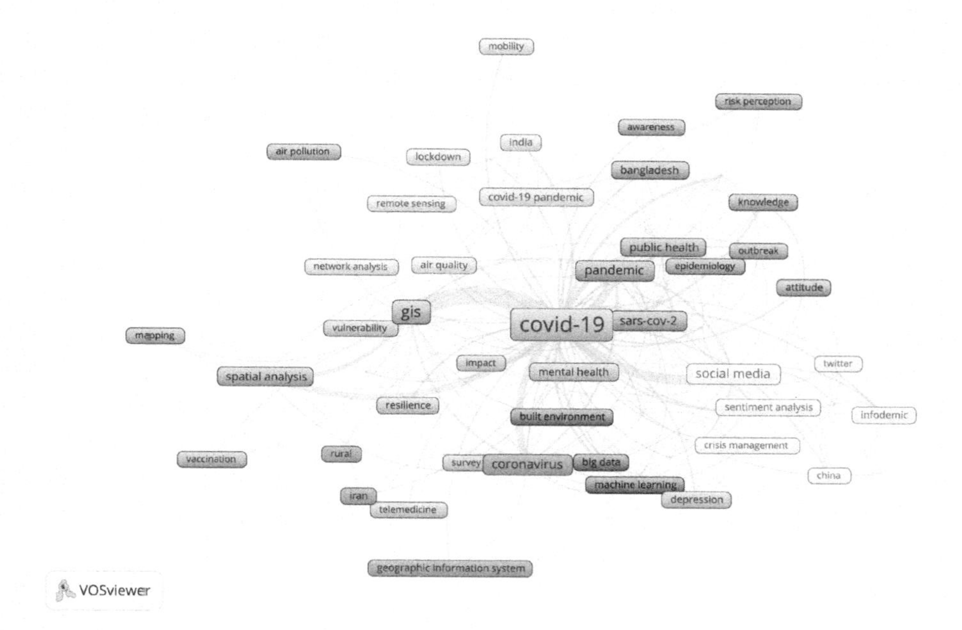

FIGURE 9.10 Co-occurrence Network—Author Keywords (DE).

In contrast to the Keyword Plus co-occurrence network, the Author Keyword network reveals how the author promotes the paper; the keywords included are those the author feels best to represent the content of their paper. The Author Keyword co-occurrence network is shown in Figure 9.10. The most frequently used keyword is COVID-19, followed by GIS; they also tend to co-occur together often as many authors included these two keywords (COVID and GIS) in their list. In the collection of Author Keywords, the words connected to GIS are *COVID-19*, *spatial analysis*, and *social media*. The collection of Author Keywords seems to embody the underlying theme or subject of the paper, whereas Keywords Plus best describes the collection's methodologies and model papers used (Figure 9.11).

Table 9.8 summarizes the top five most cited papers along with the subject category and the author and keyword plus terms. The term "GIS" is listed in Author Keywords (DE) in four of the five top papers.

9.3.5 Evolution and Dynamical Structure (Thematic Mapping)

In order to identify the underlying themes within the collection, thematic maps are built and analyzed. Comparing Keywords Plus to Author Keywords, shown in Figure 9.12, Thematic Map–Keywords Plus (Top)–Author Keywords (Bottom), a thematic map is a plot used to analyze themes according to the quadrant in which they are placed: (1) upper-right quadrant: motor themes are important themes with high centrality and density; (2) lower-right quadrant: basic themes are themes with strong centrality but low development; (3) lower-left quadrant: emerging or disappearing

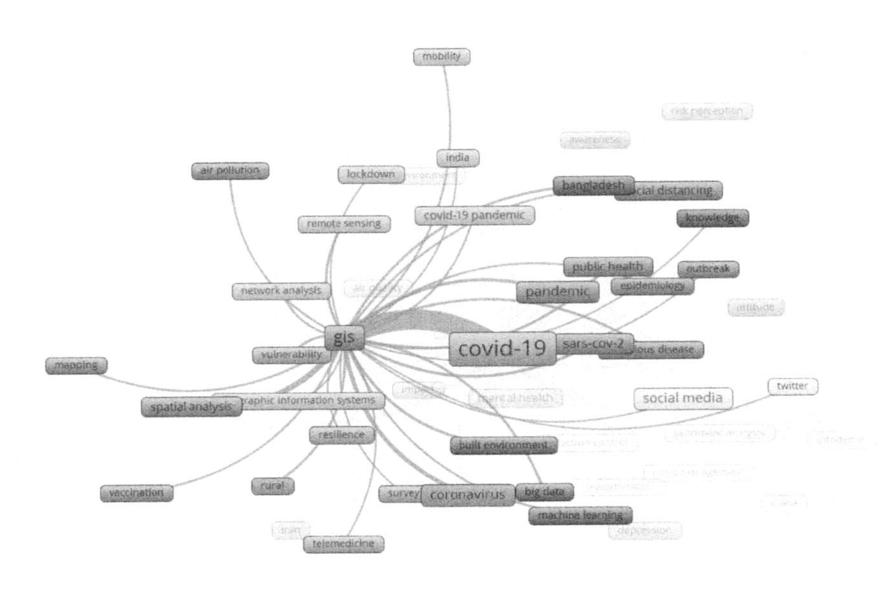

FIGURE 9.11 Co-occurrence Network–Author Keyword (DE)–GIS.

TABLE 9.8
Top 5 Cited Papers with Subject Category and Keywords

Times Cited (TC)	Cited Author(s)	Subject Category	Author Keywords (DE)	Keywords Plus (ID)
237	Kamel Boulos, M. N., & Geraghty, E. M. (2020)	Public, Environmental & Occupational Health	COVID-19; SARS-COV-2; **GIS**	None Listed
Geographical tracking and mapping of coronavirus disease COVID-19/severe acute respiratory syndrome coronavirus 2 (SARS-CoV-2) epidemic and associated events around the world: How 21st century GIS technologies are supporting the global fight against outbreaks and epidemics. International Journal of Health Geographics, 19(1), 8. https://doi.org/10.1186/s12942-020-00202-8				
217	Kakodkar, P., Kaka, N., & Baig, M. (2020).	General & Internal Medicine	COVID-19; SARS-COV-2; SEVERE ACUTE RESPIRATORY INFECTION; PANDEMIC; MRNA-1273 VACCINE; REMDESIVIR (GS-5734); CHLOROQUINE; ARDS; ACE2; LOPINAVIR AND RITONAVIR	None Listed

TABLE 9.8 (Continued)
Top 5 Cited Papers with Subject Category and Keywords

Times Cited (TC)	Cited Author(s)	Subject Category	Author Keywords (DE)	Keywords Plus (ID)
A Comprehensive Literature Review on the Clinical Presentation, and Management of the Pandemic Coronavirus Disease 2019 (COVID-19). Cureus. https://doi.org/10.7759/cureus.7560				
192	Mollalo, A., Vahedi, B., & Rivera, K. M. (2020)	Environmental Sciences & Ecology	COVID-19; **GIS**; MULTISCALE GWR; SPATIAL NON-STATIONARITY	GEOGRAPHICALLY WEIGHTED REGRESSION
GIS-based spatial modeling of COVID-19 incidence rate in the continental United States. Science of The Total Environment, 728, 138884. https://doi.org/10.1016/j.scitotenv.2020.138884				
170	Zhou et. al. (2020).	Science & Technology—Other topics; Physical Geography	COVID-19; BIG DATA; **GIS**; SPATIAL TRANSMISSION; SOCIAL MANAGEMENT	CHINA; MIGRATION; MODELS
COVID-19: Challenges to GIS with Big Data. Geography and Sustainability, 1(1), 77–87. https://doi.org/10.1016/j.geosus.2020.03.005				
144	Kim, S. J., & Bostwick, W. (2020).	Public, Environmental & Occupational Health	AFRICAN AMERICAN; EMERGENCY; GENERAL TERMS; **GIS**; HEALTH DISPARITIES; HEALTH EQUITY; NEIGHBORHOOD; PLACE; POPULATION GROUPS; RISK AND CRISIS COMMUNICATION	UNITED STATES; HEALTH; DISPARITIES
Social Vulnerability and Racial Inequality in COVID-19 Deaths in Chicago. Health Education & Behavior, 47(4), 509–513. https://doi.org/10.1177/1090198120929677				

themes represent themes that need qualitative analysis to understand whether they are emerging or losing relevance; (4) upper-left quadrant: very specialized/niche themes are clusters with high density and low centrality. The clusters were plotted in two-dimensional diagrams based on centrality (x-axis) and density (y-axis) values. The co-word analysis draws clusters of keywords. They are considered themes whose density and centrality are used to classify themes and map in a two-dimensional diagram (Cobo et al., 2011).

9.3.6 Social Structure

Collaboration networks show how authors, institutions, and countries relate to others in a specific field of research. This section examines the scientific performance

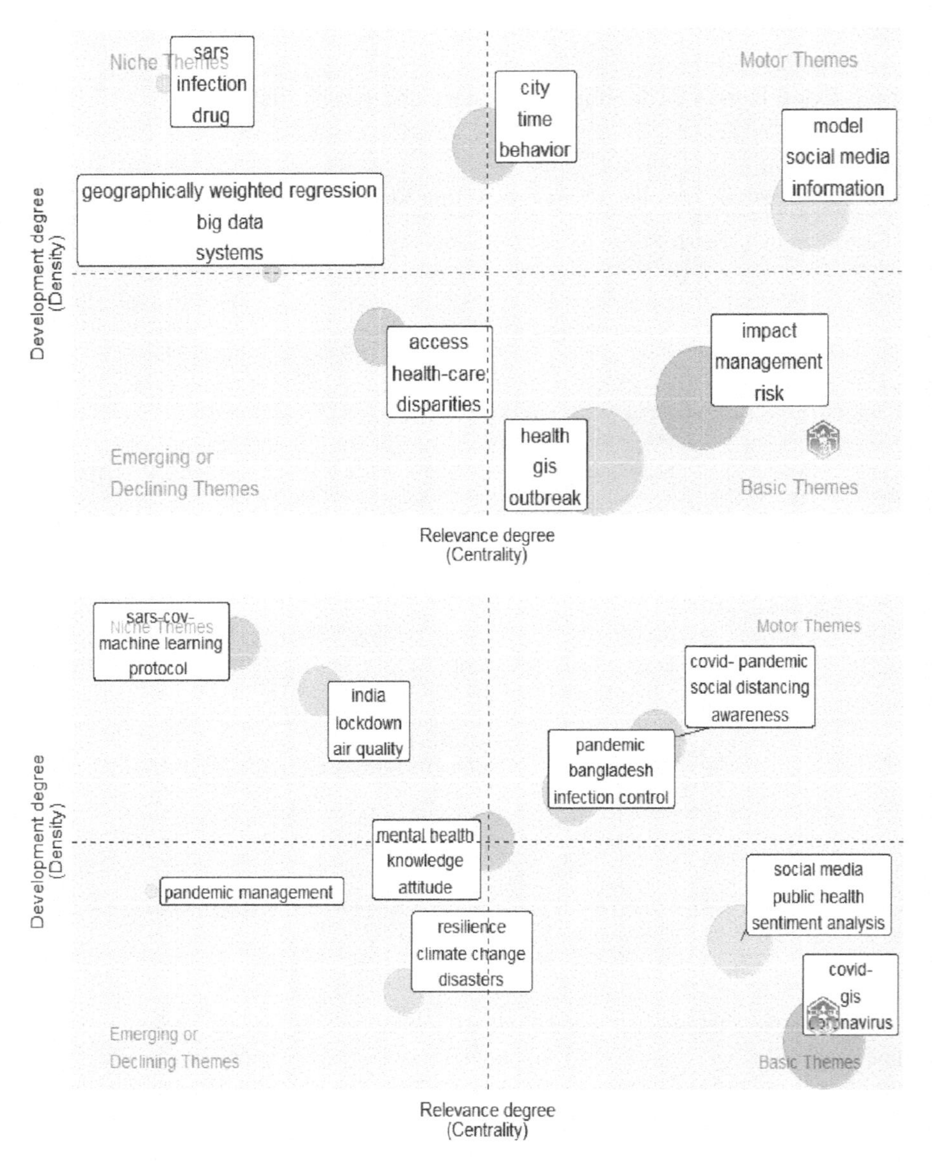

FIGURE 9.12 Thematic Map–Keywords Plus (Top)–Author Keywords (Bottom).

and collaboration patterns across three levels: author, institution, and country. The most productive countries, publications wise, are shown in Figure 9.13. With 619 articles contributing 82, or 13.2%, of the total publications, China is the most productive country. Figure 9.14 displays the country collaboration network. Figure 9.15 demonstrates a coauthor network, and it showcases regular study groups with a minimum of three coauthored documents, groups of scholars, and pivotal authors.

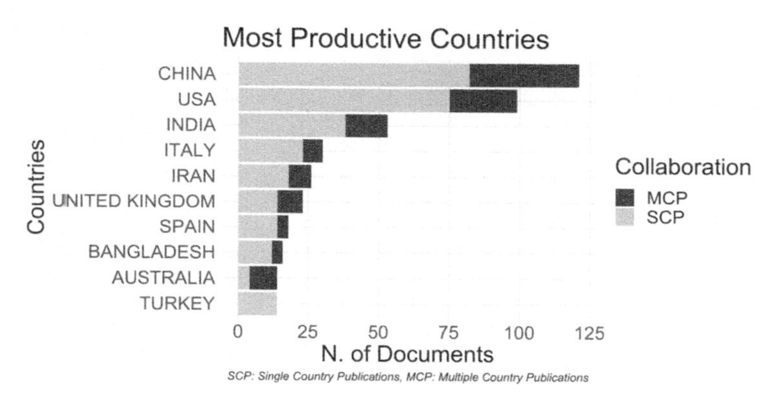

FIGURE 9.13 Most productive countries.

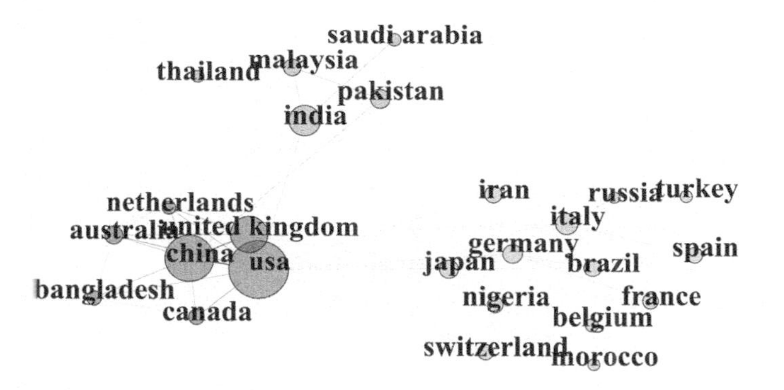

FIGURE 9.14 Country collaboration network.

Figure 9.16 illustrates the collaboration network of education or institution network, and it uncovers relevant institutions in a specific research field and their relations.

9.4 DISCUSSION AND LIMITATIONS

9.4.1 Discussion

Bibliometrics is a structured analysis of a large body of information to infer trends over time and themes researched, to identify shifts in the boundaries of the disciplines, to detect most of the prolific scholars and institutions, and to show the overall extent of research. Various methods exist to summarize the amount of scientific activity in a domain, but bibliometrics has the potential to introduce a systematic, transparent, and reproducible review process. Understanding the structure of this network is important as it can help improve research prioritization. Through science mapping analysis, this review reveals GIS's role in COVID-19 pandemic management and control. The study approached this question from five aspects: first, describing and finding trends

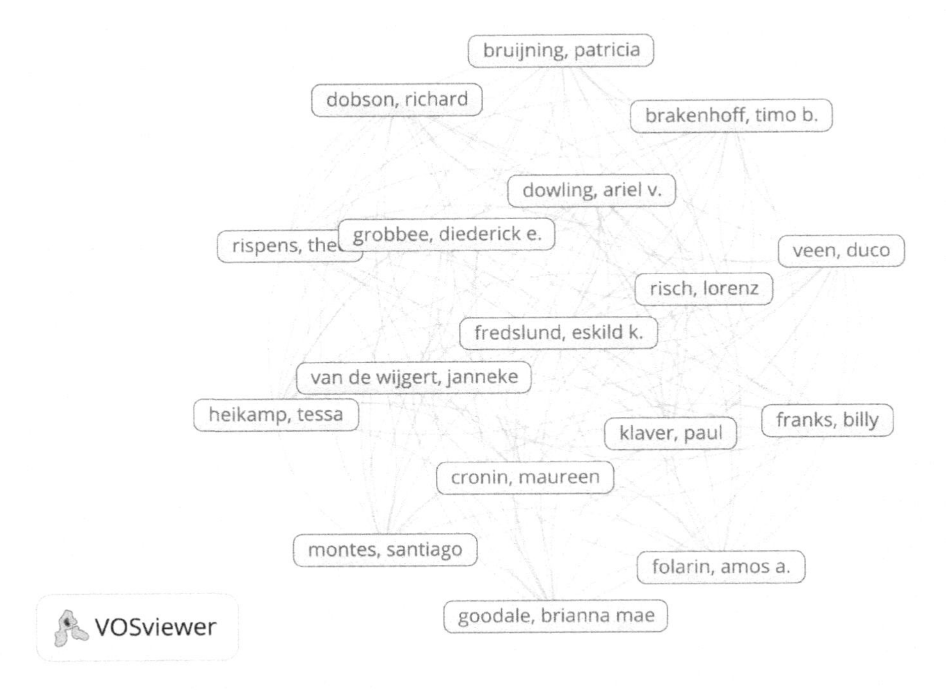

FIGURE 9.15 Author collaboration network.

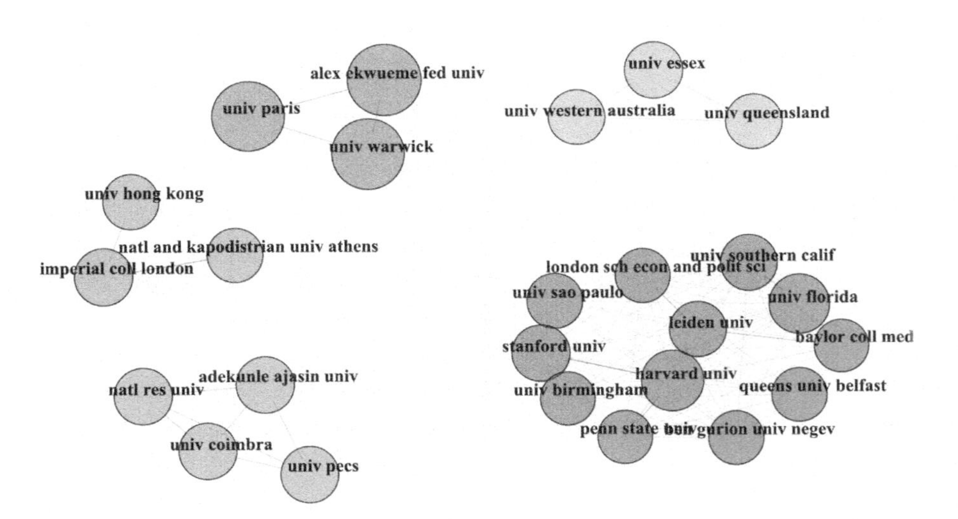

FIGURE 9.16 Institution collaboration network.

of publications within the collection, which is 2020 to 2022, followed by revealing the main themes of the scholarship through analyzing the Keywords Plus (ID) and the Author Keywords (DE). Then, visualizing the intellectual structure by mapping the author co-citation networks. Next, mapping the conceptual structure by visualizing co-occurrence networks to understand the topics covered by scholars. Finally, summarizing the collection's evolution and dynamic aspects by looking at the themes or the thematic maps of the Keywords Plus (ID) and Author Keywords (DE) in terms of their density and centrality, considering their theme type.

This bibliometric analysis of the GIS-related COVID-19 scholarship revealed worldwide research trends and performance in the subject area explored. This study used keywords (COVID, GIS, COVID CONTROL, COVID MANAGEMENT, SOCIAL MEDIA) to extract the most relevant papers from the Web of Science (WoS) Core Collection database from 2020 to 2022. Three hundred fifty-seven (357) source titles, including journals and conferences, were listed in 95 WoS subject categories with 11 different document types. The five most popular trends categories were *Environmental Science & Ecology*, *Public, Environmental & Occupational Health*, *Science & Technology*, *Health Care Sciences & Service* and *Computer Science*. The findings show that China, the United States, and India were the most productive countries that played a significant role in this collection, with the highest number of published papers globally. The analysis of authors among all and highly cited papers showed that there were 3,858 authors in all papers, and the average was 6.23, meaning that there were more than six authors for each paper. The Author Keywords (DE) analysis showed that *COIVD19*, *GIS*, *Pandemic*, *Social Media*, and *Coronavirus* are among the top five, whereas the Keywords Plus (ID) top five most frequent terms are: *Impact, Health, GIS*, *Management*, and *Outbreak*. In both the DE and ID terms, the keyword GIS has increasing records from 2020 to 2022. The analysis of research areas between all and highly cited papers was sorted by the number of records based on the total count of citations. Overall, there were 95 research areas for all paper groups and seven (7) research areas from the highly cited group. Both groups' top two research areas were *Environmental Science & Ecology* and *Public, Environmental & Occupational Health,* and GIS is centric in each of these subjects.

9.4.2 Limitations and Future Work

There are several limitations to acknowledge in this study. First, the WoS search algorithm provides only results searched in the most prominent sections of the article but not the whole article (e.g., title, abstract, keyword), so it fails to detect articles with deeper layers of meaning. Also, the bibliometrics method helps develop a holistic and objective viewpoint on the studied field; nevertheless, key findings cannot always be explained in detail. This research reveals the presence of closed research groups where collaborative research likely happens only between members. Further understanding requires the implementation of a qualitative review. In addition to showing what research has already been done, bibliometric maps indicate opportunities for future research. Because they show the existing relationships among publications within a research topic, bibliometric maps also show connections within the topic that have not yet been made through blank spaces on the map. This study teased out the Keyword Plus and the Author Keywords and then focused on the term "GIS" to determine connections to other keywords; additional work could be done to investigate further individual, cited papers, and author co-citation networks.

REFERENCES

Agbor, V. N., Simo, L. P., & Epie, T. B. (2021). Social media and management of COVID-19 in a developing country: The case of Cameroon. *Pan African Medical Journal, 38*. https://doi.org/10.11604/pamj.2021.38.344.25033

Ahasan, R., & Hossain, M. M. (2021). Leveraging GIS and spatial analysis for informed decision-making in COVID-19 pandemic. *Health Policy and Technology, 10*(1), 7–9. https://doi.org/10.1016/j.hlpt.2020.11.009

Al-Ramahi, M., Elnoshokaty, A., El-Gayar, O., Nasralah, T., & Wahbeh, A. (2021). Public discourse against masks in the COVID-19 era: Infodemiology study of Twitter data. *JMIR Public Health and Surveillance, 7*(4), e26780–e26780. https://doi.org/10.2196/26780

Aria, M., & Cuccurullo, C. (2017). Bibliometrix: An R-tool for comprehensive science mapping analysis. *Journal of Informetrics, 11*(4), 959–975. https://doi.org/10.1016/j.joi.2017.08.007

Brimblecombe, P., Hayashi, M., & Futagami, Y. (2020). Mapping climate change, natural hazards and Tokyo's built heritage. *Atmosphere, 11*(7), https://doi.org/10.3390/atmos11070680

Caspari, A., Wood, D., Campbell, A., Jefferson, D., Huynh, T., & Reddy, A. (2021). Using real-time data to detect delays and improve customer communications at New York City Transit. *Transportation Research Record: Journal of the Transportation Research Board, 2675*(7), 45–57. https://doi.org/10.1177/03611981211994115

Cobo, M. J., López-Herrera, A. G., Herrera-Viedma, E., & Herrera, F. (2011). An approach for detecting, quantifying, and visualizing the evolution of a research field: A practical application to the Fuzzy Sets Theory field. *Journal of Informetrics, 5*(1), 146–166. https://doi.org/10.1016/j.joi.2010.10.002

Cobo, M. J., Martínez, M. A., Gutiérrez-Salcedo, M., Fujita, H., & Herrera-Viedma, E. (2015). 25 years at knowledge-based systems: A bibliometric analysis. *25th Anniversary of Knowledge-Based Systems, 80*, 3–13. https://doi.org/10.1016/j.knosys.2014.12.035

Dong, E., Du, H., & Gardner, L. (2020). An interactive web-based dashboard to track COVID-19 in real time. *The Lancet. Infectious Diseases, 20*(5), 533–534. https://doi.org/10.1016/S1473-3099(20)30120-1

Esri. (n.d.). www.esri.com/en-us/home

Franch-Pardo, I., Napoletano, B. M., Rosete-Verges, F., & Billa, L. (2020). Spatial analysis and GIS in the study of COVID-19. A review. *Science of The Total Environment, 739*, 140033. https://doi.org/10.1016/j.scitotenv.2020.140033

Goodchild, M. F. (2022). Commentary: General principles and analytical frameworks in geography and GIScience. *Annals of GIS, 28*(1), 85–87. https://doi.org/10.1080/19475683.2022.2030943

Guo, J., Feng, X. L., Wang, X. H., & van IJzendoorn, M. H. (2020). Coping with COVID-19: Exposure to COVID-19 and negative impact on livelihood predict elevated mental health problems in Chinese adults. *International Journal of Environmental Research and Public Health, 17*(11). https://doi.org/10.3390/ijerph17113857

Ibrahim, M. D., Binofai, F. A., & MM Alshamsi, R. (2020). Pandemic response management framework based on efficiency of COVID-19 control and treatment. *Future Virology*, 10.2217/fvl-2020–0368. PMC. https://doi.org/10.2217/fvl-2020-0368

Jain, S., & Sharma, T. (2020). Social and travel lockdown impact considering coronavirus disease (COVID-19) on air quality in megacities of India: Present benefits, future challenges and way forward. *Aerosol and Air Quality Research*, *20*, 1222–1236. https://doi.org/10.4209/aaqr.2020.04.0171

Jiang, W., Xiong, Z., Su, Q., Long, Y., Song, X., & Sun, P. (2021). Using geotagged social media data to explore sentiment changes in tourist flow: A spatiotemporal analytical framework. *ISPRS International Journal of Geo-Information*, *10*(3). https://doi.org/10.3390/ijgi10030135

Kakodkar, P., Kaka, N., & Baig, M. (2020). A comprehensive literature review on the clinical presentation, and management of the pandemic coronavirus disease 2019 (COVID-19). *Cureus*, *12*(4), e7560. https://doi.org/10.7759/cureus.7560

Kamel Boulos, M. N., & Geraghty, E. M. (2020). Geographical tracking and mapping of coronavirus disease COVID-19/severe acute respiratory syndrome coronavirus 2 (SARS-CoV-2) epidemic and associated events around the world: How 21st century GIS technologies are supporting the global fight against outbreaks and epidemics. *International Journal of Health Geographics*, *19*(1), 8. https://doi.org/10.1186/s12942-020-00202-8

Kanga, S., Sudhanshu, Meraj, G., Farooq, M., Nathawat, M. S., & Singh, S. K. (2020). Reporting the management of COVID-19 threat in India using remote sensing and GIS based approach. *Geocarto International*, *0*(0), 1–8. https://doi.org/10.1080/10106049.2020.1778106

Kim, S. J., & Bostwick, W. (2020). Social vulnerability and racial inequality in COVID-19 deaths in Chicago. *Health Education & Behavior*, *47*(4), 509–513. https://doi.org/10.1177/1090198120929677

Koch, T. (2017). *Cartographies of disease: Maps, mapping, and medicine* (New expanded ed.). Esri Press.

Kumar, D., Batra, L., & Malik, M. T. (2020). Insights of novel coronavirus (SARS-CoV-2) disease outbreak, management and treatment. *AIMS Microbiology*, *6*(3), 183–203. https://doi.org/10.3934/microbiol.2020013

Mann, H. (2022). *Joint Statement on the Value of GIS in the Pandemic*. National States Geographic Information Council (NSGIC). www.nsgic.org/index.php?option=com_dailyplanetblog&view=entry&year=2020&month=04&day=16&id=23:joint-statement-on-the-value-of-gis-in-the-pandemic

Mapbox. (n.d.). Mapbox. www.mapbox.com/

Mollalo, A., Vahedi, B., & Rivera, K. M. (2020). GIS-based spatial modeling of COVID-19 incidence rate in the continental United States. *Science of The Total Environment*, *728*, 138884. https://doi.org/10.1016/j.scitotenv.2020.138884

National States Geographic Information Council (NSGIC). (n.d.). [.org]. National States Geographic Information Council (NSGIC). www.nsgic.org/

QGIS.org, 2022. QGIS Geographic Information System. QGIS Association. www.qgis.org

Qin, L., Sun, Q., Wang, Y., Wu, K.-F., Chen, M., Shia, B.-C., & Wu, S.-Y. (2020). Prediction of number of cases of 2019 novel coronavirus (COVID-19) using social media search index. *International Journal of Environmental Research and Public Health*, *17*(7). https://doi.org/10.3390/ijerph17072365

R Core Team. (2022). *R: A Language and Environment for Statistical Computing*. R Foundation for Statistical Computing. www.R-project.org/

Sandre, A. A., Fruehauf, A. L., Miyahara, A. A. L., Rosa, A. A., Maruyama, C. M., Locoselli, G. M., Candido, L. F., Lombardo, M. A., Coelho, M. A., Murolo, R. P., Pombo, R. M. R., Marques, T. H. N., & Pellegrino, P. R. M. (2021). Geodesign Brazil: Trees for the metropolitan area of São Paulo. In O. Gervasi, B. Murgante, S. Misra, C. Garau, I. Blečić, D. Taniar, B. O. Apduhan, A. M. A. C. Rocha, E. Tarantino, & C. M. Torre (Eds.), *Computational Science and Its Applications—ICCSA 2021* (Vol. 12954, pp. 463–475). Springer International Publishing. https://doi.org/10.1007/978-3-030-86979-3_33

Stieg, C. (2020). How this Canadian start-up spotted coronavirus before everyone else knew about it. *CNBC*. www.cnbc.com/2020/03/03/bluedot-used-artificial-intelligence-to-predict-coronavirus-spread.html

Szczepańska, A., Kaźmierczak, R., & Myszkowska, M. (2021). Virtual reality as a tool for public consultations in spatial planning and management. *Energies, 14*(19). https://doi.org/10.3390/en14196046

Van Eck, N. J., & Waltman, L. (2014). Visualizing bibliometric networks. In Y. Ding, R. Rousseau, & D. Wolfram (Eds.), *Measuring scholarly impact: Methods and practice* (pp. 285–320). Springer International Publishing. https://doi.org/10.1007/978-3-319-10377-8_13

World Health Organization. (2020). *National capacities review tool for a novel coronavirus, 10 January 2020*. World Health Organization. https://apps.who.int/iris/handle/10665/332298

Yang, S., Keller, F., & Zheng, L. (2017). *Social network analysis: Methods and examples*. SAGE.

Zhang, C. H., & Schwartz, G. G. (2020). Spatial disparities in coronavirus incidence and mortality in the United States: An ecological analysis as of May 2020. *Journal of Rural Health, 36*(3), 433–445. https://doi.org/10.1111/jrh.12476

Zhang, J., Yu, Q., Zheng, F., Long, C., Lu, Z., & Duan, Z. (2016). Comparing keywords plus of WOS and author keywords: A case study of patient adherence research. *Journal of the Association for Information Science and Technology, 67*(4), 967–972. https://doi.org/10.1002/asi.23437

Zhao, Y., Cheng, S., Yu, X., & Xu, H. (2020). Chinese public's attention to the COVID-19 epidemic on social media: Observational descriptive study. *Journal of Medical Internet Research, 22*(5), e18825. https://doi.org/10.2196/18825

Zhou, C., Su, F., Pei, T., Zhang, A., Du, Y., Luo, B., Cao, Z., Wang, J., Yuan, W., Zhu, Y., Song, C., Chen, J., Xu, J., Li, F., Ma, T., Jiang, L., Yan, F., Yi, J., Hu, Y., … Xiao, H. (2020). COVID-19: Challenges to GIS with big data. *Geography and Sustainability, 1*(1), 77–87. https://doi.org/10.1016/j.geosus.2020.03.005

10 The Use of Geographic Information Systems to Shape COVID-19 Policies Regarding Masking and Distancing with a Partially Vaccinated Population

Allison P. Plaxco, Jennifer M. Kmet, Liang Li, Rachel Rice, Chaitra Subramanya, David Sweat, Alisa Haushalter, Michelle Taylor, Li Tang, Jesse Smith, Motomi Mori, Gregory T. Armstrong, Lilian A. Nyindodo, Manoj Jain, Karen C. Johnson, and Fridtjof Thomas

CONTENTS

DOI: 10.1201/9781003227106-10

10.1 INTRODUCTION

Geographic information systems (GIS) combine location-attributed data with software tools to summarize, visualize, and interpret such data. We describe how we used GIS to support the Coronavirus Disease 2019 (COVID-19) pandemic response in Shelby County and the larger Memphis metropolitan area in Tennessee (TN), United States. We highlight successes and challenges when using GIS for this purpose and suggest desirable aspects of educational courses for GIS as well as skill sets in personnel that we came to appreciate as essential.

Shelby County is located in the southwestern corner of Tennessee, and is the largest and most populous[1] of the 95 counties in the state. The 2019 one-year American Community Survey estimated the total population at 937,166, 48% male, 52% female, 39% Whites, 54% Black or African Americans, and 7% other or multiple races.[2] In Shelby County, the annual median income for earners age 16 and older was $32,355,[3] with a median household income of $52,092, and 19% of its population lives in poverty as defined by the Census Bureau.[4, 5] The Shelby County Health Department has jurisdiction over public health issues for Shelby County, which includes the adjacent cities/towns of Arlington, Bartlett, Collierville, Germantown, Lakeland, Memphis, and Millington, which all have their own decision-making bodies. Memphis is the largest city, with a total population of 651,088.[6] The Mississippi River borders Shelby County to the west, and neighboring counties include Tipton County, TN, to the north; Fayette County, TN, to the east; Desoto County, Mississippi (MS) to the south, and Crittenden County, Arkansas (AR) to the west across the Mississippi River.

Shelby County serves as a central transportation hub in the mid-South area of the United States and is the crossroad for several interstate highways running north to south (Interstates 55 and 269), and east to west (Interstate 40). About 60,000 vehicles daily transverse the Mississippi River on the interstate bridges[7] connecting Arkansas and Memphis, TN, for long-distance as well as commuter traffic. Memphis International Airport is the second-busiest airport in the world in terms of cargo and is home to the FedEx Express global super hub.[8] Memphis also connects the Amtrak southern passenger train network to its midwestern train routes as a major stop on the New Orleans–Memphis–Chicago 19-hour train journey. Several large hospitals and the US Veterans Affairs Memphis Medical Center serve the larger mid-South population, and Memphis has the only designated Level 1 Trauma Center in a 150-mile (241 km) radius for high-risk patients from Tennessee, Mississippi, Arkansas, and Missouri.[9] As a consequence, the city of Memphis and Shelby County are not only affected by the health of their own residents but also provide resources to care for a population in the larger area in general, and hospital beds for COVID-19-related hospitalizations in particular.

10.2 BACKGROUND

10.2.1 Organizing a Regional COVID-19 Response

In March 2020, the COVID-19 pandemic had already rapidly spread worldwide for several months. In Bergamo (Italy, Europe), coffins were transported by military trucks to remote cremation sites because local morgues did not have the capability to manage the number of deaths caused by the coronavirus infections.[10,11] From New York, pictures were broadcast showing bodies wrapped in plastic being loaded onto refrigerated container trucks used as temporary morgues.[10,12] Under the impression of these disturbing happenings, all restaurants, bars, gyms, entertainment and recreational establishments, and the like were swiftly closed in Shelby County by a combination of differing mechanisms depending on the respective jurisdiction. Shelby County[13] along with local municipalities within the county issued Shelter in Place orders shortly after the first confirmed COVID-19 cases in Memphis, TN, were reported on March 8, 2020 (lockdown effective March 25, 2020, until May 4, 2020).

The City of Memphis and Shelby County assembled a broad group of stakeholders in a COVID-19 Joint Task Force under the highest level of leadership (chaired by the chief operating officer and chief administrative officer for the City of Memphis and Shelby County, respectively) that initially met daily (including Sundays and holidays), with the frequency of meetings increasing or decreasing as the pandemic evolved. Early in the pandemic, there were times when meetings took place twice per day. Later in the pandemic, meeting frequency changed to twice a week and then weekly. These 8 a.m. virtual meetings were regularly attended by over 120 individuals. The coordinating of efforts and resources, division of labor, and reporting was organized through Emergency Support Functions (ESF), including ESF-8 Public Health and Medical Services with a Data Subcommittee focused on COVID-19-related data and its interpretation for decision-making.

The Data Subcommittee initially elicited available information on the "timing of the surge" (how many citizens will fall ill and need hospital care and when). It was evident early on that an unfettered "wave" of COVID-19 infections resulting from an unmanaged community spread of the virus would overwhelm the hospitals and available resources. In the absence of vaccines and other prophylactic or therapeutic medications, the focus was initially on the relative efficiency of mitigation strategies such as masking and physical separation/social distancing measures.[14] While every part of the county (as well as the country and the world) is affected by the virus, apparently not all are affected in the same way and at the same time. Over the course of the pandemic, subcounty-level data was used to monitor trends and to tailor a response accordingly. This data was used to drive the evolution of COVID-19 policy decisions in real time, with the aim of ameliorating the pandemic and reducing transmission, morbidity, and mortality.

10.2.2 GIS and Choropleth Maps

To document and communicate the dynamics of the evolving spread of COVID-19, we mainly worked with choropleth maps using ZIP Code Tabulation Areas (ZCTAs) and use here the colloquial name "ZIP code maps." The United States Postal Service (USPS) ZIP codes are collections of mail delivery routes and do not have a clear areal representation (they do not need to be polygons, and some ZIP codes are assigned to specific companies or mailboxes within a post office (PO boxes) and have no areal extension at all). For that reason, the US Census Bureau has created ZCTAs that reflect as much as possible the geographical extent of ZIP codes and are areas that can be displayed on maps.[15]

ZIP codes remain important references, because most people know their own and neighboring ZIP codes of their residential addresses, and they can readily be extracted from street addresses provided for communication and the billing of health services. In contrast, point locations (e.g., geocoding of home street addresses) suffer from higher frequency of the misrecording of street addresses and so on, compared to ZIP codes, resulting in incomplete data sets and consequently dropped records in the analysis stage. Therefore, ZIP codes often have higher utility for public health surveillance than point locations and allow for rapidly analyzed data using more complete data sets.

10.2.3 Privacy of Health Information

In the United States, the "Health Insurance Portability and Accountability Act of 1996 (HIPAA) is a federal law that required the creation of national standards to protect sensitive patient health information from being disclosed without the patient's consent or knowledge."[16] HIPAA considers all information relating to health that is from human subjects and relates to medical data (doctor visits, lab tests, etc.) or medical billing data. There are 18 HIPAA identifiers, and every single one of them turns an individual record-level data set into "identifiable" data and protected health information (PHI). Among those are any dates (except year-only information) relating to age, admissions, lab tests, and the like, and any geographical subdivisions smaller than a state (with an exception that combined three-digit ZIP codes can be provided if specific small ZIP codes are displayed collectively as 000). Thus, in the United States, information about COVID-19 tests is protected under HIPAA and must not be publicly displayed due to confidentiality concerns. Protecting the privacy of individuals in ever-increasing use of data is an area of active research.[17, 18]

In accordance with HIPAA, records for small geographical areas such as ZCTAs should not be shared if their populations make up fewer than 20,000 people (HHS HIPAA Section 164.514(b)(2)[19]). Shelby County, TN, has a total of 33 ZCTAs, several of which are small enough to fall under that regulation. This does affect mapmaking in the specific circumstances as detailed in the following example: To address the issue of low population areas, ZCTAs that are similar in demography and geography were merged to obtain populations above 20,000. One of the merges included three ZCTAs (38103+38104+38105) that differ in important aspects: ZCTA 38103 is considered a high socioeconomic area, 38104 is considered a middle-class area, and 38105 is considered a low socioeconomic area. Thus, the necessary merge effectively erased a trait (socioeconomic status) that we would ideally maintain for analysis purposes.

There are at least three general approaches to guard against loss of confidentiality for individuals. First, no identifiable information is contained in the individual displayed records. However, because HIPAA does not allow for individual records to contain any geographical subdivisions smaller than states, it is apparent that otherwise well-established geo-masking approaches by jittering (adding noise) to geographical coordinates will not suffice with respect to health records and medical testing information, because the location is not disguised "enough."

A second approach to preventing identification of individuals uses aggregated records, and it must be ensured that individual cell counts in tables or on maps adhere to the required specifications/aggregation level. The Shelby County Health Department sets the minimal number to ten for cell counts; therefore, cell counts no smaller than these can be displayed in tables or on maps.

A somewhat different approach is that cases/counts are not displayed at all, but instead, computational results are displayed that do not allow for backtracking of individual cases. We used that technique to show COVID-19 density maps (Figure 10.1): Here each address location listed for a COVID-19 case is individually geocoded, and a smoothing algorithm is used to create density estimates in longitudinal (x-axis) and latitudinal (y-axis) directions. Note that a simple rate together with the population count of an area allows for easy back-calculation of recorded cases, whereas this smoothing approach together with the resulting representation does not suffer from that drawback. Kernel density analysis was used to create density maps such as that shown in Figure 10.1. This method was used to represent COVID-19 case data as a collection of geographical address points (per square mile). The density analysis smooths the data points and puts them within a color ramp, which is easy to interpret. For example, the red areas denote higher concentration of cases when compared to the green areas where the cases are less concentrated and more spread out. The same data was used internally for identifying clusters and to initiate prompt and targeted interventions to control and limit COVID-infections. From this data, concentration of cases was identified in a variety of settings such as mobile homes, assisted living facilities, and household clusters. In the early months of the pandemic, this information was used to inform early interventions and policy decisions aimed at containing clusters. When many cases are reported and rampant community spread is evident, these density plots mostly resemble the population distribution across the larger area. However, geocoding of addresses was also used to inform various aspects of the COVID-19 response in several other key ways. For example, geocoding of testing sites and vaccination sites (both public and private) was used in assessing availability and lack of availability of resources, which informed targeted responses to address lack of access to facilities in specific areas so identified.

10.2.4 Health Disparities

There are well-documented differences in health-related outcomes between the racial-ethnic groups in the United States. Not only is life expectancy at birth for racial-ethnic groups markedly different in the United States, but it also is not uniform within each

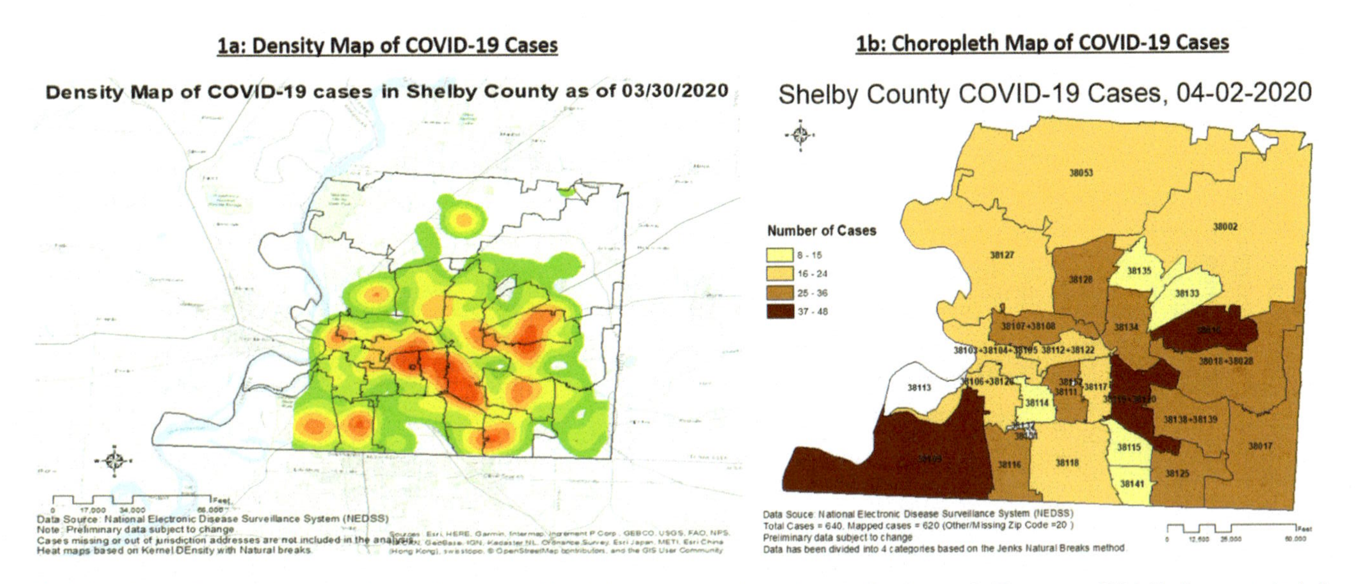

FIGURE 10.1 The jurisdiction of the Shelby County Health Department is limited to Shelby County in Tennessee, USA (1a larger encircled area), with the City of Memphis being part of that county. Borders are to other counties in TN in the north and east, and to the states of Mississippi (MS) in the south and Arkansas (AR) to the west (separated by the Mississippi River, blue line). The county includes the City of Memphis. 1b: The choropleth map gives a similar indication of where most cases occur. Note how the density map (1a) accumulates case density according to where individually geocoded home addresses of cases were located, with red areas denoting a higher case load. Maps displaying the actual geocoded point locations of cases were deemed inappropriate out of a concern for protecting the privacy of the individuals; see 10.2.3 for details.

group across the country, and relative improvements within each group from 2000 to 2019 also vary substantially. What is more, *differences* between racial-ethnic groups differ across the country.[20] These are important observations, as they document that health differences between groups are not "fixed" but change over time. Thus, they cannot all be attributed to natural, biological variation. While some health outcomes might be inevitable or unavoidable, others are potentially unnecessary.[21] Modifiable risk factors include access to and utilization of health-care services, socioeconomic status, the physical environment, literacy levels, and legislative policies that are all aspects of social determinants of health that materialize in an individual's as well as a population's health.

Components of these determinants of health that often have a strong association with "the part of town or country" people live in, include access to high-quality education; nutritious food; decent and safe housing free from overcrowding; affordable, reliable public transportation; culturally sensitive health-care providers; health insurance; clean water and nonpolluted air.[22] In addition to race and ethnicity, health disparities are recognized between other groups identified by sex, gender identity and sexual orientation, disability status or special care needs, and geographical location (rural vs. urban). General demographic factors, life expectancy, rates of illnesses, and the extent of chronic conditions are just a few aspects that can be attributed to these groups. COVID-19 was early known to spread by close contact through the severe acute respiratory syndrome coronavirus 2 (SARS-CoV-2) in aerosols, and additional factors were therefore of importance, such as multiple-generation households that are more prevalent in low-income areas with greater crowding and less ability to separate different generations from each other.

Though maps are useful tools to communicate differences in metrics like case rates, test rates, and vaccination rates across locations, they cannot simultaneously represent known factors and health disparities that influence the health status of the individuals in the respective populations. It is desirable to consider many confounders simultaneously when interpreting geographical variation in the rates seen on a map. The possibly easiest to adjust for is population size in an area. In principle, additional aspects such as the age/race/gender composition in each area can be used in such standardized rates, but these distributions are much less readily available as the multivariate joint distribution is required for this purpose and the respective marginal distributions do not suffice for this standardization. Such confounders in epidemiology[23, 24] are by definition traits or factors that are associated with the "exposure" as well as the "outcome," and can potentially lead to erroneous associations in naive, unadjusted analyses. When analyses are not adjusted for these potential confounders, observed associations/correlations can even change from positive to negative or vice versa (referred to as Simpson's paradox[25, 26]). Maps often show aggregated measures, like means or proportions, but the problem of unadjusted confounders is well known to extend to ecologic studies[27] using such aggregated measures as well as environmental measures/physical characteristics of the places where members of each group live. While existing combined measures like the hardship index[28] can reflect several factors simultaneously to some degree, for the study of the etiology of diseases and actionable factors, it is desirable to include all known potential confounders as adjustments in more complex models or to use them for standardized rates[29] that are subsequently included in analyses. Maps cannot substitute for such models and either display simple rates and so on that are prone to the above issues of unadjusted confounding factors, or results from complex epidemiological models that are often difficult to communicate effectively to policy makers and/or the public. These more advanced spatial modeling strategies that can be used to adjust for confounders were not used in our monitoring and reporting of the COVID-19 pandemic and are beyond the scope of this text.[30–35]

In the context of COVID-19, which is caused by the virus SARS-CoV-2, preexisting health conditions and the ability to protect oneself from infection within someone's own home or at work, must be assumed to differ substantially between different individuals and subpopulations. Besides health-care workers, workers in numerous industries and occupations were classified as essential to maintain critical infrastructure and continue critical services and functions.[36–38] These workers were required to maintain "normal work schedules" to a greater extent than others despite the ongoing COVID-19 pandemic, and many had to carry out work-related duties that needed to be performed on-site and involved being near coworkers or the public. Despite effective transmission mitigation strategies widely utilized by many workplaces including masking, social distancing, staggered breaks, and later, vaccination requirements in some cases, workers in essential/critical positions were put at higher risk for infection, if they continued to report to work, than workers in positions that could be shifted to be completed entirely remotely. Many workers in such essential/critical positions risked becoming unemployed if they decided not to continue to report to work. Areas across the county differ in their composition of workers, often related to education level and income: workers who perform manual labor are often paid by the hour or on a piecework basis ("blue-collar workers"), whereas administrative and management positions are often held by higher-income individuals with annual salaries and health insurance ("white-collar" workers). These populations were different in their relative risk of getting COVID-19 during work and this difference may have extended into the infection risk associated with their different residential addresses. Consequently, worker types ("blue collar" vs. "white collar") and residential addresses each meet the definition of a confounder,[23, 24] and will induce or distort any association observed in a direct/naive comparison of ZIP code areas with COVID-19 infections. Likewise, every variable closely associated with these work positions and, for example, social vulnerability index,[39] their resulting purchasing powers as expressed by median house values, offered internet speeds by internet providers, urban blight indicators relating to the absence of maintenance of properties and so forth, are also possible confounders of that relationship. This is not to say that these factors cannot possibly be responsible for differences in health outcomes, but it is important to note that associations cannot be fully understood without proper adjustment of their confounding components.

Nonetheless, from the perspective of public health, it was important to determine in which areas the absolute or relative numbers of COVID-19-infected individuals were higher, and no adjustment for confounders is needed for these facts. For example, to plan for hospital resources and other resources that needed to be allocated, unadjusted maps were a more appropriate tool, even though they did not speak to *why* the differences across larger regions exist. In such cases of immediate response, it is most important to identify *where* resources are most needed, and *why* the differences exist is not as important. In contrast, whenever the *cause* of these different outcomes is considered and actionable aspects are to be identified, careful attention must be given to confounding variables. Because COVID-19 is an acquired infection by a virus with known medical risk factors for severe symptoms,[40] the list of potential confounders is long, and many are not readily accessible for the different subpopulations of interest. Therefore, it is considerably easier to furnish maps with raw counts or simple standardized rates, accounting for relative population sizes, to facilitate a solid understanding where disease burden is relatively higher. It is considerably more difficult to furnish such maps to guide an understanding of *why* infection rates differ across an area.

A separate difficulty that we noticed during the ongoing COVID-19 pandemic was a temporal pattern such that different areas regularly had very comparable dynamics in COVID-19 cases but offset by a few days or weeks. As a result, comparative maps might have visually suggested specific patterns across an area at a specific time point that were entirely muted once the other areas had "caught up." This was especially noticeable with respect to rural/urban classifications that also tend to come with different political ideologies, household earnings, distance to supermarkets, access to health-care facilities, and the like. A snapshot of one point in time does not capture such dynamics, and map patterns that seemingly support a specific narration could often be found by picking the "suitable" point in time during an otherwise very similar dynamic development.

10.3 FACILITATING POLICY MAKING WITH GIS

10.3.1 What Information Is Needed?

The choropleth maps showed, by type and intensity of the color, an aggregate number or a geographical characteristic with spatial enumeration as ZIP codes. This type of map has been used by the Shelby County Health Department to track many metrics from the time that the COVID-19 pandemic reached Shelby County. In the early phases of the pandemic, counts and rates of cumulative cases, cumulative tests administered, and overall test positivity rate by ZIP code were topics prioritized for monitoring graphically with maps (Figures 10.1–10.3). As the pandemic continued and case counts continued to rise, cumulative case and test counts held less meaning, and the focus shifted to monitoring 14-day trends of topics including test counts and rates, case counts and rates, and 14-day test positivity rates by ZIP code (Figure 10.4). This approach allowed for monitoring the shifting areas of priority over time, monitoring areas that heavily utilized testing resources and areas where access might be lacking, and prioritizing intervention efforts based on areas of greatest potential impact.

As the vaccine campaign began, and more of the population became vaccinated, we used ZIP code maps to monitor vaccination rates and the percent of the population vaccinated with at least one dose by ZIP code. Over time, as new virus variants were introduced into the population, it became more important to be fully vaccinated,

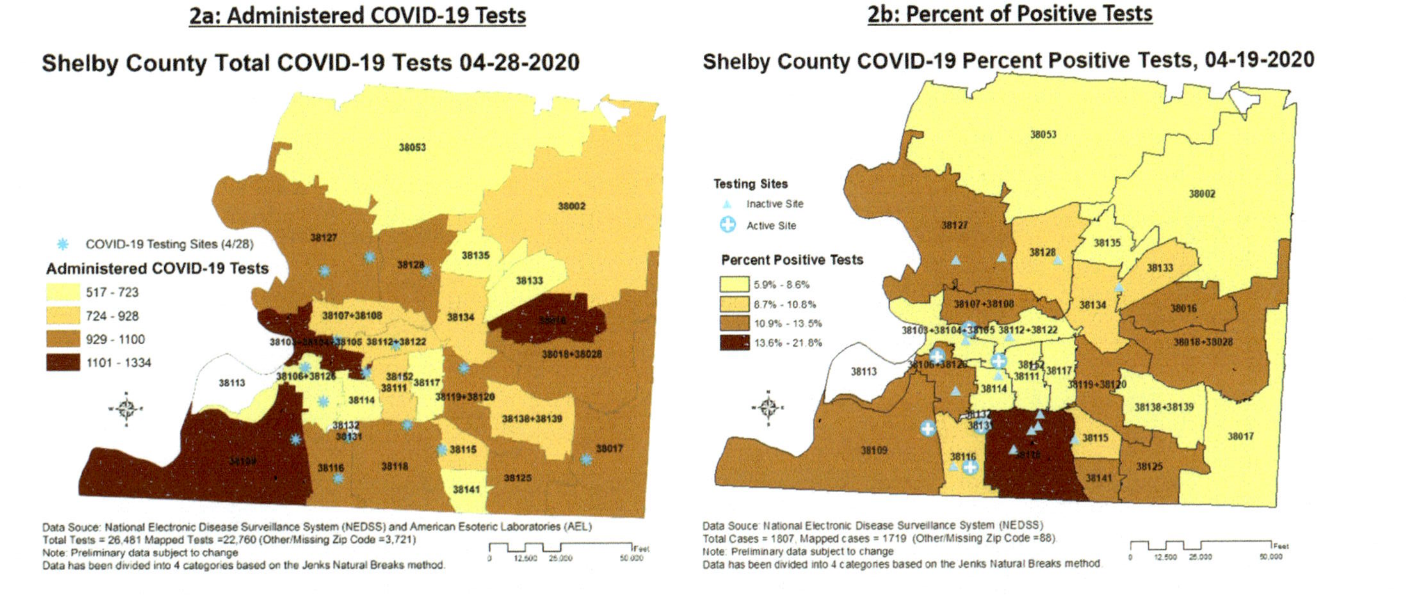

FIGURE 10.2 Examples of maps used to monitor testing counts (2a) and test positivity (2b) in the first months of the COVID-19 pandemic in Shelby County. In these early days of the pandemic, community spread was beginning to accelerate, but daily case counts were still relatively low.

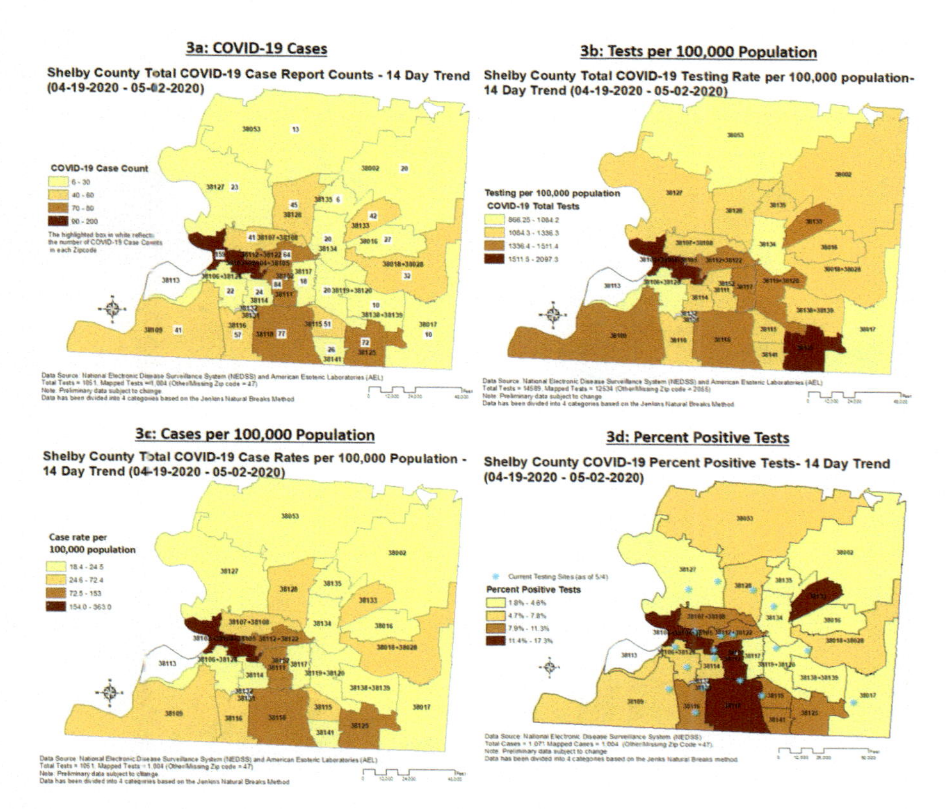

FIGURE 10.3 Examples of early iterations of 14-day trend maps used to monitor case counts (3a), testing relative to population size (3b), case rates (3c), and proportion of positive tests (3d) at the subcounty level.

and cases among children began rising. This led to new priority topics that needed to be monitored and the introduction of new ZIP code maps tracking the 14-day rate of pediatric cases and the percent of the population fully vaccinated by ZIP code (Figure 10.4). These additional maps allowed us to monitor the pediatric population that was at increased risk of disease in the context of the Delta variant and allowed us to change the method of monitoring population protected by vaccination as the protection afforded by suboptimal vaccination shifted in the context of newly emerging variants.

To assess and inform intervention efforts made in areas of high priority, we began monitoring geographical disparities in vaccination, case, testing, and death rates on a quarterly basis (Figure 10.5). These maps proved particularly useful in combination with other approaches and were, for example, used to identify areas with higher burden of disease, followed by a closer analysis of the demographics in those areas. The Shelby County Health Department used that approach to subsequently contract field-based observational studies to assess masking with an emphasis on diverse populations within each area. When mandatory masking was implemented, follow-up

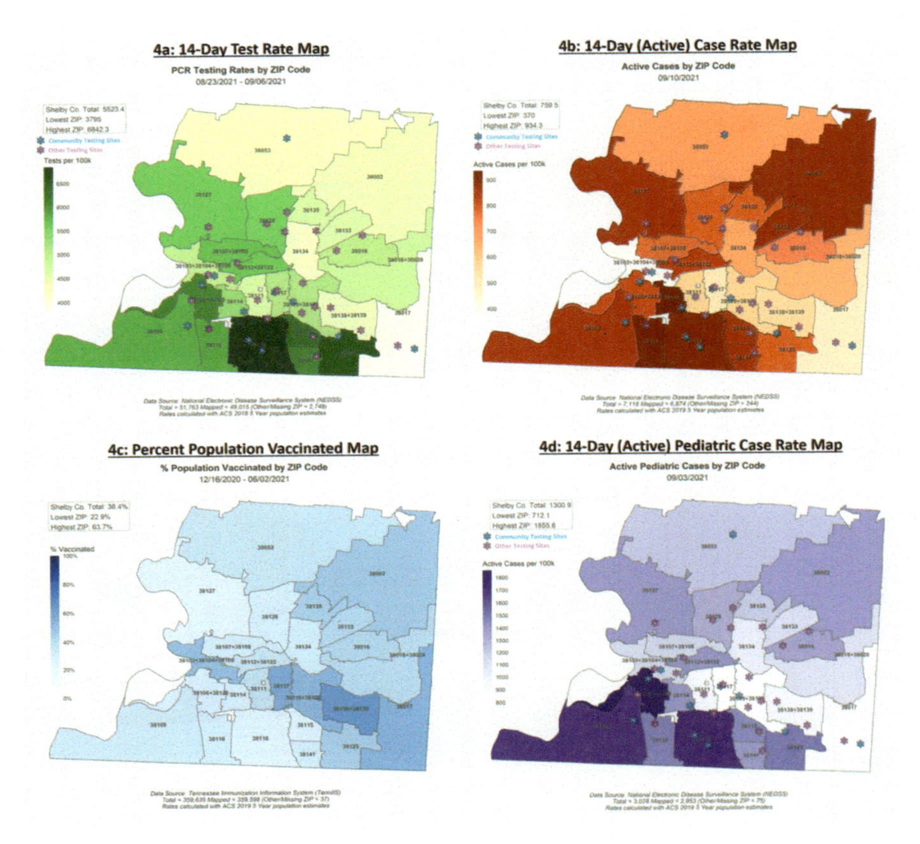

FIGURE 10.4 Examples of four critical maps which were automated and produced daily to support monitoring near real-time data. We decided that 14-day testing (4a) and case rates (4b) along with the percent of population vaccinated with at least one dose of COVID-19 vaccine (4c) were critical topics to monitor and communicate frequently.

observational studies and continued surveillance allowed a better understanding of messaging to diverse populations. Thus, the GIS data was used in a meaningful and timely way to respond to the pandemic locally.

10.3.2 Social and Policy Environment Around Introduction of the Maps

Examples of early maps are given in Figure 10.2, and these were used to monitor testing counts (10.2a) and test positivity (10.2b) in the first months of the COVID-19 pandemic in Shelby County. Continual monitoring of testing utilization geographically throughout the pandemic allowed for identification of potential gaps in access and influenced communication strategies and targeted messages in the community.

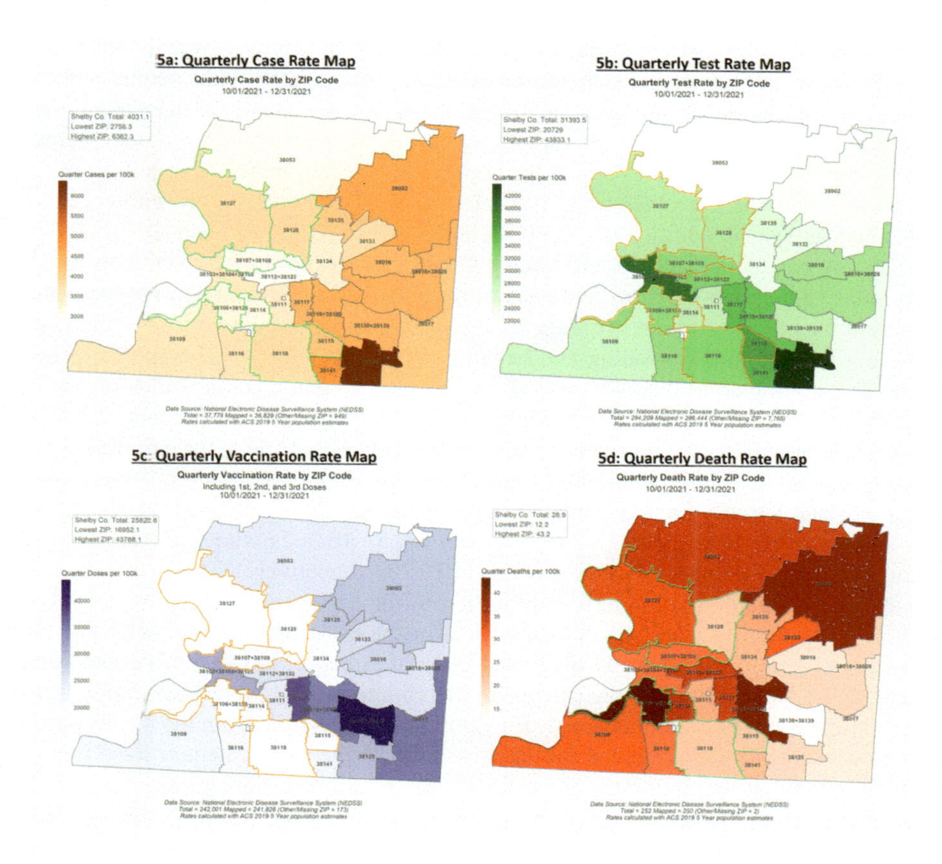

FIGURE 10.5 Example of maps created to visually display quarterly case (5a), testing (5b), vaccination (5d), and death (5d) rates by ZIP code. The highlighted ZIP codes (emphasized borders) identify areas of potential for high impact from intervention efforts due to continued high case rates and low vaccination uptake (5c) as of June 2021, compared to the rest of Shelby County.

In these early days of the pandemic, community spread was beginning to accelerate, but daily case counts were still relatively low. Test sites and monitoring infrastructure were still being created, and the initial focus was on where cases were detected. Under the assumption that COVID-19 symptoms and/or known exposure to cases would drive the demand for testing, one may have expected that the proportion of positive tests (COVID-19 is present) correlates strongly with the number of tests administered, something that did not appear to be the case in 10.2a and 10.2b. However, such a comparison is hampered by at least three drawbacks: First, the ZIP code areas have different population sizes. Second, individuals may have sought multiple tests following the same exposure or symptom onset and may have continued to test positive in the 90 days following initial infection, even after they had recovered. Third, access to testing was not equal for every resident. In the beginning of the pandemic, access for asymptomatic individuals was restricted to primarily first responders and essential workers. Especially the latter category is associated with income, education level, age, gender, and race, and these traits are not equally distributed across all ZIP code areas. This calls for computing and displaying standardized rates, something that is readily done to adjust for population size differences, but not as easily accomplished for the other traits (see also 10.2.4) that presumably influence the demand for tests in addition to COVID-related symptoms and exposure.

As the pandemic continued in Shelby County, the focus shifted from monitoring cumulative numbers of cases and tests to monitoring case and testing numbers and rates in the previous 14 days (Figure 10.3). At this point, cumulative counts and rates did not hold as much meaning and began to mask the shifting geographical priorities that continued to fluctuate periodically as the pandemic continued. Monitoring cases and tests on a 14-day interval allowed for more specific assessment of where cases were rising and falling over time, identification of hot spots, for monitoring for clusters in areas where community transmission was continually high or increasing quickly, and for easy comparison between time periods.

Note the identical shading of the maps in Figure 10.3 showing the total number and the relative number of cases (10.3a and 10.3c, respectively). This could suggest that there is small variation in the proportion of the population affected by COVID-19 at this point but could also be an artifact of the used default approach to create the four categories known as the Jenks Natural Breaks Method.[41] This approach determines break points for the categories by minimizing the average deviation from the class mean while simultaneously maximizing the deviation from the means of the other groups. All that is needed for the areas to be in the same category for different measures is that break points can be found that satisfy the above optimization criteria, even if there is potentially substantial "movement" within each category. This demonstrates the need to take control over default settings in software when processing maps. The standardization of measures to the population size (10.3b and 10.3c) allowed a direct comparison of relative test frequency with relative number of cases and showed that it is not true that cases were merely detected where tests were conducted, something which would result in a large agreement of shading in the two maps and that was a frequently encountered opinion in the early days of the pandemic. There were clearly differences in the relative number of cases that were not "explained" by the number of tests requested. In addition, 10.3d shows how the proportion of positive tests was at best only partially "explained" by the relative number of cases or tests in an area. Other factors must have impacted this percentage as well. Note how it became "visually bewildering" to use the same color scheme for a growing collection of different metrics. This motivated us to reserve different color schemes for different aspects/metrics (see Figure 10.4).

The vaccination campaign in Shelby County began in mid-December 2020. Upon the introduction of COVID-19 vaccinations in the community, the Shelby County Health Department began monitoring vaccination trends in the community by ZIP code. At the beginning of this process, when access to vaccines was restricted based on age group and high-risk occupation, cumulative vaccination per 100,000 population by ZIP code was monitored using maps created in ArcMAP/ArcGIS (by

Esri). As access restrictions to the early vaccines were eased and vaccination uptake increased across the county and some residents became eligible for their second dose, the Shelby County Health Department began monitoring the percent of the population vaccinated with at least one dose by ZIP code instead of all vaccinations, beginning in mid-May 2021. This decision to focus on those vaccinated with at least one dose was made for multiple reasons. First, in the earlier stages of the pandemic, when historical data on the nature of immunity from COVID-19 and vaccination were lacking, the hope at the time was that immunization would be persistent for the duration of this "initial" pandemic stage. Before the availability of COVID-19 vaccinations and early in the initial vaccination campaign, cumulative case counts were monitored and considered to represent the proportion of the population that had substantial protection from COVID-19. In these early pandemic phases, reinfections were known to be a possibility, but they were relatively uncommon. Additionally, at the time, the common assumption was that herd immunity could be achieved at about 70%–80% vaccinations/immunity. Because receipt of one COVID-19 dose indicated that an individual does not have principled objections toward the vaccine, the percent of the population who has received at least one dose is a good indication of the proportion that is willing to get vaccinated. Finally, early evidence supported the prioritization of monitoring percent of population vaccinated with at least one dose, as a substantial level of protection was shown to be conferred by even one dose of a two-dose series against the wild type of COVID-19 and against variants that caused earlier surges internationally[42–44] (e.g., the Alpha variant). Figure 10.4 shows examples of maps used to monitor COVID-19 vaccination levels in Shelby County at the ZIP code level.

At this time, the Shelby County Health Department Epidemiology Team was manually producing many maps, including those monitoring testing numbers and rates, case numbers and rates, vaccine rates, and percent of population vaccinated with at least one dose by ZIP code, multiple times per week. As cases continued to fluctuate in Shelby County, and data about cases in near real-time were needed by leadership and community partners to make decisions related to containment policies, it became necessary to select the most meaningful map related to each topic and find a way to automate them. We decided to automate three critical maps, which would be produced daily, including a 14-day test rate map (10.4a), a 14-day case rate map (10.4b), and a map showing the percent of population vaccinated with at least one dose by ZIP code (10.4c). Having the daily map production automated using the programming language R (The R Foundation for Statistical Computing) increased the team's ability to provide current analyses based on the most current data and increased the bandwidth for other data requests and ad-hoc analyses.

Automatization also facilitated consistent use of color schemes, which helped all users of these maps quickly locate earlier versions of these maps in their material. It also enhanced discussions during meetings as maps could be referred to by their color scheme instead of often lengthy names of the displayed metrics.

In the context of the Delta virus variant, the pediatric population was experiencing rates of infection that had not been seen previously during the pandemic. To monitor activity in this population, we created a new ZIP code map that shows the rate of 14-day pediatric cases across the county (10.4d). Additionally, in the context of the Omicron variant,[45] which along with its subvariants, was prevalent in Shelby County beginning in December 2021 through the time that this work was submitted for publication (October 2022), protection conferred by COVID-19 vaccination was attenuated, so we created maps to monitor the percent of population fully vaccinated by ZIP code (not shown) in addition to the percent of population vaccinated with at least one dose by ZIP code (10.4c).

10.3.3 Maps to Monitor and Address Health Disparities

Over the course of the pandemic, ZIP codes with continually high case rates and low vaccine uptake were identified and became the focus of an initiative to reduce health disparities with respect to COVID-19 cases and vaccination. Nine ZIP codes with high case rates and low vaccination uptake as of June 2021 were identified by assessing the ZIP code maps (Figure 10.5). The ZIP code maps helped inform the need for additional targeted intervention and supported applications for additional grant funding. At the onset of this project, maps were adapted to visually compare the nine targeted ZIP codes with the other ZIP codes in the county, as shown in Figure 10.5 by the highlighted area borders. Additionally, new maps were developed to monitor quarterly vaccination, testing, case, and death rates in the targeted ZIP codes and in other ZIP codes across the county, to monitor progress in addressing the existing health disparities. In the scope of this project, targeted interventions were developed within the nine targeted ZIP codes, including health fairs, community events, and direct contact with individuals (door-to-door canvasing) to provide education about COVID-19 and COVID-19 vaccination (at the home of the residents by trained teams). These efforts were designed to increase access to and uptake of testing and vaccination in those areas. Such door-to-door and health fair programs required funding, and these maps also facilitated the required quarterly reporting for specific grants for the targeted efforts. ZIP code maps were very useful in identifying areas of greatest potential impact for such interventions and in monitoring progress toward the goals of improving vaccination and case rates in the targeted areas.

10.3.4 The Visual Aspects of the Maps

To monitor trends over time, it was critical to produce maps frequently enough and in a consistent format to make comparisons. When we began producing maps automatically in R (The R Foundation for Statistical Computing), we dedicated a distinguishable color palette to each topic monitored using maps. For example, when designing the original maps generated in R, we chose to use green to denote testing rates, orange to denote active cases, and blue to denote percent of population vaccinated.

Each map represented a snapshot in time and looking at one map by itself was not necessarily helpful from a monitoring perspective. What was helpful, from a monitoring perspective, was the comparison over time and between different areas of the county. Because we were interested in monitoring the relative situation across the ZIP codes, we chose to make the testing rate maps and the active case maps using variable scales instead of fixed scales. This was helpful, as it made relative positions very clear at any point in time, it was simple to identify where the highest and lowest rates were across the county at any given time, and it demonstrated how transmission levels were changing across ZIP codes as case rates increased or declined (Figure 10.6). However, it added an additional layer of complexity when comparing maps over different periods of time because the maps might look very similar at first glance, but the scale might have changed dramatically from one period to the next.

Figure 10.6 illustrates this point, with a comparison of the active/14-day case rate map immediately prior to the Omicron surge (10.6a) and during the peak of the Omicron surge (10.6b), four weeks later. At first glance, it might look like cases decreased dramatically in the northern part of the county from the first time point to the second time point, but when considering the change of scale, it becomes apparent that cases increased sharply across the entire county. Regardless, the rates in these areas were relatively different when compared to the situation in other parts of the county, between the two maps. Of course, this problem was a result of the limited number of color shades that can be meaningfully displayed and/or discerned by the human eye for comparison across different figures. One might be tempted to furnish additional maps applying different scales for different purposes such as longitudinal comparisons of ZIP codes versus comparisons of ZIP codes at the same point in time, but we found that there is a limit to the number of figures that can be presented and communicated, and we therefore settled on regularly communicating a standard map for each topic and a consistent message about what these maps showed (and what they did not show).

10.4 DISCUSSION

10.4.1 Aspects of Interorganizational Communication of These Maps and Importance of a Dashboard in Disseminating Information to the Public

ZIP code maps played a critical role in communicating information to key stakeholders in Shelby County across all levels from public, private, government, or nongovernment sectors. At multiple levels, these maps served as an essential tool for decision-making and resource allocation. When used as a communication tool or visual aid to presentations or status reports, these maps provided an opportunity to educate the public and key stakeholders about the epidemiology of the disease.

In the context of the COVID-19 pandemic, the sharing of data and data sources was a crucial element. The epidemiology team at the Shelby County Health Department created data dashboards (graphical user interfaces with key performance indicators) and continually shared information, including the maps, through them. This was critical to the ability to pass timely information about the ever-evolving pandemic to the public. This information and the public's ability to access it was important in changing perceptions and behaviors such as masking and vaccination uptake. In developing ZIP code maps and other informational tools, it was critically important to also develop a plan for sharing this information with the public frequently and in a timely manner. The ZIP code maps were an integral tool in communicating findings, but they would

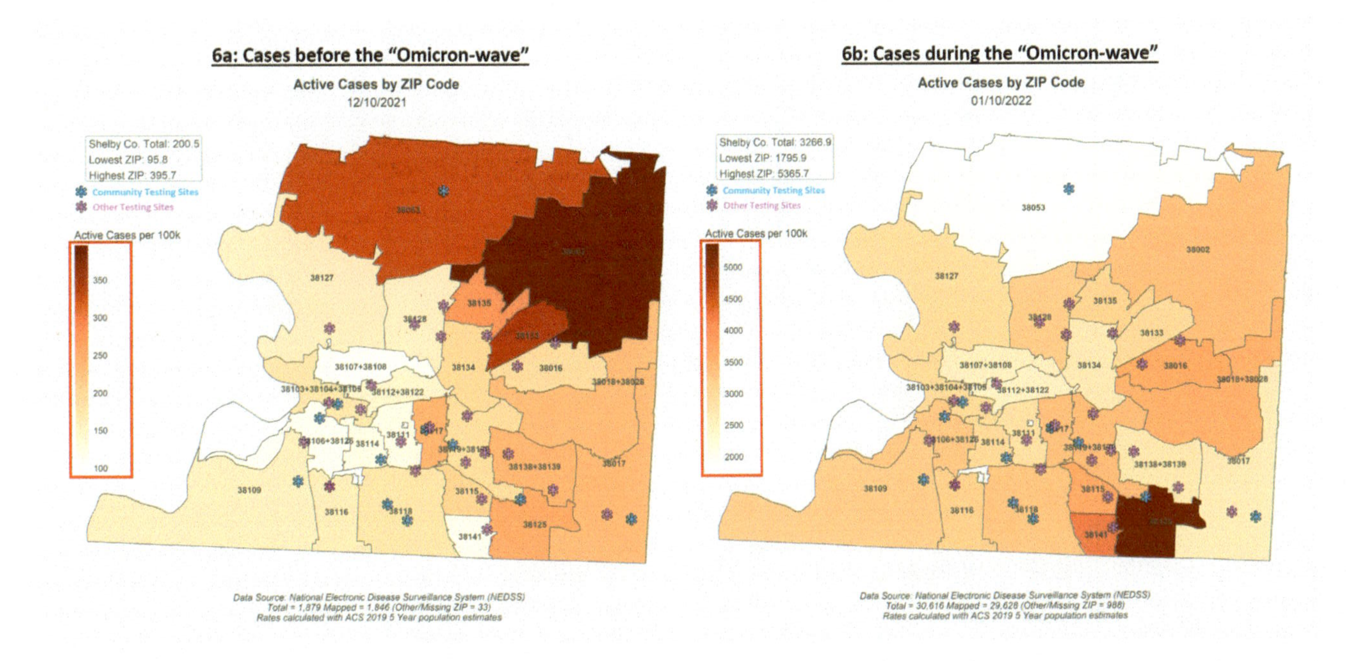

FIGURE 10.6 Illustration of a key consideration of implementing maps with changing scales: Active case rate maps immediately before the Omicron surge (6a) and during the peak of the Omicron surge (6b). At first glance, one might incorrectly conclude that cases have decreased dramatically in the northern part of the county (dark colors in 6a), but a detailed look at the scales is required to understand that cases have increased sharply across the entire county. If the scale in 6a would be applied four weeks later (6b), the map would be uniformly in the darkest color, thus carrying very little information.

have been ineffective without a plan to disseminate this information broadly in a continual and standardized way through the data dashboards. Additionally, the data dashboards were critically important in reducing the burden of standard tasks on the Epidemiology team and allow more bandwidth for nonstandard data requests about COVID-19 data.

10.4.2 General Lessons Learned about the Use of and Process for Creating the Maps

Epidemiology traditionally uses statistical tools and analyses to determine and describe meaningful differences in observed "patterns." Findings yielded from these approaches are often difficult to convey to the public in ways that communicate useful information to people who need to take actions to protect themselves and their families. Our team found that visualizing key data elements using maps was an effective way to share information and increase transparency, which also helped build trust and credibility for the pandemic response effort overall. For public health leaders and those developing strategies to respond to the pandemic, monitoring mechanisms such as the ZIP code maps were crucial to the decision-making process. Human resources and financial resources were limited, and using them most effectively was vital to the mission of reducing mortality and preserving the capacity of the health-care system to function despite the extraordinary workload caused by the pandemic. Our strategy of automatically mapping key data elements in near-real time facilitated shifting deployment of resources to places and populations most in need and helped leadership evaluate the effectiveness of the pandemic response efforts as the pandemic evolved.

Good data visualization must clearly present a story or a point of view in a truthful manner. Our team continually evaluated aspects of the visualizations we produced and refined them. One example of this was the simple shift in the ZIP code maps to a continuous scale for the respective rates and proportions, which allowed us to align the visual intensity of the map with the intensity of the data in a way that the initially used algorithm to create a limited number of groups for display did not. Conversely, creating fixed "high impact" criteria for highlighting ZIP codes allows for pointing out transitions/changes in space or time that align meaningfully with real-world action, without obscuring information. These are considerations that must be made according to the evolving nature of the pandemic and to the purpose that the map is intended to serve. The material presented here is in part intended to help the reader reduce the number of iterations needed to reach a useful analysis product, but some form of change is inevitable, particularly in fast-moving scenarios such as a pandemic. It is therefore advisable to create all maps and analyses with the mindset that these will change and evolve over time. When making such changes, both continuity of established elements as well as a clear explanation what has changed and why are indispensable. We sometimes showed the "new" version alongside an earlier version for some time before retiring the latter; we nearly always needed to invest significant resources in communicating the meaning of even minor changes. Communicating the purpose and context of such changes skillfully and carefully is all-important in avoiding misinterpretation and misguided decisions. It is equally important to appreciate that decision makers can legitimately arrive at different decisions/conclusions based on the same correctly interpreted information as a decision is inevitably based on the factual information (in part provided by us) as well as preferences including moral and ethical considerations that can differ despite a shared understanding of the "facts on the ground."

Over the course of the pandemic, our team has used R (by The R Foundation for Statistical Computing) and ArcMAP/ArcGIS (by Esri) to generate many different types of maps as the local situation evolved and changed. We used both geocoding and choropleth mapping strategies intermittently and as demand and data needs fluctuated. Benefits of geocoding strategies include potential for higher precision when data (addresses) are complete and highly accurate. Choropleth maps by ZIP code proved more useful in the context of visualizing our general surveillance data, where addresses had a high potential to be coded incorrectly or missing details altogether. A major advantage of producing these maps in R was that they were able to be automatically generated daily. However, geocoded heat maps or kernel density maps were very useful in certain situations where address data were of high quality and reasonably reliable, such as early in the pandemic for understanding the location of early cases and for some specific use cases such as cluster/outbreak investigation.

If the location information in the source data is of high quality, census tract-level choropleth maps are alternatives to ZIP code maps that display additional granularity to the data and could be used to inform more targeted interventions. In these cases, concerns for protection of health information can be alleviated by aggregating data over a longer period of time. For example, in the context of COVID-19 and with the current virus variants, it seems that immunity lasts about 90 days and that reinfections become more common thereafter.[46] Thus, aggregating infection-related data over a 90-day period using standardized population as a denominator will provide more stable and useful information for decision- and policy making in the context of creating maps at the sub-ZIP code level. This is also an example for how maps should be adjusted depending on the context: aggregation over 90-day periods is not useful for monitoring a dynamic development of ongoing COVID-19 community spread, but can very well be adequate to monitor vaccination intervention effectiveness.

The COVID-19 pandemic has served as a learning situation for epidemiological mapping and the importance of accurate and reliable data. We learned several important lessons that can be applied to other data sources in the future. If a disease is highly dynamic and evolves quickly, cumulative summary patterns or summary patterns for longer than a 14- or 30-day time frame should be avoided; otherwise, misleading conclusions may be reached, which could bias the decision-making process. ZIP code maps served as a successful communication tool that could be disseminated to advisers (academic partners and the data team), to deciders (politicians and health department personnel), and to the public (informed and unaware residents). These maps proved useful as starting points for a more nuanced discussion as well. In the context of the COVID-19 pandemic, we produced key ZIP code maps daily for internal monitoring purposes. We communicated these findings to key stakeholders and to the public multiple times per week throughout most of the pandemic. These mapping strategies did not allow for assessment of clustering at the sub-ZIP code level, but they did allow us to monitor and assess the shifting impact across ZIP codes, and this information was critical to informing policy, messaging, and to timely intervention and allocation of resources.

10.4.3 Strengths and Weaknesses of the Maps

These unadjusted or minimally adjusted (for population size), frequently produced ZIP code maps were very useful in the context of the COVID-19 pandemic, where information about what specific areas are being most affected over time was very important in shaping the response strategies and recommendations. This information needed to be communicated frequently in a way that would be understandable to both policy- and decision makers as well as the public. These maps were a great communication tool to address this need. Key maps were able to be automatically produced daily, which allowed for the most up-to-date information to be used for community monitoring and allocation of limited resources to where they would be most impactful.

Because ZIP code-level maps often depict topics like crude rates, there are several limitations to their interpretations. One should be cautious when interpreting map-based illustrations when known or suspected confounders are not adjusted for. Such confounders can include, but are not limited to, socioeconomic patterns, age distributions, and access to testing sites. Although many confounders can usually be controlled in a multivariate model or with confounder-standardized metrics, this is not always feasible, for example, when the number of confounders is too large, or when there are many unmeasured confounders. Further, finer adjustment may not be possible due to lack of availability of individual data on potential confounders in the analytical data. Therefore, such graphical illustrations are most useful when raw or minimally adjusted rates and the like are sufficient and needed for policy decisions or when a few aspects/levels of an important confounder can be used to show multiple figures in a stratified analysis.

10.4.4 Toward Better Education and Work-Force Development

The seriousness of the rapidly unfolding COVID-19 pandemic required us to work intimately together despite diverse backgrounds. The need to access, transfer, process, visualize, and communicate data in this high-tempo operation resulted in an amalgamation of these aspects that illustrates the importance of "data science" par excellence. While specialization in any of these aspects is useful, we consider it essential that all involved individuals are exposed to all aspects to some extent. For education and training purposes, this suggests to us that statistical programs are well advised to expose their students to the technical aspects of accessing and transferring data and devote more time to teaching methods for making data "analysis ready," including processing of data that are not in "rectangular" form. For computing and data science-oriented programs, we advocate including methods-oriented coursework and exercises that leverage the substantial body of knowledge from epidemiology and applied statistics especially with respect to confounding variables and selection biases arising in observational data.

However, while the above are necessary skills, there are additional traits we have valued throughout our work and that we find important to look for in new hires as well as to develop in current team members. Collaboration, teamwork, and communication skills are, of course, required abilities. Understanding of emergency response and crisis management is something that also needs to be developed. Effective contributors need to stay flexible and have the ability to adjust based on the evolution of the emergency and the needs of key decision makers. Forethought—the ability to anticipate what will be beneficial to decision makers—is very important in a high-tempo operation to generate results in a timely way. This includes understanding of various data sets and how/where to access information long before the need arises to actually process said data. Likewise, a fundamental understanding of the analytical methods and their limitations should be a long-term goal in preparation for any crisis response. Resilience and stamina are required when only long work hours can produce the required responses.

10.5 CONCLUSIONS

Besides the technical challenges described here of accessing data, geocoding addresses, automating mapmaking, and so forth, the greatest challenge for us during the COVID-19 pandemic in the work presented here was communication. At every level of government, communication between those designing the pandemic response and the communities impacted by decisions that were made often determined whether the response was effective or was doomed to failure. In Shelby County, the regular production and the sharing of even simple maps established a shared understanding of how COVID-19 was impacting the county and formed a basis of understanding between all parties involved about why decisions were made. Even when there were disagreements on policy decisions and approaches, the common shared understanding that the maps helped establish made the discussions and debates more useful and productive. Local health departments must develop and ensure a capability to produce maps and perform basic geospatial analyses of both social determinants of health as well as disease patterns in the population. Public health workforces must have a basic core competency in mapmaking and decision-making grounded in geospatial analysis to be effective in preventing disease, protecting people, and promoting their best health outcomes. GIS mapping and access to real-time local data were key to our local COVID-19 response. Forming a data team that consists of a variety of individuals with diverse skills was critical. We engaged decision makers, epidemiologists, biostatisticians, and clinicians in continuous discussion to develop our data visualization resources, and particularly our ZIP code maps.

Through our experience, we have found that ZIP code maps are a good tool to illustrate the spatial distribution of disease, and that they are a tool that can be easily used to communicate the current situation and the areas of greatest need to policy makers. Overall, GIS information is a powerful, informative, and efficient tool in facilitating health policy making and infectious diseases monitoring. Correct interpretations with recognition of applicable limitations are key to offering timely, intuitive, and close-to-unbiased guidance. To better consider the effect of important and well-known confounders, standardized or relative metrics are recommended, along with stratified illustrations, if feasible. Using residential ZIP codes as a GIS indicator is convenient and often reliable. However, one should always avoid overinterpreting the results as ZIP codes do not capture all underlying information. Any "pattern" should always be interpreted at the population level rather than the individual level. While the ZIP code-level maps hold a personal level of interpretation, as most people know their ZIP code, ZIP codes are relatively large, which limits the interpretation for custom intervention in sub-ZIP code areas.

GIS data communicated through choropleth maps can be highly effective for communicating simple statistics across relatively large geographical levels but can become challenging to interpret when significant background information is required, or very granular spatial information is used. Trade-offs in mapmaking are inevitable; in our case, a particularly stark trade-off was often necessary between temporal consistency and visual informativeness. Such trade-offs need to be made with both faithfulness to the data and the needs of the users in mind, ideally while in communication with said users.

Maps have helped convert data into something visually appealing and meaningful, analyze trends and risks in our population for COVID-19, allow comparison across ZIP codes for targeted interventions, identify health disparities and resource allocation, and last but not the least, convey data to the lay audience in a way most people can make sense of. ZIP code maps made communicating COVID-19 data to the external and internal partners easier and more effective. On several occasions this approach helped facilitate data-driven decisions around masking, vaccinations, and isolation and quarantine guidelines. The COVID-19 pandemic has identified gaps in data and gaps in visualization of data. As we head past this pandemic, the lessons learned from COVID-19 can be applied to other chronic and infectious diseases that afflict the community to drive policy change.

ACKNOWLEDGMENTS

All authors have served on the Data Subcommittee of the City of Memphis and Shelby County COVID Joint Task Force Emergency Support Function #8–Public Health and Medical Services and gratefully acknowledge the steadfast support they have received from the leadership of their respective organizations, which made this support possible. The views here expressed are those of the authors and should not be construed to represent those of the Shelby County Health Department or the City of Memphis and Shelby County COVID Joint Task Force.

REFERENCES

1. Data.census.gov. B01003|Total Population 2022 [Internet]. [August 15, 2022]. Available from: https://data.census.gov/cedsci/table?q=total%20population&g=0400000US47%240500000&tid=ACSDT1Y2019.B01003.
2. Data.census.gov. DP05|ACS Demographic and Housing Estimates 2022 [Internet]. [August 15, 2022]. Available from: https://data.census.gov/cedsci/table?q=demographics&g=0500000US47157&t.id=ACSDP1Y2019.DP05.

3. Data.census.gov. S2001|Earnings om the Past 12 Months (In 2019 Inflation-Adjusted Dollars) 2022 [Internet] [August 15, 2022]. Available from: https://data.census.gov/cedsci/table?q=median%20income&g=0500000US47157&tid=ACSST1Y2019.S2001.
4. census.gov. QuickFacts Shelby County, Tennessee 2022 [Internet]. [September 3, 2022]. Available from: www.census.gov/quickfacts/shelbycountytennessee.
5. census.gov. How the Census Bureau Measures Poverty 2021 [Internet]. [September 3, 2022]. Available from: www.census.gov/topics/income-poverty/poverty/guidance/poverty-measures.html.
6. Data.census.gov. S0101|Age and Sex 2022 [August 15, 2022]. Available from: https://data.census.gov/cedsci/table?q=arlington&g=1600000US4701740,4703440,4716420,4728960,4740350,4748000,4749060&tid=ACSST1Y2019.S0101.
7. TN.gov. Interstate 40 Hernando DeSoto Bridge 2022 [Internet]. [August 15, 2022]. Available from: www.tn.gov/tdot/projects/region-4/i-40-hernando-desoto-bridge.html.
8. International Airport Review. The top 10 busiest airports in the world by cargo handled 2022 [Internet]. [September 3, 2022]. Available from: www.internationalairportreview.com/article/107921/top-10-busiest-airports-world-cargo/.
9. Tha.com. Find a Hospital Near You 2022 [Internet]. [August 15, 2022]. Available from: https://tha.com/hospitals/.
10. Insider. Sobering photos reveal how countries are dealing with the dead left by the coronavirus pandemic 2020 [Internet]. [March 23, 2021]. Available from: www.businessinsider.com/coronavirus-covid-19-victims-bodies-burials-morgues-cemeteries-photos-2020-4#italy-is-undoubtedly-one-of-the-hardest-hit-countries-the-mediterranean-nation-has-one-of-the-highest-death-tolls-to-date-at-least-19900-since-its-first-confirmed-case-on-january-31-4
11. Horowitz J, Bucciarelli F. The lost days that made Bergamo a coronavirus tragedy. New York Times. 2020 Nov 29.
12. New York Post. Hundreds of COVID victims' bodies still in refrigerated trucks on NYC waterfront 2021 [Internet]. [September 12, 2022]. Available from: https://nypost.com/2021/05/07/many-covid-victims-bodies-still-in-refrigerated-trucks-on-nyc-waterfront/.
13. Shelby County Health Department. Formal Issuance of Health Directive: Public Health Announcement on COVID-19 Response March 25, 2020 [Internet]. [2022-10-13]. Available from: www.shelbycountytn.gov/DocumentCenter/View/36527/SCHD-Covid-19-Health-Directive.
14. Olney AM, Smith J, Sen S, Thomas F, Unwin HJT. Estimating the effect of social distancing interventions on COVID-19 in the United States. Am J of Epidemiol. 2021;190(8): 1504–1509. https://doi.org/10.1093/aje/kwaa293.
15. United States Census Bureau. ZIP Code Tabulation Areas (ZCTAs) [Internet]. 2022 [2022-07-30]. Available from: www.census.gov/programs-surveys/geography/guidance/geo-areas/zctas.html.
16. CDC. Health Insurance Portability and Accountability Act of 1996 (HIPAA). [Internet]. Centers for Disease Control and Prevention; 2022 [2022-05-09]. Available from: www.cdc.gov/phlp/publications/topic/hipaa.html.
17. Beenstock M, Felsenstein D. Freedom of information and personal confidentiality in spatial COVID-19 Data. Journal of Official Statistics. 2021;37(4):791–809. https://doi.org/10.2478/jos-2021-0035.
18. Bowen CM. The art of data privacy. Significance. 2022;19(1):14–9. https://doi.org/10.1111/1740-9713.01608
19. US Department of Health & Human Services. Guidance Regarding Methods for De-identification of Protected Health Information in Accordance with the Health Insurance

Portability and Accountability Act (HIPAA) Privacy Rule 2012 [Internet]. [2022-10-13]. Available from: www.hhs.gov/sites/default/files/ocr/privacy/hipaa/understanding/coveredentities/De-identification/hhs_deid_guidance.pdf.
20. Dwyer-Lindgren L, Kendrick P, Kelly YO, Sylte DO, Schmidt C, Blacker BF, Daoud F, Abdi AA, Baumann M, Mouhanna F, Kahn E, Hay SI, Mensah GA, Nápoles AM, Pérez-Stable EJ, Shiels M, Freedman N, Arias E, George SA, Murray DM, Phillips JWR, Spittel ML, Murray CJL, Mokdad AH. Life expectancy by county, race, and ethnicity in the USA, 2000–19: a systematic analysis of health disparities. The Lancet. 2022. https://doi.org/10.1016/S0140-6736(22)00876-5.
21. Whitehead M. The concepts and principles of equity and health. International Journal of Health Services. 1992;22(3):429–45.
22. US HHS ODPHP. Disparities: US Department of Health and Human Services, Office of Disease Prevention and Health Promotion. (Internet]. 2022 [2022-07-30]. Available from: www.healthypeople.gov/2020/about/foundation-health-measures/Disparities.
23. Greenland S, Morgenstern H. Confounding in health research. Annu Rev Public Health. 2001;22(1):189–212.
24. Rothman KJ, Greenland S, Lash TL. Modern Epidemiology. Third ed. Philadelphia: Lippincott Williams & Wilkins; 2008.
25. Norton HJ, Divine G. Simpson's paradox … and how to avoid it. Significance. 2015;12(4):40–3. https://doi.org/10.1111/j.1740-9713.2015.00844.x.
26. Pearl J. Understanding Simpson's Paradox (Technical Report R-414). University of California, Los Angeles, Computer Science Department, 2013.
27. Morgenstern H. Ecologic studies. In: Rothman KJ, Greenland S, Lash TL, editors. Modern epidemiology. Third ed. Philadelphia: Lippincott Williams & Wilkins; 2008. p. 511–31.
28. Nathan RP, Adams C. Understanding central city hardship. Political Science Quarterly. 1976;91(1):47–62. https://doi.org/10.2307/2149158.
29. Greenland S, Rothman KJ. Measures of occurrence. In: Rothman KJ, Greenland S, Lash TL, editors. Modern epidemiology. Philadelphia: Lippincott Williams & Wilkins; 2008. p. 32–50.
30. Bivand RS, Pebesma E, Gómez-Rubio V. Applied spatial data analysis with R. Second ed. New York: Springer; 2013.
31. Lawson AB. Statistical Methods in Spatial Epidemiology. Second ed. Chichester: Wiley; 2006.
32. Lawson AB. Bayesian disease mapping. Second ed. Boca Raton, FL: CRC Press; 2013.
33. Lawson AB, Kim J. Space-time Covid-19 Bayesian SIR modeling in South Carolina. PLoS One. 2021;16(3):e0242777. https://doi.org/10.1371/journal.pone.0242777.
34. Sartorius B, Lawson AB, Pullan RL. Modelling and predicting the spatio-temporal spread of COVID-19, associated deaths and impact of key risk factors in England. Sci Rep. 2021;11(1):5378. Epub 2021/03/10. https://doi.org/ 10.1038/s41598-021-83780-2. PubMed PMID: 33686125; PMCID: PMC7940626.
35. Blangiardo M, Cameletti M. Spatial and spatio-temporal bayesian models with R-INLA. Chichester, UK: Wiley; 2015.
36. CDC. Interim List of Categories of Essential Workers Mapped to Standardized Industry Codes and Titles 2021 [Internet]. [updated March 29, 2021September 12, 2022]. Available from: www.cdc.gov/vaccines/covid-19/categories-essential-workers.html.
37. US Department of Homeland Security—Cybersecurity & Infrastructure Security Agency (CISA). Identifying Critical Infrastructure during COVID-19 2020 [Internet]. [September 12, 2022]. Available from: www.cisa.gov/identifying-critical-infrastructure-during-covid-19.

38. US Department of Homeland Security–Cybersecurity & Infrastructure Security Agency (CISA). Advisory Memorandum on Identification of Essential Critical Infrastructure Workers during COVID-19 Response 2020. [Internet]. Available from: www.cisa.gov/sites/default/files/publications/CISA_Guidance_on_the_Essential_Critical_Infrastructure_Workforce_Version_2.0_1.pdf.
39. Agency for Toxic Substances and Disease Registry. CDC/ATSDR Social Vulnerability Index 2022 [Internet]. [2022-10-13]. Available from: www.atsdr.cdc.gov/placeandhealth/svi/index.html.
40. CDC. Underlying Medical Conditions Associated with Higher Risk for Severe COVID-19: Information for Healthcare Professionals 2022 [Internet]. [September 12, 2022]. Available from: www.cdc.gov/coronavirus/2019-ncov/hcp/clinical-care/underlyingconditions.html.
41. Jenks GF, editor. The Data Model Concept in Statistical Mapping1967.
42. CDC. Interim Estimates of Vaccine Effectiveness of Pfizer-BioNTech and Moderna COVID-19 Vaccines Among Health Care Personnel—33 US Sites, January–March 2021 [Internet]. [September 3, 2022]. Available from: www.cdc.gov/mmwr/volumes/70/wr/mm7020e2.htm.
43. Vasileiou E, Simpson CR, Shi T, Kerr S, Agrawal U, Akbari A, Bedston S, Beggs J, Bradley D, Chuter A, de Lusignan S, Docherty AB, Ford D, Hobbs FDR, Joy M, Katikireddi SV, Marple J, McCowan C, McGagh D, McMenamin J, Moore E, Murray JLK, Pan J, Ritchie L, Shah SA, Stock S, Torabi F, Tsang RSM, Wood R, Woolhouse M, Robertson C, Sheikh A. Interim findings from first-dose mass COVID-19 vaccination roll-out and COVID-19 hospital admissions in Scotland: a national prospective cohort study. The Lancet. 2021;397(10285):1646–57. https://doi.org/10.1016/S0140-6736(21)00677-2.
44. Hall VJ, Foulkes S, Saei A, Andrews N, Oguti B, Charlett A, Wellington E, Stowe J, Gillson N, Atti A, Islam J, Karagiannis I, Munro K, Khawam J, Chand MA, Brown CS, Ramsay M, Lopez-Bernal J, Hopkins S, SIREN Study Group. COVID-19 vaccine coverage in health-care workers in England and effectiveness of BNT162b2 mRNA vaccine against infection (SIREN): a prospective, multicentre, cohort study. The Lancet. 2021;397(10286):1725–35. https://doi.org/10.1016/S0140-6736(21)00790-X.
45. Plaxco A, Kmet J, Nolan VG, Taylor M, Smeltzer MP. Real-world association between mRNA vaccination and infection from the Omicron strain of SARS-CoV-2: A population-level analysis. AJPM Focus. 2022:100010. https://doi.org/10.1016/j.focus.2022.100010.
46. Centers for Disease Control and Prevention. Science Brief: SARS-CoV-2 Infection-induced and Vaccine-induced Immunity 2021 [Internet]. [2022-10-13]. Available from: www.cdc.gov/coronavirus/2019-ncov/science/science-briefs/vaccine-induced-immunity.html.

11 Using GIS to Map Women's Health, Well-Being, and Economic Opportunities in the Context of COVID-19

Ginette Azcona, Antra Bhatt, and Julia Brauchle

CONTENTS

11.1 INTRODUCTION

Perhaps unlike any other crisis in recent memory, the COVID-19 pandemic has had a devastating impact on the lives and well-being of people everywhere, from women in small rural villages of Southeast Asia, to residents, rich and poor, of large metropolitan cities. Millions have lost loved ones, and been forced to socially distance and isolate, including from their family and friends. Women especially, have faced other risks, including higher threats of violence and harassment, inadequate access

DOI: 10.1201/9781003227106-11

to sexual and reproductive health services, and greater prevalence of anxiety and depression. In a survey of 13 countries, 49% of women in urban areas said they feel less safe walking alone at night since COVID-19. The prevalence of anxiety and depressive disorders also increased, among women reaching 28% and 30%, respectively. Women from low-income households, already least likely to receive adequate care before the pandemic, faced even greater risks. In Canada, women who were pregnant during the pandemic were twice as likely to show symptoms of depression and anxiety, and the rate was even higher among those from low-income households (UN Women, 2022b).

Apart from the health-related impacts, the pandemic has also exacerbated economic inequalities. Before the COVID-19 pandemic, the share of people living on less than $1.90 a day had fallen from 11.2% in 2013 to 8.6% in 2018. COVID-19 derailed this progress, with the rate expected to rise to around 9% in 2022. Hunger is also on the rise, with an increasing number of men and women facing challenges in putting food on the table for their families. Moderate or severe food insecurity among adult women, for instance, rose during the pandemic from 27.5% in 2019 to 31.9% in 2021. Among men, it increased from 25.7% to 27.6%, enlarging the gender gap from 1.8 to 4.3 percentage points. Women in locations ravaged by war and conflict are among the most at risk of hunger. 60% of the world's undernourished people live in conflict-affected areas. In these settings, female-headed households experience moderate or severe food insecurity at much higher rates than male-headed households, 37.5% compared to 20.5% among male-headed households (UN Women, 2022b).

Gender differences are often compounded with other forms of inequalities. Going beyond the national aggregate to more granular levels of analysis that enable an assessment of inequality in outcomes by sex, location, and the intersection of these is essential for understanding the divergent impacts of the pandemic and for informing an evidenced-based policy response. For instance, combining data sets that contain information on geographical access to services alongside granular information on outcomes (e.g., by sex and location), can bring to light how deprivations overlap, where they overlap, and for whom. These insights are essential for designing policy interventions that aim to address the needs of those hardest to reach and most in need, including women and girls from marginalized and vulnerable communities. And yet, while these types of data and approaches are increasingly possible, they are sorely underutilized.

In this chapter, we combine spatial data with sex-disaggregated survey data collected in the first year of the pandemic in three countries, Mexico, Kenya, and the United States, to better understand the spatially differentiated impact of the pandemic from a gender perspective. Section 11.2 of the chapter begins by providing a summary of the existing evidence base on use of geographic information system (GIS) analysis in mapping the prevalence of COVID-19 as well as on its use in understanding the impact of the pandemic at the local level and from a gender perspective. Section 11.3 presents the data sources for each of the case studies as well the methodology. In section 11.4 we highlight the key findings from the three case studies on Mexico, Kenya, and the United States. Section 11.5 concludes and discusses policy implications.

11.2 LITERATURE REVIEW

The overlay of GIS data and analysis with individual- and household-level data on health and other socioeconomic data is particularly relevant to understanding the COVID-19 virus, its geographical spread, and the secondary impacts on the economy. In the early days of COVID-19, researchers, governments, and media outlets used GIS techniques to map the prevalence of the virus and predict the spread of the virus. The earliest studies came from China and explored the distribution of cases by province (Guan et al., 2020) and their correlation with the migration of the Wuhan population (Chen et al., 2020). The mapping of COVID-19 outbreaks by subregional boundaries quickly become popular across other countries and across large media organizations including *The New York Times*, *El Pais*, *Le Monde*, and the BBC (Pardo et al., 2020).

One of the strengths of conducting a geospatial analysis is the ability to place subjects within the context of their localized interactions. This has the potential to highlight influencing environmental factors that otherwise may not be easily knowable (Mclafferty, 2020). This can be especially useful within the context of tracking the trajectory of health and disease, but also labor and migration. For example, previous geospatial analyses have uncovered the effects of access to roads on women's labor force participation in Nepal (Brown, 2003), and epidemiological research studies frequently employ geospatial dimensions in their analyses of health inequality (Ozdenerol, 2017). Yet, studies that look at COVID-19 and gender through a GIS lens are sparse. In a review of 67 scientific articles on geospatial and spatial-statistical analysis of the geographical dimensions of COVID-19, published during 2020, only two papers appear to include a sex variable in their analysis (Pardo et al., 2020).

The majority of gender-related research has focused on indicators directly related to the risk and impact of contracting the SARS-CoV-2 virus (Flor et al., 2022). Few have taken a gender and GIS approach. The National Aeronautics and Space Administration's Socioeconomic Data and Applications Center (SEDAC) Global COVID-19 Viewer maps national population estimates by sex and age to identify countries with a higher risk of deaths based on the size of the population of older males, as more deaths were reported among older men than among younger people and females (SEDAC, 2020). In another rare example of research analyzing gendered data at the subnational level, Bertocchi and Dimico (2021) show that Black women in and around Chicago, Illinois, and living in neighborhoods subject to historical redlining were more likely to contract the virus than Black women living in other neighborhoods, likely due to poverty, occupational segregation in the health care and transportation sectors, and higher reliance on public transportation (Bertocchi & Dimico, 2021).

Beyond the direct impacts of contracting the virus, indirect impacts on health and non-health-related well-being must also be considered. From a gender perspective, researchers have detailed the negative impact of the pandemic on women's and girls' economic opportunity, increases in gender-based violence, and reduced access to education (UN Women, 2020). Other studies, for example on Bangladesh, have looked at student suicide, insomnia, and fear of COVID-19 during the pandemic, mapping these health outcomes across district and by sex. The findings suggest that female students were more likely to die by suicide than male students across Bangladesh. Female students were also more likely to experience insomnia during COVID-19 and more likely to be afraid of contracting COVID-19 (Mamun et al., 2021; Mamun et al., 2022; Mamun, 2021).

Outside of COVID-19, gendered spatial analyses are more frequent. The current authors' own research previously employed GIS techniques to show how gender intersects with ethnic and geographical disparities in Pakistan, finding that Pashtun women living in northeastern regions, for example, were more deprived of basic needs including access to water, sanitation, fuel, and health services than other groups of women in other parts of the country (Azcona & Bhatt, 2021).

Despite the large amount of information collected during the pandemic, persistent gender- and sex-disaggregated data gaps persevere at all levels. Countries have a dearth of data with both gender and geospatial components. Subnational data required to pinpoint where women are most disadvantaged and aid in delivering resources where they are most needed is often unavailable by sex. Moreover, use of GIS in health-care planning and health information systems remains limited (Dotse-Gborgbortsi, Wardrop, Adewole, Thomas, & Wright, 2018). The present chapter uses a case study approach, piecing together available data across three countries: Mexico, Kenya, and the United States to better understand the gendered impacts of the COVID-19 pandemic, including the divergent impact at the national and the subnational level.

11.3 DATA SOURCES AND METHODOLOGY

For the case study on Mexico, we make use of a telephone survey on COVID-19 and the pandemic's impact on labor markets and labor market outcomes. The Encuesta Telefónica sobre COVID-19 y Mercado Laboral (ECOVID-ML) survey was conducted by El Instituto Nacional de Estadística y Geografía (INEGI) in July 2020. The survey collected subnational data on the effect on the pandemic on the labor market to understand which groups of individuals have been temporarily absent from work, have been working fewer hours, and have lost their jobs because of COVID-19. The survey covered more than 50,000 people between April and July 2020 aged 18 and over and is nationally[1] and regionally representative for all 32 regions. However, while generally representative, it is important to note that for some rural areas the survey may not be fully representative given the limited access to telephones in those parts of the country. The survey outcomes tabulations were weighted to adjust for nonresponse, regional population that has access to telephones, and other key population characteristics (INEGI, 2020b).

ECOVID-ML not only collected information on basic labor market statistics but also asked questions about the specific reason due to which jobs were lost (care reasons, suspension of activity, lockdowns, etc.), help received from employers, and the situation of self-employed professionals (INEGI, 2020a). This information was then triangulated using other indicators on the rates of informal and vulnerable employment by region to understand which groups of individuals were most impacted. These additional indicators were sourced from the Sistema Nacional de Informacion Estadistica y Geografica, El Catálogo Nacional de Indicadores, Mexico.[2]

For Kenya, the key data source used in this chapter is the 2020 Kenya Malaria Indicator Survey (MIS), which focuses on households with women and girls between the ages of 15 and 49. The Survey was originally planned to be conducted in June and July 2020. However, data collection was delayed due to the pandemic and did not take place until November and December 2020. Although the survey was not specifically designed with the pandemic in mind, data collection coincided with some of the worst periods of the health crisis, thereby enabling the survey to provide important insights about the pandemic's impact on women's access to essential health services, including antenatal services.

The survey used a stratified sampling approach and adjusted for possible differences in response rates using sampling weights (DNMP & ICF, 2021). The survey also included a GIS module that allows for a spatial analysis of the data on outcomes. The outcomes data from the MIS survey were combined with additional indicators from the Kenya National Bureau of Statistics, "2019 Kenya Population and Housing Census Volume I: Population by County and Sub-County," on COVID-19, including case counts as well as information on lockdowns/closures.

For the United States case study, only one data set has been used: the United States Census Bureau Household Pulse Survey 2020. This is an online survey that studied how the COVID-19 pandemic has impacted households across the country from a social and economic standpoint. The survey was launched in April 2020 and data were collected throughout the year. The survey is nationally representative and is well equipped to produce estimates at the state level for each of the 50 states and the District of Columbia. Several measures were applied to ensure national and regional representation including providing weights, which can be used for household nonresponse adjustment, number of adults within the housing unit, and for population adjustment for ethnicity, race, sex, and age distributions (Fields et al., 2020). The United States Census Bureau Household Pulse Survey 2020 is unique in the sense that it also captured information on mental health issues—a range of questions on nervousness, anxiousness, depression, and the like were included in the survey.[3] The survey also allowed for disaggregation of outcomes by sex, age, and race/ethnicity as well as the presence of children in the household.

The data on access to mental health services has also been sourced from the Pulse survey. Using geospatial mapping, mental health outcomes and usage of health services from the US household Pulse 2020 survey were overlaid side by side, to understand how inequalities in access compare to mental health outcomes for women across the 50 states in the United States. The geospatial analysis could only be broken down by sex and location simultaneously and not by race/ethnicity or presence of children. This is because the data on access to mental health services is more restrictive and is only available at the regional level by sex and no additional dimensions such as race/ethnicity or presence of children are available.

11.4 ANALYSIS AND FINDINGS

For each case study, we make use of data collected in the first year of the pandemic, 2020, to explore the impact of the crisis on subnational inequalities for different subgroups of women. In Mexico, we highlight the disproportionate impact of the economic fallout on women's employment and access to resources. In Kenya, we show how lockdowns impacted women's access to maternal care services, both in urban and rural areas. In the United States, we explore the association between pre-existing spatial inequalities in women's access to mental health care and spatial patterns with regard to reported increases in depression and anxiety outcomes during COVID-19.

11.4.1 COVID-19's Impact on Women's Employment: A Case Study from Mexico

The COVID-19 pandemic devasted the Mexican economy—the country's gross domestic product (GDP) declined by 8.7% in 2020, the largest contraction since the 1994 Tequila Crisis. Total employment dropped by 3.7% from March 2020 to December 2020. The decline, however, was highest for those in informal employment (4.5%), compared to those in formal employment (2.6%). Women, disproportionately represented in the informal sector, faced more job losses than men. Employment for women dropped by 7.6%, compared to 2.4% for men (Cañas & Smith, 2021).

Women were more likely than men to say they were forced to suspend their paid work due to COVID-19 and COVID-19 restrictions, 28.3% compared to 23.8%, respectively. Additionally, women were less likely to receive any economic help from their employer due to the pandemic (6.7% of women reported receiving help from their employer versus 8.7% of men). Among self-employed workers, women similarly reported facing business-related problems at a higher rate than men (77% for women versus 65% for men) (INEGI, 2020a).

Apart from the fact that women were overrepresented in hard-hit sectors, including personal care, leisure and hospitality, the increase demand for unpaid care work during the pandemic took a major toll on women's labor market participation as well as their physical and mental well-being (Azcona, Bhatt, Davies, Harman, Smith, & Wenham, 2020; ILO, 2022). Schools and day-care centers were largely closed, leaving families with few options for childcare. Emerging data during the pandemic showed that in Mexico, women and men both increased their unpaid care hours, but women's estimated increase in weekly hours on childcare was 13.2 hours compared to 6.3 hours for men (Care Burden COVID-19 UN Women, 2020). The increased care burden, for many, meant less time in paid work. Partnered women with children in Mexico experienced sharper pandemic-related drops in labor force participation (LFP) than partnered men with children—and these were most pronounced for women living with children under six for whom LFP stood at 41.6% at the end of Q1 2020 and decreased to 29.1% by Q2 2020 (UN Women, 2020).

The analysis at the subnational level unpacks these findings further. In 20 of the 32 regions in Mexico, a larger share of women, as compared to men, said they were prevented from working due to COVID-19-related work closures, suspensions, and dismissals (see Figure 11.1a and b).

The most impacted regions from women's employment perspective are the southern regions of Chiapas, Guanajuato (GUA), Guerrero (GRO), Oaxaca (OAX), Puebla (PUE) and Veracruz (VER); Sonora (SON) and Sinaloa (SIN) in the northwest; and Chihuahua (CHH) in the north. In Sinaloa and Sonora respectively, 65.2%

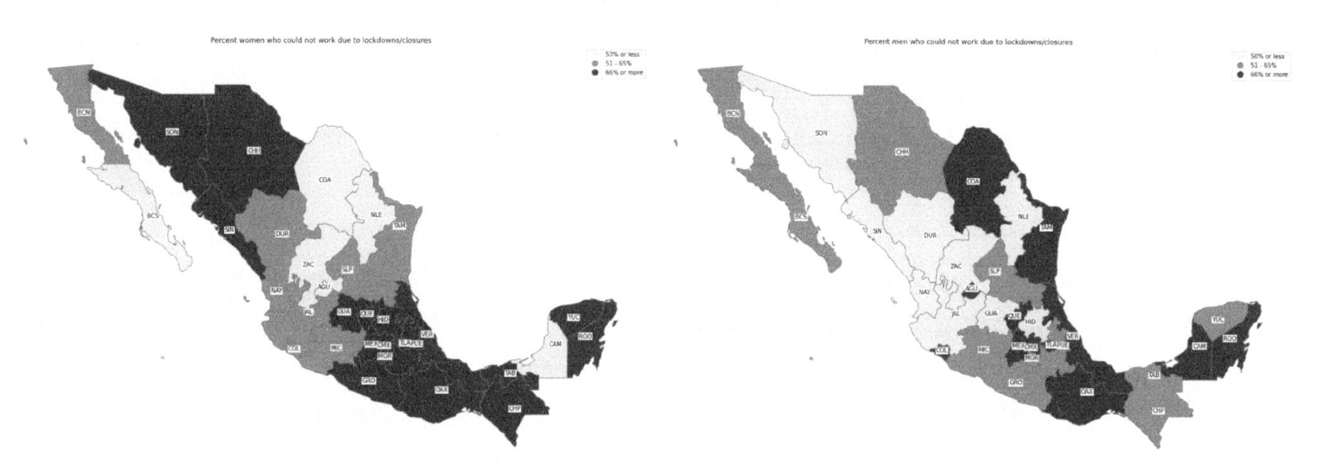

FIGURE 11.1A AND B Individuals who could not work due to COVID-19 related closures, suspensions, or dismissals, by sex, %.

Note: The question about being able to work focused on the "last week," i.e., seven days preceding the survey. (Source: Author's calculations from INEGI. Encuesta Telefónica sobre COVID-19 y Mercado Laboral (ECOVID ML). Julio 2020.)

and 75.2% of women in comparison to 17.5% and 36.1% of men reported not being able to work due to COVID-19-related closures/suspensions or dismissals. Similarly, in Guanajuato (GUA), the corresponding figures for women were 83% compared to 48.3% for men.

When viewed side by side, these spatial patterns in barriers to paid work during COVID-19 mirror the regional distribution of women in vulnerable jobs, which similarly have a spatial dimension (SNIEG, 2022).[4] At the end of 2019, 57.5% of employed women held vulnerable jobs versus 55.1% of men, but with substantial regional variation. In southern regions such as Puebla (PUE), 76.8% of employed women versus 72.2% of employed men were in vulnerable jobs before the pandemic began (see Figure 11.2 below). In Puebla, 80.7% of women reported not being able to work in the last week due to COVID-19 related closures, dismissals, or suspensions versus 64.7% of men.

The other geographical region where women reported much larger shares of COVID-19-related exits from paid work is the northwest: Sonora (SON) and Sinaloa (SIN) and Chihuahua (CHH) in the north. Evidence from 2020 shows that Sinaloa lost 9.8% of its jobs—the third-largest number of job losses in the country (Casado Izquierdo, 2021). Trade and supply chain disruptions could be responsible for this as the state exports large amounts of produce to the United States. Women in agriculture already faced precarious working conditions pre-pandemic, and the pandemic only heightened these perilous working conditions. Meanwhile, in Chihuahua, the primary economic sector is the service industry (led by retail trade, general services,

FIGURE 11.2 Share of women in informal employment, by region. (Source: Author's calculations using Catalogo Nacional de Indicatores, Mexico.)

and food and hospitality). Women tend to be a larger share of employees in these sectors (Data Mexico, 2022), often paid less than men and with less access to social protection, and fewer opportunities for management positions. This sector also faced major disruptions during the periods of COVID-19-related lockdowns, negatively impacting women's employment.

The findings, mapped by geographical region and overlaid with information on quality of work and sex distribution reveal a dire situation for women's work, particularly the impact on women working in vulnerable employment with limited access to social protection. With economic activity at a halt during the pandemic, women and men, both faced challenges and declines in their capacity to earn a living, but women in specific regions of the country, working in specific sectors, faced the biggest challenges. These insights are critical for targeting policy interventions that reach those most harmed by the economic fallout of the pandemic.

11.4.2 Disruption in Sexual and Reproductive Health Services: The Case of Antenatal Care in Kenya

Antenatal care (ANC) is key to ensuring healthy pregnancies and reducing maternal and neonatal deaths, with one study finding that close to 40% of neonatal deaths in Kenya could be attributed to lack of ANC (Global Health Action, 2017). Kenya had made progress in recent years to increase access to ANC, but these gains have been reversed in the wake of COVID-19. In the two years preceding COVID-19, the proportion of women getting at least four antenatal checks, the minimum recommended by the World Health Organization, stood at 61.7%, up from 47.1% in 2008–09 and 57.6% in 2014 (DHS StatCompiler). In the first year of the pandemic, the figure dropped to 55.6%, a rate lower than that achieved in 2014.

Spatial dimensions of inequality in key health outcomes improved during the pandemic, but this was the result of outcomes worsening among groups that had higher outcomes previously, and thus a leveling down of health outcomes and not by any means a leveling up. Analysis at the subnational level shows the disheartening reality that inequalities in ANC rates between provinces were reduced in 2019–20, but only due to declines in provinces with a previously higher average number of visits. Pre-COVID-19, most women living in more urban provinces, including Nairobi and the Central province, were receiving the minimum recommended number of antenatal care visits. Between March and December 2020, provinces that had previously maintained an average of four or more antenatal care visits per birth saw their averages drop during the height of the COVID-19 pandemic. In three provinces (Nairobi, Central province, and the Coast), the mean visits, previously at four or above, dropped to below the WHO recommended minimum (Figure 11.3). These regions now report averages similar to those seen in more rural provinces such as Rift Valley, Northeastern and Eastern, where long distances to health services often prevent access to essential care (Joseph, 2022).

Spatial analysis shows that decreases in ANC coincided with higher population density, more stringent lockdown measures, and higher COVID-19 caseloads (Figures 11.4a and 11.4b). While the entire country was under a dusk-to-dawn curfew

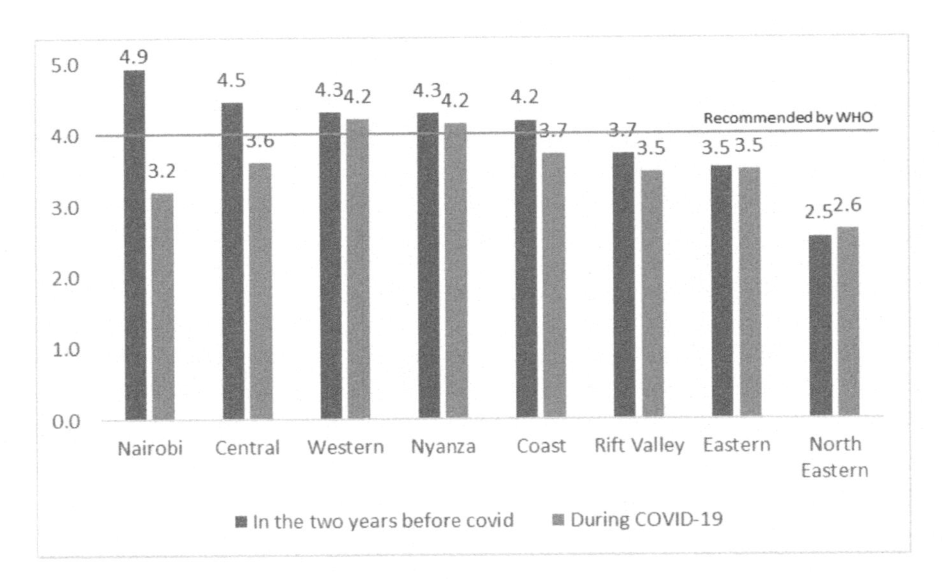

FIGURE 11.3 Mean antenatal visits per birth, by region, pre- and during COVID-19. (Source: Author's calculations using Kenya MIS 2019-2020 ICF International.)

Note: There were 490 actual births recorded in Kenya via the MIS 2019–2020 between April 1, 2020, and December 8, 2020. At least six births were reported per region during this time period.

for large periods of 2020, in certain counties with the highest coronavirus cases, additional movement bans were also imposed (see Figure 11.4b).[5]

In Nairobi, with the highest reported caseload in the country, restrictions on movement and harshly enforced curfews (Human Rights Watch, 2020) made it the epicenter for reduced antenatal care (Humanitarian Response, 2020). The proportion of women attending at least four antenatal care visits dropped from 63.5% in the two years before the pandemic to a striking 16.7% between March and December 2020, the lowest of any province. In the Coast and Western provinces, more stringent lockdown measures and higher caseloads also coincided with decreases in the proportion of women receiving at least four antenatal checks of 10.6% and 7.2%, respectively.

Surveys conducted during the pandemic in urban settings have found that fear and stigma around contracting COVID-19, deprioritization of services, and economic constraints reduced patients' demand for maternal health care (Oluoch-Aridi, 2020). On the supply side, facility closures, reduced staffing, and confusion surrounding what constituted essential care contributed to reductions, especially in the early days of the pandemic.[6]

To mitigate adversely impacting broader health outcomes, the Kenyan Ministry of Health developed and rolled out guidelines for accessing maternal health care during the pandemic, which included substituting in-person visits with telemedicine. Findings from this case study, however, suggest that promotion and uptake of

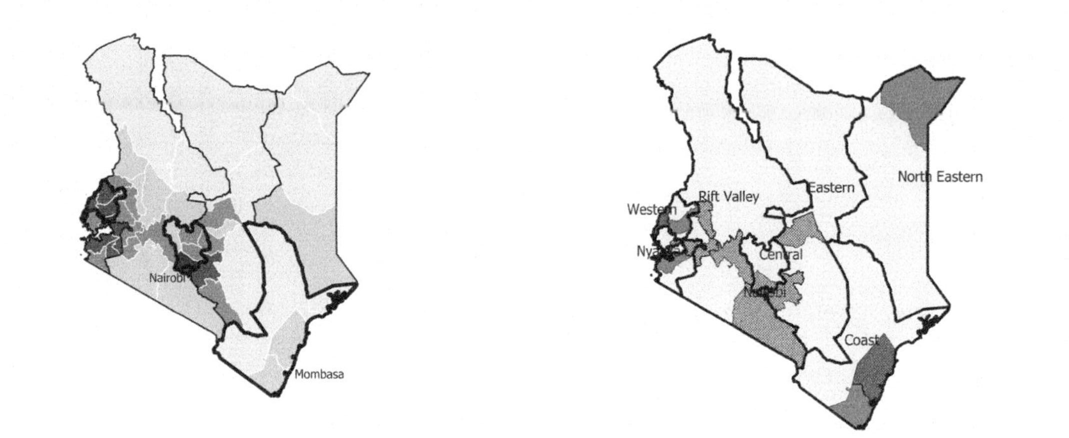

FIGURE 11.4A Decreases in ante-netal care (ANC) and population density, by province.

FIGURE 11.4B Presence of severe COVID-19 disruptions, by county. (Sources: Author's calculations using Kenya MIS 2019-2020. ICF International; OCHA, "Kenya: COVID-19 Operations Dashboard." Accessed April 15, 2022; Human Rights Watch, "Kenya: Police Brutality during Curfew," 2020; Kenya National Bureau of Statistics, "2019 Kenya Population and Housing Census Volume I: Population by County and Sub- County.")

such measures was in all likelihood quite low and insufficient to stave off declines in women's access to health care, including access to ANC services (Kenya Ministry of Health, 2020).

11.4.3 COVID-19 Impact on Mental Health: Spatial Inequalities in Women's Access to Mental Health Services in the United States

Amid lockdowns, job losses, and shifting uncertainties surrounding the COVID-19 crisis, mental disorders, including depressive and anxiety disorders, have soared. In the United States, the proportion of individuals reporting symptoms of depressive or anxiety disorder[7] rose more than threefold from 10.8% in 2019 to 37.4% in 2020.[8] A higher share of women (41.3%), compared to men (33.2%), reported experiencing symptoms of depressive or anxiety disorder (CDC, 2022).

When disaggregating the data by sex and race/ethnicity, further disparities are revealed. Hispanic women reported the largest increase in symptoms of anxiety or depression, jumping from 11.5% in 2019 to almost half (46.5%) in 2020, followed by Black and Asian women, each reporting a 30 percentage point increase in symptoms.[9] Additional research is needed to understand these disparities, but disproportionately high caseloads and death counts (Hill, 2022), lower rates of health insurance coverage (Vasquez Reyes, 2020), higher employment losses (Gemelas et a ., 2021), and fewer resources to cope with economic shocks (Bhutta et al., 2020) have likely negatively impacted the mental health of Hispanic, Black, and Asian women.

Rates of anxiety and depressive disorder also vary by age group. The largest gender gaps are seen among those between 18 and 25 years of age, with 58.7% of young women reporting symptoms of either disorder compared to 44.6% of young men, a gender gap of 14 percentage points (Figure 11.5). Other studies have attributed this gender gap among younger adults to women's larger social networks prior the pandemic and increased loneliness during it (Etheridge & Spantig, 2020).

Household composition by sex is also a revealing stratifier of the data on mental disorders. Throughout 2020 and 2021, women with children reported the highest rates of anxiety and depression, followed by women without children, men with children, and finally men without children (Figure 11.6). The pressures of taking on more unpaid care at home, including of school-age children, and balancing this additional time burden with paid work, has had a major impact on women's health, as well as women's economic prospects. More than two million moms left the labor force in 2020 according to new global estimates covering 189 countries and territories (UN Women, 2022a).

Women residing in the southern and western regions of the United States reported the highest levels of anxiety and/or depressive symptoms. These geographical differences mirror the inequalities in care provisioning. States with lower access to health care (i.e., with a larger share of women and men saying they did not have access to care) were more likely to report high levels of anxiety and depression (Figure 11.7a and 11.7b). The state with the highest rate of lack in care provisioning, New Mexico (35.8%), also had the fourth-highest rate of anxiety and depression, 44.8%.

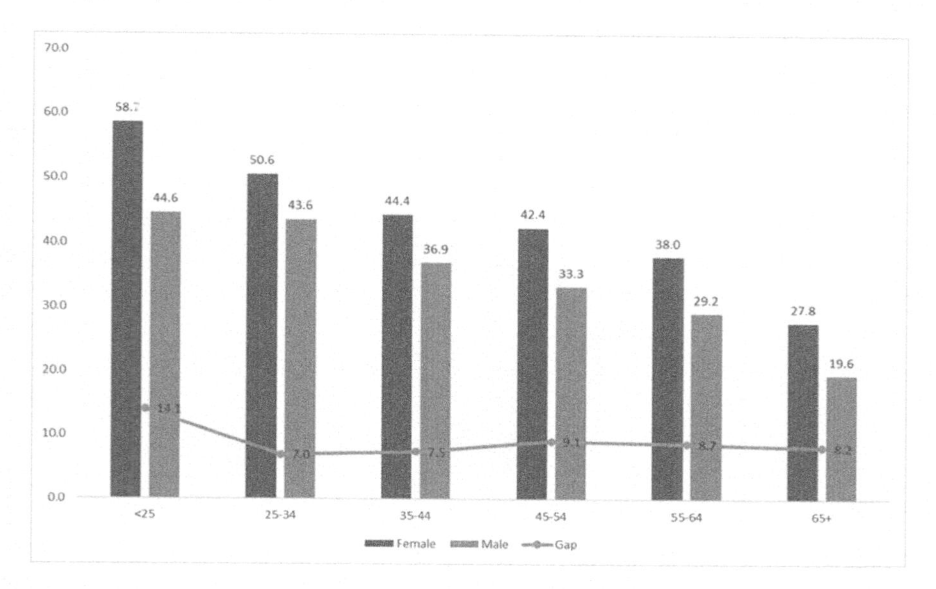

FIGURE 11.5 Individuals reporting symptoms of depressive disorder and/or anxiety, by sex, 2020, percentage. (Source: Author's calculations using US Census Bureau Household Pulse Survey, 2020.)

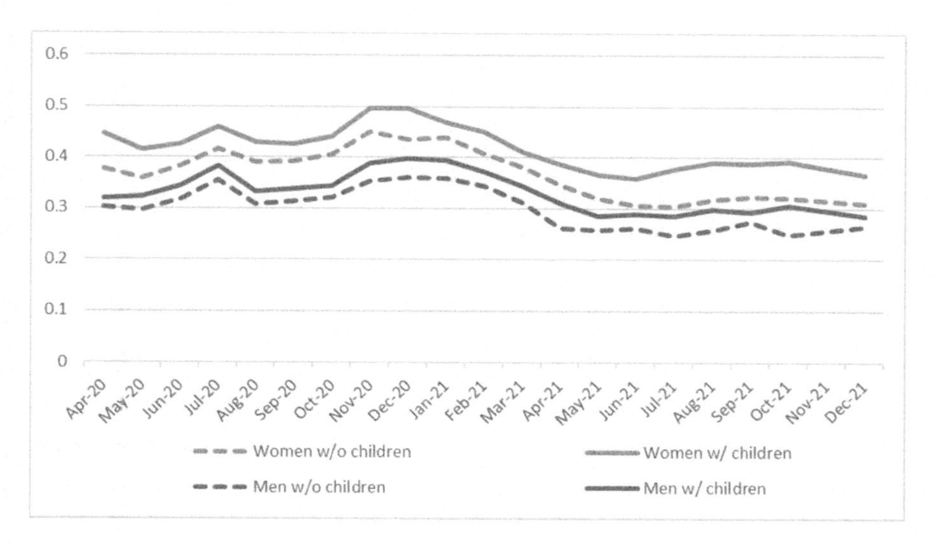

FIGURE 11.6 Individuals reporting symptoms of depressive disorder and/or anxiety, by sex and presence of children under 18, April 2020 to December 2021, percentage. (Source: Author's calculations using US Census Bureau Household Pulse Survey, 2020–2021.)

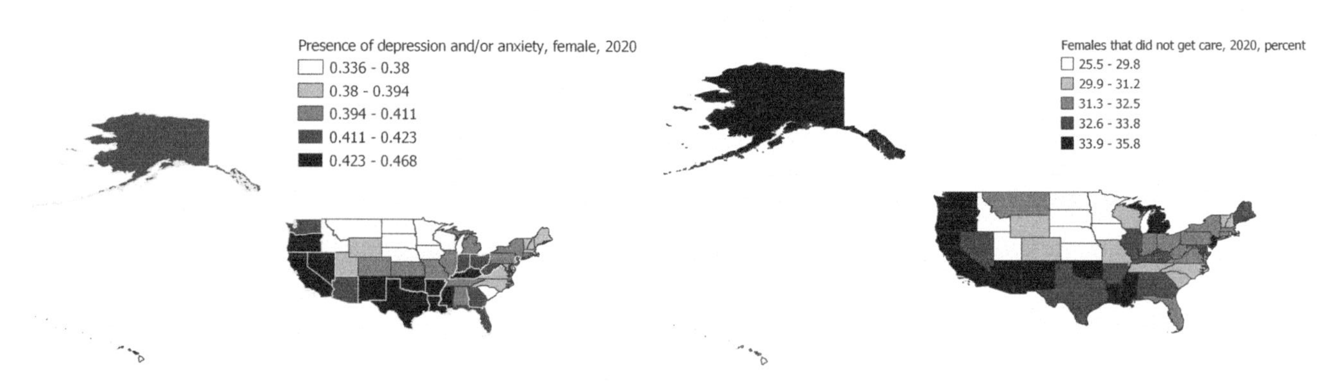

FIGURE 11.7 Depression and/or anxiety rates in the United States for females by location, and females who did not get care during COVID-19 by location, 2020. (Source: Author's calculations using US Census Bureau Household Pulse Survey, 2020.)

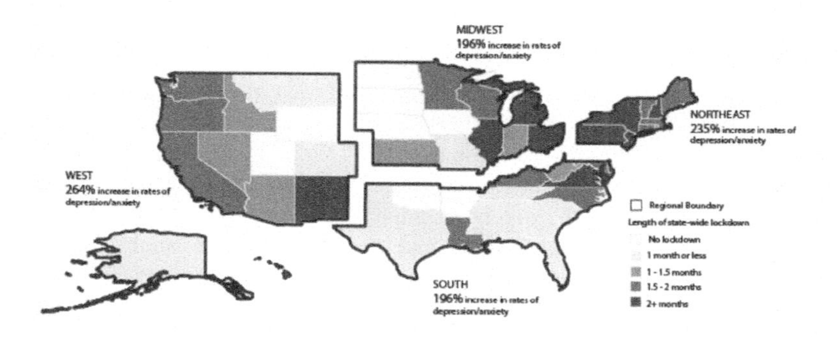

FIGURE 11.8 Lockdown length by state and percent change in rates of female depression or anxiety during COVID-19, 2020. (Source: Author's calculations using US Census Bureau Household Pulse Survey, 2020; United States Centers for Disease Control and Prevention, National Health Interview Survey2019; Delbert "States that have been locked down the longest," 2020.)

Lack of pre-COVID data on mental health prevalence limits the ability to compare changes at the state level. However, regional results from 2019 and 2020 show the largest increases in reports of anxiety or depressive disorder were found in the Northeast and the West, two areas that experienced long lockdowns during 2020 (Figure 11.8). In 2021, female rates of depression and/or anxiety reached 37.1 % nationally, well above the pre-pandemic 2019 levels of 10.8%.

11.5 CONCLUSION AND POLICY IMPLICATIONS

Findings from these three country case studies suggest that pandemic response, recovery, and preparedness efforts would benefit from a granular multilevel analysis of the barriers faced by different groups and subgroups, including women, many of whom have seen a dramatic decline in their health and well-being. While the pandemic has touched everyone in one way or the other, groups facing systemic inequality pre-pandemic are likely to have borne the biggest shock and often received the least amount of support. Women in precarious work situations pre-pandemic, for example, faced more challenges than other groups of women and men as shown in the Mexico case study. Moreover, while all countries struggled with the strain the pandemic placed on their health systems, countries with poorly resourced health-care systems, and little option for telehealth faced even more challenges. Women's health and access to health care suffered as a result, as shown in the Kenya case study. Spatial inequalities, which often meant women in rural and isolated areas faced acute barriers to accessing health care, have only compounded the problem.

In other cases, the disruptions to women's health care have reversed years of progress and made outcomes worse in regions and spatial locations that had previously done much better. The real danger of massive reversals in women's health is corroborated by other country-level data showing a worrisome trend of higher rates of maternal mortality, lower access to family planning services, and higher rates of adolescent births (UN Women, 2022). Moving forward, greater efforts are needed to document and study the divergent impact of crises and their tendency to exacerbate gender inequalities across areas and dimensions. As the analysis in this chapter shows, GIS and gender approaches to the study of crisis impact can provide a rich understanding of the hot spots, regions, and areas in a country that require more attention, but also groups that need greater support.

ACKNOWLEDGMENTS

The authors would like to acknowledge Asha Meagher, Yunjoo Park, and Kathleen Clements for excellent research assistance.

DISCLAIMER

Views presented in this chapter are of the authors only and do not represent the views of UN Women. The designations employed and the presentation of material on the maps do not imply the expression of any opinion whatsoever on the part of the United Nations or the authors concerning the legal status of any country, territory, city or area, or of its authorities, or concerning the delimitation of its frontiers or boundaries.

NOTES

1 en.www.inegi.org.mx/contenidos/investigacion/ecovidml/2020/doc/ecovid_ml_diseno_muestral.pdf
2 www.snieg.mx/cni/escenario.aspx?idOrden=1.1&ind=6200093953&gen=2229&d=s
3 US 2020 COVID-19 Household Pulse Survey www2.census.gov/programs-surveys/demo/technical-documentation/hhp/Phase_1_Questionnaire_06_11_20_English.pdf
4 In Mexico, several indicators for informal employment are available. The indicator used here looks at the proportion of the employed population that is working in non-registered economic units (informal sector); paid domestic service workers who do not have social security; those employed on their own account in subsistence agriculture; unpaid workers; subordinate and paid workers whose services are used by registered economic units and work without social security protection. The rate is calculated based on the total employed population. For more information see www.snieg.mx/cni/escenario.aspx?idOrden=1.1&ind=6200093953&gen=2229&d=s
5 The first COVID-19 case in Kenya was reported on March 13, 2020
6 www.scribd.com/document/548599491/Kenya-RMNCH-Deep-Dive-brief
7 Depression variable constructed using two questions from the Pulse survey: Over the last 7 days, how often have you been bothered by: 1) having little interest or pleasure in doing things? 2) feeling down, depressed, or hopeless? Anxiety variable constructed using two questions from the Pulse survey: Over the last 7 days, how often have you been bothered by the following problems: 1) Feeling nervous, anxious, or on edge? 2) Not being able to stop or control worrying? Would you say not at all, several days, more than half the days, or nearly every day? Select only one answer. For each question, respondents choose not at all, several days, more than half the days, or nearly every day. For each scale, the answers are assigned a numerical value: not at all = 0, several days = 1, more than half the days = 2, and nearly every day = 3. The two responses for each scale are added together. A sum equal to three or greater has been shown to be associated

with diagnoses of major depressive or anxiety disorder. For adults with scores of three or greater, further evaluation by a clinician or other health professional is generally recommended. See www.cdc.gov/nchs/covid19/pulse/mental-health.htm

8 UN Women calculations using US Census Bureau Household Pulse Survey, 2020

9 UN Women calculations using US Census Bureau Household Pulse Survey, 2020

REFERENCES

AfricaNews. (2020, October 5). Kenya coronavirus: Updates from March–April 2020. Africanews. www.africanews.com/2020/05/10/enforcement-of-coronavirus-lockdown-turns-violent-in-parts-of-africa//

Al Jazeera. (2020, April 6). COVID-19: Kenya bans travel in and out of Nairobi, other areas. www.aljazeera.com/news/2020/4/6/covid-19-kenya-bans-travel-in-and-out-of-nairobi-other-areas

Azcona, G., & Bhatt, A. (2021). How to undertake an inequality, gender and sustainable development analysis. In E. Ozdenerol (Ed.), *Gender Inequalities* (pp. 165–192).Boca Raton, FL: Taylor & Francis. https://doi.org/10.1201/9780429196584-8

Azcona, G., Bhatt, A., Davies, S. E., Harman, S., Smith, J., & Wenham, C. (2020). Spotlight on gender, COVID-19 and the SDGs: Will the pandemic derail hard-won progress on gender equality? *Spotlight on the SDGs Paper Series*. www.unwomen.org/sites/default/files/Headquarters/Attachments/Sections/Library/Publications/2020/Spotlight-on-gender-COVID-19-and-the-SDGs-en.pdf

Bhutta, N., Chang, A. C., Dettling, L. J., & Hsu, J. W. (2020, September 28). Disparities in wealth by race and ethnicity in the 2019 Survey of Consumer Finances. Federalreserve.gov. www.federalreserve.gov/econres/notes/feds-notes/disparities-in-wealth-by-race-and-ethnicity-in-the-2019-survey-of-consumer-finances-20200928.htm

Brown, S. (2009). Spatial analysis of socioeconomic issues: Gender and GIS in Nepal. *Mountain Research and Development, 23*(4), 338–344. https://doi.org/10.1659/0276-4741(2003)023[0338:SAOSIG]2.0.CO;2

Cañas, J., & Smith, C. N. (2021). COVID-19 poses stubborn challenge to economic growth in Mexico. Federal Reserve Bank of Dallas. www.dallasfed.org/research/swe/2021/swe2101/swe2101c.aspx

Casado Izquierdo, J. M. (2021). De Crisis Sanitaria a Crisis Económica Y Laboral: Patrones Espaciales Del Impacto de La COVID-19 En El Empleo Formal de México. *Investigaciones Geográficas*, 104 (January). https://doi.org/10.14350/rig.60212

Catálogo Nacional de Indicadores. (2022). Snieg.mx. 2022. www.snieg.mx/cni/escenario.aspx?idOrden=1.1&ind=6200093953&gen=2229&d=s

Centers for Disease Control and Prevention (CDC) (2022). National Center for Health Statistics. COVID-19 Data from NCHS. Anxiety and Depression. Household Pulse Survey. www.cdc.gov/nchs/covid19/pulse/mental-health.htm

Chen, Z. L., Zhang, Q., Lu, Y., Guo, Z. M., Zhang, X., Zhang, W. J., Guo, C., Liao, C. H., Li, Q. L., Hanm, X. H., & Lu, J.H. (2020). Distribution of the COVID-19 epidemic and correlation with population emigration from Wuhan, China. *Chinese Medical Journal, 133*(9), 1044–1050. https://doi.org/10.1097/CM9.0000000000000782

Chihuahua: Economy, Employment, Equity, Quality of Life, Education, Health and Public Safety. (2020). Data México. https://datamexico.org/en/profile/geo/chihuahua-ch

The COVID-19 pandemic has increased the care burden, but by how much? (2020, December 3). UN Women Data Hub. https://data.unwomen.org/features/covid-19-pandemic-has-increased-care-burden-how-much-0

Delbert, C. (2020). States that have been locked down the longest. Stacker. https://stacker.com/stories/4218/states-have-been-locked-down-longest

Division of National Malaria Programme (DNMP) [Kenya] and ICF. (2021). Kenya Malaria Indicator Survey 2020. Nairobi, Kenya and Rockville, Maryland, USA: DNMP and ICF.

Dotse-Gborgbortsi, W., Wardrop, N., Adewole, A. et al. (2018). A cross-sectional ecological analysis of international and sub-national health inequalities in commercial geospatial resource availability. *International Journal of Health Geographics, 17*(14). https://doi.org/10.1186/s12942-018-0134-z

Effectiveness of antenatal care services in reducing neonatal mortality in Kenya: Analysis of national survey data. (2017). *Global Health Action, 10*(1). www.tandfonline.com/doi/full/10.1080/16549716.2017.1328796

Etheridge, B., & Spantig, L. (2020). The gender gap in mental well-being during the Covid-19 outbreak: Evidence from the UK. Econstor.eu. ISER Working Paper Series No. 2020-08. http://hdl.handle.net/10419/227789

Fallout of COVID-19: Working moms are being squeezed out of the labour force. (2020, November 25). UN Women Data Hub. https://data.unwomen.org/features/fallout-covid-19-working-moms-are-being-squeezed-out-labour-force

Fields, J. F., Hunter-Childs, J., Tersine, A., Sisson, J., Parker, E., Velkoff, V., Logan, C., & Shin, H. (Forthcoming). Design and operation of the 2020 Household Pulse Survey, 2020. US Census Bureau. www2.census.gov/programs-surveys/demo/technical-documentation/hhp/2020_HPS_Background.pdf

Flor, L. S., Friedman, J., Spencer, C. N., Cagney, J., Arrieta, A., Herbert, M. E., Stein, C. et al. (2022). Quantifying the effects of the COVID-19 pandemic on gender equality on health, social, and economic indicators: A comprehensive review of data from March, 2020, to September, 2021. *The Lancet, 399*(10344), 2381–2397. https://doi.org/10.1016/s0140-6736(22)00008-3

Franch-Pardo, I., Desjardins, M. R., Barea-Navarro, I., & Cerdà, A. (2021). A review of GIS methodologies to analyze the dynamics of COVID-19 in the second half of 2020. *Trans GIS, 25*(5), 2191–2239. https://doi.org/10.1111/tgis.12792

Gemelas, J., Davison, J., Keltner, C., & Ing, S. (2021). Inequities in employment by race, ethnicity, and sector during COVID-19. *Journal of Racial and Ethnic Health Disparities, 9*(1), 350–355. https://doi.org/10.1007/s40615-021-00963-3

Graziella, B., & Dimico, A. (2021). COVID-19, race, and gender. EIEF Working Papers Series 2106. Einaudi Institute for Economics and Finance (EIEF), revised March 2021.

Guan, W. J., Nim, Z. Y., Hu, Y., Liang, W. H., Ou, C. Q., He, J. X., & Du, B. (2020). Clinical characteristics of coronavirus disease 2019 in China. *New England Journal of Medicine, 382*(18), 1708–1720. https://doi.org/10.1056/NEJMoa2002032

Hill, L., & Artiga, S. (2022, February 22). COVID-19 cases and deaths by race/ethnicity: current data and changes over time. KFF. www.kff.org/coronavirus-covid-19/issue-brief/covid-19-cases-and-deaths-by-race-ethnicity-current-data-and-changes-over-time/

Instituto Nacional de Estadística y Geografía (INEGI). (2020a). Encuesta Telefónica Sobre COVID-19 Y Mercado Laboral (ECOVID-ML). Inegi.org.mx. www.inegi.org.mx/investigacion/ecovidml/2020/

Instituto Nacional de Estadística y Geografía (INEGI). (2020b). Encuesta Telefónica Sobre COVID-19 Y Mercado Laboral (ECOVID-ML). Inegi.org.mx. http://en.www.inegi.org.mx/contenidos/investigacion/ecovidml/2020/doc/ecovid_ml_diseno_muestral.pdf

International Labour Organization (ILO). (2022, May 23). ILO Monitor on the World of Work. Ninth edition. Ilo.org. www.ilo.org/wcmsp5/groups/public/---dgreports/---dcomm/---publ/documents/publication/wcms_845642.pdf

Joseph, N. K., Macharia, P. M., Ouma, P. O., Mumo, J., Jalang'o, R., Wagacha, P. W., Achieng V. O. et al. (2020). Spatial access inequities and childhood immunisation uptake in Kenya. *BMC Public Health, 20*(1). https://doi.org/10.1186/s12889-020-09486-8

Kenya. Ministry of Health. (2020). *Kenya COVID19 RMNH guidelines: A Kenya practical guide for continuity of reproductive, maternal, newborn and family planning care and services in the background of COVID19 pandemic*. www.health.go.ke/wp-content/uploads/2020/04/KENYA-COVID19-RMNH.pdf

Kenya: COVID-19 Operations Dashboard. HumanitarianResponse. (2020). Humanitarianresponse.info. www.humanitarianresponse.info/en/operations/southern-eastern-africa/kenya-covid-19-operations-dashboard

Kenya: Police brutality during curfew. (2020, April 22). Human Rights Watch. www.hrw.org/news/2020/04/22/kenya-police-brutality-during-curfew

La COVID-19 Y Su Impacto En Las Mujeres En México. (2020). Inegi.org.mx. www.inegi.org.mx/tablerosestadisticos/mujeres/

Mamun, F., Gozal, D., Hosen, I., Misti, J. M., & Mamun, M. A. (2021). Predictive factors of insomnia during the COVID-19 pandemic in Bangladesh: a GIS-based nationwide distribution. *Sleep Medicine, 91*, 219–225. https://doi.org/10.1016/j.sleep.2021.04.025

Mamun, M. (2021). Exploring factors in fear of COVID-19 and its GIS-based nationwide distribution: The case of Bangladesh. *BJPsych Open, 7*(5), E150. https://doi.org/10.1192/bjo.2021.984

Mamun, M. A., Al Mamun, M., Hosen, I., Ahmed, T., Rayhan, I., & al-Mamun, F. (2022, December). Trend and gender-based association of the Bangladeshi student suicide during the COVID-19 pandemic: A GIS-based nationwide distribution. *International Journal of Social Psychiatry*. https://doi.org/10.1177/00207640211065670

Mclafferty, S. (2002). Mapping women's worlds: Knowledge, power and the bounds of GIS. *Gender. Place and Culture: A Journal of Feminist Geography, 9*(3), 263–269. https://doi.org/10.1080/0966369022000003879

Oluoch-Aridi, J., Chelagat, T., Nyikuri, M. M., Onyango, Guzman, D., Makanga, C., Miller-Graff, L., & Dowd, R. 2020. COVID-19 effect on access to maternal health services in Kenya. *Frontiers in Global Women's Health, 26* (November). https://doi.org/10.3389/fgwh.2020.599267

Ozdenerol, E. (2017). *Spatial health inequalities: Adapting GIS tools and data analysis*. Boca Raton, FL: CRC Press.

PATH. (2021, February). RMNCAH-N services during COVID-19: A spotlight on Kenya's policy responses to maintain and adapt essential health services. www.scribd.com/document/548599491/Kenya-RMNCH-Deep-Dive-brief

SEDAC. Global COVID-19 Viewer: Population Estimates by Age Group and Sex. (2020). Columbia.edu. https://sedac.ciesin.columbia.edu/mapping/popest/covid-19/

UBS Editorial Team. (2021). Mexico: COVID and the impact on women in the labour force. Investment Bank. UBS. www.ubs.com/global/en/investment-bank/in-focus/covid-19/2021/women-employment-rates.html

UN Women. (2020, November). Fallout of COVID-19: Working moms are being squeezed out of the labour force. UN Women–Headquarters. https://data.unwomen.org/features/fallout-covid-19-working-moms-are-being-squeezed-out-labour-force

UN Women. (2021, September 17). From insights to action: Gender equality in the wake of COVID-19. UN Women–Headquarters. www.unwomen.org/en/digital-library/publications/2020/09/gender-equality-in-the-wake-of-covid-19

UN Women. (2022a, February 21). More than 2 million moms left the labour force in 2020 according to New Global Estimates. UN Women Data Hub. https://data.unwomen.org/features/more-2-million-moms-left-labour-force-2020-according-new-global-estimates

UN Women. (2022b). Progress on the Sustainable Development Goals: The Gender Snapshot 2022. UN Women–Headquarters. September 2022. www.unwomen.org/sites/default/files/2022-09/Progress-on-the-sustainable-development-goals-the-gender-snapshot-2022-en_0.pdf

US Census Bureau. (2022, July 20). Household Pulse Survey Public Use File (PUF). Census.gov. www.census.gov/programs-surveys/household-pulse-survey/datasets.html

US Centers for Disease Control and Prevention. (2019). National Health Interview Survey 2019. www.cdc.gov/nchs/nhis/2019nhis.htm.

Vasquez Reyes, M. (2020). The disproportional impact of COVID-19 on African Americans. *Health and Human Rights, 22*(2), 299–307. www.ncbi.nlm.nih.gov/pmc/articles/PMC7762908/

12 The Effects of Lifestyle on COVID-19 Vaccine Hesitancy in the United States

An Analysis of Market Segmentation

Esra Ozdenerol and Jacob Daniel Seboly

CONTENTS

12.1 INTRODUCTION

Since the US Food and Drug Administration (FDA) authorized the first COVID-19 vaccines, more than one hundred million people in the United States have

DOI: 10.1201/9781003227106-12

been vaccinated. The initial US COVID-19 vaccination effort potentially reduced COVID-19 infections, hospitalizations, and deaths in the United States (CDC, 2021). However, many people remain hesitant to receive the vaccine. Our research and market segmentation tools can help identify the lifestyle traits associated with vaccine hesitancy, enabling vaccine uptake to be predicted at local (e.g., zip code, census tract, block group) levels (Ozdenerol & Seboly, 2021). Our targeted approach enables key stakeholders to use these tools in their interventions and focus on these specific households, which are exhibiting patterns of vaccination hesitancy but are likely to see significant benefit from COVID-19 vaccinations.

Our prior work has explored associations between lifestyle factors and several health outcomes (e.g., COVID-19 infections, Lyme disease) (Ozdenerol & Seboly, 2021; Ozdenerol et al., 2021). Associated lifestyle factors and market segments with the relevant health outcomes (e.g., COVID-19) and health-care decisions (e.g., vaccination rates) could potentially become an integral step for the implementation of coordinated vaccination interventions (Ozdenerol et al., 2021). The ability to target a population in terms of its vaccination characteristics (e.g., vaccination acceptance, vaccination hesitancy) and its vaccination needs is becoming increasingly critical for the successful management and control of COVID-19. Vaccine hesitancy refers to the delay in acceptance or refusal of vaccines despite the availability of vaccine services (Salmon et al., 2015). This phenomenon is complex and context specific, varying across time, place, and vaccines. It is influenced by factors such as complacency, convenience, and confidence, which are all related to lifestyle attributes of individual households and communities—consumer behaviors, civic engagement, income, education, dietary preferences, and so forth (Salmon et al., 2015; McClure, 2018).

We demonstrated the impact of lifestyle on COVID-19 vaccination by applying the CDC's vaccination rates at the US county level. We utilized ESRI's lifestyle segmentation system, which classifies all US households into 14 LifeMode categories, and identified segments that are associated with higher rates of vaccine hesitancy (Dubé, 2013). To effectively analyze the impact of vaccination on American households by their lifestyle characteristics, we specifically tested three research questions: "Which LifeModes have vaccination rates that are statistically higher/lower than average?", "How has vaccine uptake/hesitancy changed over time among the different LifeModes?", and "How well can vaccination rates be predicted at subcounty levels using lifestyle segmentation data?" We focused on comparing each LifeMode's mean to the national mean to ascertain spatial and temporal patterns of high-risk households and the effects of lifestyle on the vaccination rates in the United States.

12.2 VACCINE HESITANCY IN THE UNITED STATES

In the United States, the COVID-19 vaccination rate has been influenced by vaccine hesitancy, which is related to a proportion of individuals who remain unwilling or uncertain about vaccination (Salmon et al., 2015; McClure, 2018; Hamel et al., 2020). Psychographics, demographics, and geography play a major role in COVID-19 vaccine hesitancy, indicating a need for targeted messaging to high-hesitancy groups.

Vaccination acceptance indicates a proportion of people who have already received or are willing to receive the vaccine (Salmon et al., 2015; McClure, 2018; Hamel et al., 2020). A national survey by Mejia and King assessed time trends and how each hesitancy group's outlook changed regarding vaccination (Salmon et al., 2015). Analyzing the data by race, education, US region, and support for Donald Trump in the 2020 election, they found COVID-19 vaccine hesitancy was higher among young people (ages 18–24), non-Asian people, and less educated (high school diploma) adults with a record of a positive COVID-19 test, who were not concerned about the severity of COVID-19 and living in regions that supported Trump.

Between January and May 2020, vaccine hesitancy decreased in the population with a high school education or less but stayed constant among those who had obtained higher levels of education. Although all racial groups observed a decrease in vaccination hesitancy in May, Black people and Pacific Islanders had the largest decrease, joining Hispanics and Asians. Higher hesitancy is observed among residents of counties with higher Trump support in the 2020 presidential election, who neither trusted the vaccine nor the government. The difference in hesitancy between high and low Trump-elected counties increased over the period studied. Less hesitant groups were interested to wait and see if the vaccine was safe. Researchers also found that the lack of change in those with strong feelings about the vaccine were not likely to change easily and that reaching that group with messages and incentives was crucial to increasing vaccine acceptance (King et al., 2021a; King et al., 2021b). Researchers suggest targeted campaigns may have increased vaccine acceptance and access, since decreases in disparities by race and educational attainment were found.

King et al. investigated vaccine hesitancy by race and age subgroups. They found younger Black people were more hesitant than younger white people in May, while the reverse was true in older populations (King et al., 2021a; King et al., 2021b). This type of detailed analysis could help policy makers identify vaccine-resistant pockets. Survey participants had greater vaccine uptake with higher education levels compared to the general population. Due to this limitation, the percentage of people who refuse the COVID-19 vaccine is likely higher.

A cohort study by Siegler et al. (2021) found that despite vaccine hesitancy having weakened between late 2020 and early 2021, inequities remained. Higher vaccine willingness converted into successfully delivered vaccinations; vaccine hesitancy converted into vaccine willingness (37%) and initial hesitancy of being vaccinated at follow-up (32%).

Callaghan et al.'s study (2021) analyzed the influence of demographics, political beliefs, and COVID-19 experiences on COVID-19 vaccine hesitancy by conducting a survey of 5,009 American adults collected from May 28–June 8, 2020. A total of 31% of participants did not pursue getting vaccinated when the United States began COVID-19 vaccinations. The two most mentioned reasons for vaccine refusal were concerns related to vaccine safety and effectiveness. Among subpopulations, women were most likely to be hesitant based on concerns about vaccine safety and efficacy. Black people were more likely to be hesitant than white people. In addition to concerns about safety and efficacy, Black people lacked the needed financial resources or health insurance, and they had already contracted COVID-19.

In the context of the global vaccine distribution, vaccine nationalism presents a challenge to the equitable access to vaccines. Even though governments have the right and duty to ensure their citizens priority access to a COVID-19 vaccine, some governments have heavily prioritized the vaccine manufactured in their own country to their citizens (Bolcato et al., 2021). The issue of equal access to vaccination led to equity-based initiatives such as the World Health Organization-devised COVAX equitable vaccine access plan. By accelerating the development and manufacturing of vaccines as well as facilitating the equitable distribution of them, COVAX ensures said vaccines reach poor countries (Stein, 2021). However, the success of the COVAX with respect to equitable access to the vaccine is likely to be limited, primarily because of the prevalence of vaccine nationalism (Eccleston-Turner & Upton, 2021).

We identified lifestyle segments that have a high propensity for vaccine uptake. High-risk lifestyle segments are clearly the areas where in the United States the public might benefit from getting vaccinated. Outreach and education campaigns targeted to these segments can reduce access and logistical barriers to vaccination. Based on their lifestyle preferences, these at-risk households could be patient communities in clinical trials' campaigns. We mapped when and where each LifeMode had above/below the overall mean vaccination rate and how COVID-19 vaccination impacted different households. Our results recommend prevention and control policies to be implemented to those specific households.

12.3 MATERIALS AND METHODS

12.3.1 Data

We integrated data from multiple sources into geographic information systems (GIS) to create a map exhibiting the concentration of the high-risk lifestyle segments (i.e., the Tapestry is used to correlate lifestyle segments with COVID-19 vaccination rates). As a first step, we examine what is ESRI's Tapestry segmentation and of what the Tapestry data set is comprised. Then, we provide information on the COVID-19 vaccination data sets we used for the analysis and further explain our study design and methods applied in detail.

12.3.1.1 An Introduction to ESRI Tapestry Segmentation

We have chosen ESRI Tapestry data for our analysis because of ESRI's authorized leadership as a location intelligence platform provider. According to the Forrester report, in their 30-criterion evaluation of location intelligence platform providers, ESRI is identified as one of the leaders among the nine most significant ones—ESRI, CARTO, Google, Hexagon, MapLarge, Microsoft, Oracle, Salesforce, and Syncsort (McGormick, 2020). Esri's Tapestry Market Segmentation is also one of the most widely used geodemographic systems that provides an accurate, detailed description of American households and categorizes them into 67 distinctive consumer market segments based on socioeconomic, demographic, and psychographic data (ESRI-Tapestry, 2017). The ESRI Tapestry Segmentation system employs Experian's Consumer View database (Experian, 2017), the Survey of the American Consumer from GfK MRI (Survey of the American Consumer, 2020), and the US Census American Community Survey (ACS, 2018). ESRI Tapestry Segmentation compiles these databases specifically to understand consumers' lifestyle choices—what they buy and how they spend their free time, their beliefs, and life patterns based on geography, behavior, demographics, and psychographics—their income and demographic parameters such as race, gender, age groups, and marital status (ACS, 2018; Mosaic USA, 2017; Acorn, 2017; Claritas, 2017; People2People & Places, 2017; Harris et al., 2005). ESRI Tapestry helps gain insights into the markets, and which markets are being underserved, ultimately improving the performance of existing locations, finding optimal site locations, and investing resources wisely.

ESRI Tapestry segments are grouped into 14 LifeMode Groups with names such as "Rustic Outposts," "Affluent Estates," and "Family Landscapes," which have commonalities based on lifestyle and life stages (ESRI-Tapestry, 2017). We downloaded ESRI Tapestry segmentation data from ESRI that contained the dominant LifeMode within each US county and the number of households and percentages of the households in the county belonging to each LifeMode (Esri Data, 2019). Table 12.1 provides a detailed description of the LifeModes.

TABLE 12.1
Description of LifeModes

Life Mode
LifeMode1 Affluent Estates
• Established wealth—educated, well-traveled married couples • Accustomed to "more": less than 10% of all households, with 20% of household income • Homeowners (almost 90%), with mortgages (65.2%) • Married couple families with children ranging from grade school to college • Expect quality; invest in time-saving services • Participate actively in their communities • Active in sports and enthusiastic travelers
LifeMode 2 Upscale Avenues
• Prosperous married couples living in older suburban enclaves • Ambitious and hard-working • Homeowners (70%), prefer denser, more urban settings with older homes and a large share of townhomes • A more diverse population, primarily married couples, many with older children • Financially responsible, but still indulge in casino gambling and lotto tickets • Serious shoppers, from Nordstrom's to Marshalls or DSW, that appreciate quality and bargains • Active in fitness pursuits like bicycling, jogging, yoga, and hiking • Subscribe to premium movie channels like HBO and Starz
LifeMode 3 Uptown Individuals
• Young, successful singles in the city • Intelligent, hard-working, and averse to traditional commitments of marriage and home ownership • Urban denizens, partial to city life, high-rise apartments, and uptown neighborhoods • Prefer credit cards over debit cards, while paying down student loans • Green and generous to environmental, cultural, and political organizations • Internet dependent, from social connections to shopping for fashion, tracking investments • Adventurous and open to new experiences and places

(continued)

TABLE 12.1 (Continued)
Description of LifeModes

Life Mode
LifeMode 5 GenXurban
• Gen X in middle age; families with fewer kids and a mortgage • Second-largest Tapestry group, comprised of Gen X married couples, and retirees • About a fifth of residents are 65 or older; about a fourth of households have retirement income • Own older single-family homes in urban areas, with one or two vehicles • Live and work in the same county, creating shorter commute times • Invest wisely, well-insured, comfortable banking online or in person • News junkies (read a daily newspaper, watch news on television, and go online for news) • Enjoy reading, renting movies, playing board games, doing puzzles, going to museums and rock concerts
LifeMode 6 Cozy Country Living
• Empty nesters in bucolic settings • Largest Tapestry group, almost half of households located in the Midwest • Homeowners, in single-family dwellings in rural areas; 30% have three or more vehicles • Politically conservative and believe in the importance of buying American • Own domestic trucks, motorcycles, and ATVs/UTVs • Prefer to eat at home, shop at discount retail stores (especially Walmart), bank in person • Own every tool and piece of equipment imaginable to maintain their homes, vehicles, gardens • Listen to country music, watch auto racing on television, and play the lottery; enjoy outdoor activities
LifeMode 7 Ethnic Enclaves
• Established diversity—young, Hispanic homeowners with families • Multilingual households and multigenerational Hispanic families • Neighborhoods feature single-family, owner-occupied homes built at city's edge, built after 1980 • Hard-working, residents aged 25 years or older, have a high school diploma or some college • Shopping and leisure also focus on their children—baby and children's products • Residents favor Hispanic programs on radio or television; children enjoy playing video games • Many households have dogs for domestic pets
LifeMode 8 Middle Ground
• Lifestyles of thirtysomethings • Millennials in the middle: single/married, renters/homeowners, middle class/working class • Urban market mix of single-family, townhome, and multiunit dwellings • Majority of residents attended college or attained a college degree • Householders have ditched their landlines for cell phones, which they use to listen to music, news • Online all the time: use the Internet for entertainment, social media, search for employment • Leisure includes nightlife (clubbing, movies), going to the beach, some travel and hiking

TABLE 12.1 (Continued)
Description of LifeModes

Life Mode
LifeMode 9 Senior Styles
• Senior lifestyles reveal the effects of saving for retirement • Households of married empty nesters or singles living alone; single-family or high-rise • More affluent seniors travel and relocate to warmer climates; less affluent, settled seniors • Cell phones are popular, but so are landlines • Many still prefer print to digital media: Avid readers of newspapers, to stay current • Subscribe to cable television to watch channels like Fox News, CNN, and The Weather Channel • Residents prefer vitamins to increase their mileage and a regular exercise regimen
LifeMode 10 Rustic Outposts
• Country life with older families in older homes • Rustic Outposts, engaged manufacturing, retail and health care, with mining and agricultural jobs • Low labor force participation in skilled and service occupations • Own affordable, older single-family or mobile homes; vehicle ownership, a must • Residents live within their means, shop at discount stores and maintain their own vehicles • Outdoor enthusiasts, who grow their own vegetables, love their pets and hunting and fishing • Technology is cost prohibitive and complicated. Pay bills in person, use the yellow pages.
LifeMode 11 Midtown Singles
• Millennials on the move—single, diverse, urban • Millennials seeking affordable rents in apartment buildings • Work in service and unskilled positions, usually close to home or public transportation • Single parents depend on their paycheck to buy supplies for their very young children • Midtown Singles embrace the Internet, for social networking and downloading content • From music and movies to soaps and sports, radio and television fill their lives • Brand-savvy shoppers select budget-friendly stores
LifeMode 12 Hometown
• Growing up and staying close to home; single householders • Close-knit urban communities of young singles (many with children) • Owners of old, single-family houses, or renters in small multiunit buildings • Religion is the cornerstone of many of these communities • Visit discount stores and clip coupons, frequently play the lottery at convenience stores • Canned, packaged, and frozen foods help make ends meet • Purchase used vehicles to get them to and from nearby jobs
LifeMode 13 Next Wave
• Urban denizens, young, diverse, hard-working families • Extremely diverse with a Hispanic majority, the highest among LifeMode groups • A large share is foreign born and speak only their native language • Young, or multigenerational, families with children are typical • Most are renters in older multiunit structures, built in the 1960s or earlier • Hard-working with long commutes to jobs, often utilizing public transit to commute to work • Spending reflects the youth of these consumers, focus on children and personal appearance • Also, a top market for moviegoers (second only to college students) and fast food • Partial to soccer and basketball

(*continued*)

TABLE 12.1 (Continued)
Description of LifeModes

Life Mode
LifeMode 14 Scholars and Patriots
• College and military populations that share many traits • Highly mobile, recently moved to attend school or serve in military • The youngest market group, with a majority in the 15- to 24-year-old range • Renters with roommates in nonfamily households • For many, no vehicle is necessary as they live close to campus, military base, or jobs • Fast-growing group with most living in apartments • Part-time jobs help supplement active lifestyles • Millennials are tethered to their electronic devices, typically spending over five hours online • Purchases aimed at fitness, fashion, technology, and the necessities of moving • Highly social, free time is spent enjoying music, being out with friends, seeing movies • Try to eat healthy, but often succumb to fast food

12.3.1.2 COVID-19 Vaccination Rates

We downloaded COVID-19 vaccination rates from the CDC's vaccination website that were updated daily on a county-by-county basis (CDC, 2022). This data set contains many variables, as there are many facets to vaccination status. For our analysis, the variable series_complete_pct was used (CDC Vaccination Rates, 2022). This represents the percentage of people in the county considered "fully vaccinated," meaning that they received two doses of the Pfizer or Moderna vaccine or one dose of the Johnson & Johnson vaccine. As the criteria for vaccine eligibility changed several times throughout 2021, we maintain temporal consistency by considering only the percentage of the entire population vaccinated, without when/who was eligible. Another consideration is that late in 2021, the FDA approved the distribution of booster doses for people more than six months removed from their most recent shot (FDA, 2022). Our analysis does not account for boosters, as eligibility for the booster depends on the time at which the initial vaccination was received, making it difficult to compare across counties. Moreover, this study is principally focused on identifying populations who refuse the initial vaccination, so analyzing the booster uptake is not relevant to our goals. The CDC data set also contains raw counts of vaccinated people as well as population data by county. These two variables are also utilized for our analysis. Figure 12.1 shows the percentage that is fully vaccinated by county on January 1, 2022.

On January 1, 2022, 58.5% of the US population was considered fully vaccinated. However, the mean vaccination percentage for all US counties (not weighted for population) on January 1, 2022, was 49%. This indicates that counties with high population (urban) disproportionately had higher vaccination rates, while counties with low population (rural) disproportionately had lower vaccination rates. Vaccination rates appeared to be normally distributed. A histogram of county-by-county vaccination rates is provided in the Supplementary Materials. The maximum vaccination rate of

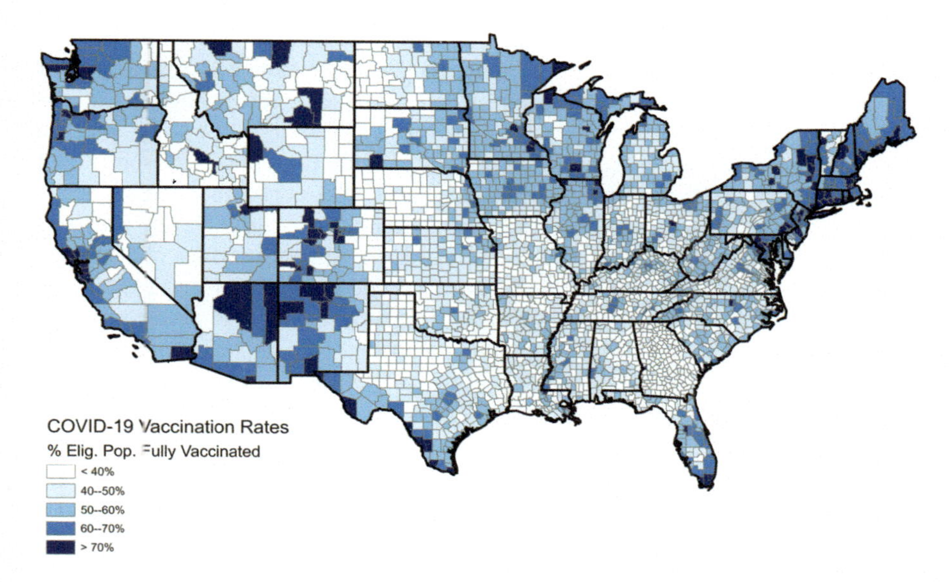

FIGURE 12.1 County-by-county vaccination rates for January 1, 2022.

95% was found in Santa Cruz County, Arizona, where most of the population is older. Vaccine uptake appears to be highest in the urban corridors of the Northeast, the West Coast, and in the retirement communities of Arizona and Florida, and lowest in the rural areas of the Great Plains and the Southeast.

12.3.2 Methods

This study is a population-based observational study that explores associations between vaccination uptake (i.e., high/low vaccine uptake) and LifeModes. We further describe in detail the methods we applied.

12.3.2.1 Temporal Analysis of High and Low Vaccine Uptake LifeModes

This section of the methods is devoted to discovering which LifeModes are associated with statistically high/low vaccine uptake. Counties are grouped by their dominant LifeMode, and the mean vaccination rates for each LifeMode are calculated. A 95% confidence interval for the mean vaccination rate for each LifeMode is generated using bootstrapping with 1,000 replications. The limits of this interval can be compared to the nationwide statistics to determine whether each LifeMode's vaccine uptake differs significantly from the nationwide average. The process is repeated using the median instead of the mean. If a LifeMode's mean vaccination rate confidence interval lies completely above the nationwide mean and the median vaccination rate confidence interval lies completely above the nationwide median, that LifeMode is designated as a "High" vaccination status. If a LifeMode's mean vaccination rate confidence interval lies completely below the nationwide mean and the median vaccination rate confidence interval lies completely below the nationwide median, that LifeMode is designated as a "Low" vaccination status. If neither of the above criteria is met, the LifeMode is designated as "Neutral." Figure S3 presents mean vaccination rate confidence interval and Figure S4 represents median vaccination rate confidence interval in Supplementary Materials.

This analysis is repeated quarterly, using vaccination rates obtained from July 1, 2021; October 1, 2021; and January 1, 2022. This enables the comparison of vaccine uptake among LifeModes across different temporal periods. July 2021 is chosen as the first quarterly date to examine because it is the first quarter in which all American adults would have had both access to the vaccines and sufficient time to reach "fully vaccinated" status. Earlier in 2021, vaccines were only available to senior and those with vulnerability due to underlying health conditions.

12.3.2.2 Principal Component Analysis

The next component of our study is the prediction of vaccination rates using ESRI's Tapestry Segmentation and other demographic variables. To reduce the complexity of the data set, a principal component analysis is applied prior to modeling. The following variables were included in the analysis: the percentage of households in the county belonging to each LifeMode (14 variables), the percentage of non-white persons, the percentage with a bachelor's degree or higher, the median income, the median age, and the percentage of the vote received by Trump in the 2020 presidential election. The demographic variables are all obtained from the 2015–2019 American Community Survey (ACS) (ESRI ACS Data, 2017), except for the voting data, which was downloaded from the GitHub under US County Election results (data originally obtained from the *New York Times*) (*New York Times*, 2021). These specific demographic variables were chosen because the King et al. studies (King et al., 2021a; King et al., 2021b) indicated that they were effective predictors of vaccination rates. This resulted in a total of 19 variables aggregated at the county level. The PCA generated 19 principal components. The percentage of variance explained by each of the first ten components is displayed in the Supplementary Materials. The first seven components were chosen for use in the model as they all had an eigenvalue greater than 1 and there was a reasonable cutoff in variance explained between components 7 and 8. Therefore, components 1 through 7 were exported to a new matrix and used as the predictor variables in the logistic regression analysis.

12.3.2.3 Logistic Regression Analysis

Regression analysis is used to model vaccination rates using lifestyle segments as predictor variables. The resulting model may then be leveraged to predict vaccination rates at subcounty levels (e.g., census tracts, block groups) based on the lifestyles found there. Because the response variable (vaccination rates) represents a binary choice (fully vaccinated or not fully vaccinated), logistic regression is the appropriate model.

The assumptions for logistic regression (Tolles & Meurer, 2016) are mostly satisfied, albeit with a few concerns. The response variable is binary. There is no multicollinearity among the predictors since they come from principal components. The sample size is sufficiently large. Unfortunately, there is the possibility of some spatial autocorrelation in the observations. This will be examined using mapping software after the model is completed. There are also some outliers in the data, but these tend to be the very populous and diverse counties and there is no reason to believe this is due to incorrect reporting. Thus, they are left in the data.

It was also noted from the maps in Figure 12.1 that there seem to be systemic issues with the reporting of vaccine percentages in Georgia, as the vaccination rates within the state are unusually low and an area of unusually low rates correlates perfectly with the state boundary. Therefore, the Georgia counties are discarded from the model. These issues were not present in the predictor variables (demographic and LifeMode data), so the model can still be used to make predictions for Georgia counties.

To choose the best fit, the 10-fold cross-validation method is used. The final model is chosen according to the initial model with the lowest root mean square error (RMSE). There was a significant degree of spatial autocorrelation (Clif & Ord, 1981) in the model residuals. Specifically, the data was spatially clustered (Moran's $I = 0.336$, $z = 56.9$, $p < 0.001$). Using a non-parametric modeling method can reduce the problems associated with this autocorrelation. Therefore, a random forest model is also generated, and its effectiveness is compared with the logistic regression model.

12.3.2.4 Random Forest

Due to the issues associated with this data set regarding logistic regression (i.e., spatial autocorrelation, outliers), a nonparametric modeling method is also attempted. The random forest model was the nonparametric method chosen for this analysis (Liaw & Wiener, 2002). This model consists of building a collection of decision trees, with each individual tree based on a subset of the variables in the data set and a random sample of the observations obtained through bootstrapping. In our case, 1,000 trees will be built, each containing five of the 19 possible variables. In a random forest model, the predictor variables need not be normally distributed or uncorrelated. This means that we can return to using the original variables, not the principal components. There are 19 predictor variables: the percentage of households in the county belonging to each LifeMode (14 variables), the percentage of non-white persons, the percentage with a bachelor's degree or higher, the median income, the median age, and the percentage of the vote received by Trump in the 2020 presidential election. The response variable is the percentage of the entire population considered fully vaccinated. Once again, the observations from Georgia are not included in the analysis due to an apparent bias.

12.4 RESULTS

12.4.1 Temporal Analysis of High and Low Vaccine Uptake LifeModes

We compared the mean vaccination rates for each LifeMode to the nationwide mean. Several clusters become apparent. LifeModes 1 (Affluent Estates), 2 (Upscale Avenues), 3 (Uptown Individuals), and 13 (Next Wave) seem to have the highest vaccination rates. A middle cluster includes LifeModes 4 (Family Landscapes), 5 (GenXUrban), 6 (Cozy Country Living), 7 (Sprouting Explorers), 8 (Middle Ground), 9 (Senior Styles), 11 (Midtown Singles), 12 (Hometown), and 14 (Scholars and Patriots). Finally, LifeMode 10 (Rustic Outposts) has the lowest vaccination rates. Figures of 95% confidence intervals for both the mean and median vaccination rate on January 1, 2022, for each of the 14 LifeModes, are provided in the Supplementary Materials. The medians show roughly the same pattern: three clusters of LifeModes. Most LifeModes have a mean below the nationwide mean, while most LifeModes have a median above the nationwide median. Thus, a LifeMode is only awarded high vaccination status if its mean and median are both above the respective nationwide mean and median. This results in LifeModes 1 (Affluent Estates), 2 (Upscale Avenues), and 13 (Next Wave) receiving high vaccination status. Similarly, a LifeMode is awarded low vaccination status if its mean and median are both below the respective nationwide mean and median. The only LifeMode meeting these criteria is 10 (Rustic Outposts). The remainder of the LifeModes are designated "neutral" regarding their vaccination status. The same process is then repeated for data from July 1, 2021, and October 1, 2021, to see if the vaccination propensity of the various LifeModes might have changed over time. Table 12.1 displays the vaccination status of each LifeMode on each date:

Table 12.2 reveals that only one change in vaccination status occurred between the July 2021 and January 2022 data: Upscale Avenues started out as neutral in July and shifted to high in October and January. Affluent Estates and Next Wave remained consistently high across the time period; Rustic Outposts remained consistently low. Figure 12.2 yields more insight by looking at the precise changes in the mean vaccination rates over time.

For all the groups, the vaccination rate increased more between July and October than between October and January. This is to be expected because there were fewer unvaccinated people remaining in the second period compared to the first. Between

TABLE 12.2
Vaccination Status for Each LifeMode on July 1, 2021; October 1, 2021; and January 1, 2022

LifeMode	July 1, 2021	October 1, 2021	January 1, 2022
[1] Affluent Estates	High	High	High
[2] Upscale Avenues	Neutral	High	High
[3] Uptown Individuals	Neutral	Neutral	Neutral
[4] Family Landscapes	Neutral	Neutral	Neutral
[5] GenXUrban	Neutral	Neutral	Neutral
[6] Cozy Country Living	Neutral	Neutral	Neutral
[7] Sprouting Explorers	Neutral	Neutral	Neutral
[8] Middle Ground	Neutral	Neutral	Neutral
[9] Senior Styles	Neutral	Neutral	Neutral
[10] Rustic Outposts	Low	Low	Low
[11] Midtown Singles	Neutral	Neutral	Neutral
[12] Hometown	Neutral	Neutral	Neutral
[13] Next Wave	Neutral	Neutral	Neutral
[14] Scholars and Patriots	Neutral	Neutral	Neutral

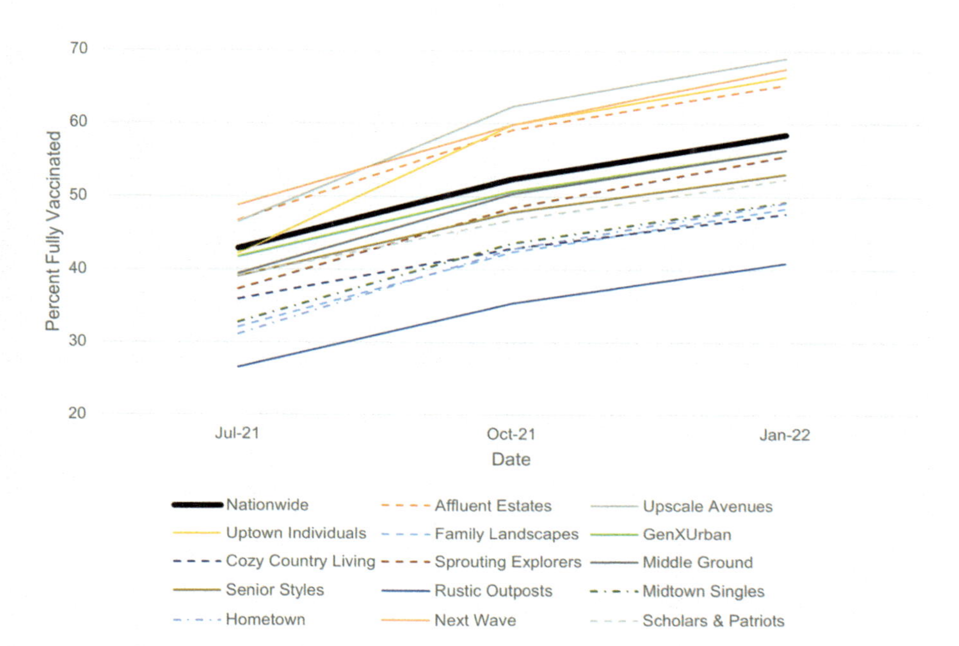

FIGURE 12.2 Mean vaccination rates for each of the 14 LifeModes versus the nationwide mean plotted over time.

July and October, Uptown Individuals saw the largest increase in vaccination rate (+17.6%), while Cozy Country Living (+7.0%) had the smallest increase (nationwide: +9.5%). Between October and January, Next Wave (+7.7%) had the largest increase in vaccination rate, while Cozy Country Living (+4.7%) once again had the smallest increase (nationwide: +6.1%). Figure 12.3 provides a nationwide map showing the locations of the LifeModes with high, low, and neutral vaccination status. By mapping the LifeModes according to their vaccination status, the neighborhoods where vaccine hesitancy is high can be predicted and interventions can be considered.

12.4.2 Logistic Regression Analysis

Table 12.3 shows the coefficients and significance levels for each of the seven predictor variables. Table 12.4 shows the measures of fit for the model.

The first six predictors (PC1 through PC6) are significant in the model; PC7 is not. The model's performance is not impressive with an R-squared of 0.401. The maps in Figures 12.4 and 12.5 show the model predictions by county and the residuals. Some spatial autocorrelation (clustering) in the residuals is clear and may be the reason for the low R-squared value.

Figure 12.5 indicates some spatial clustering in the model residuals. There is a cluster of positive residuals (modeled vaccination rates higher than reality) in areas from the northern Great Plains; these are extremely isolated rural areas that are mostly white and politically conservative. Because of the sparse, isolated nature of

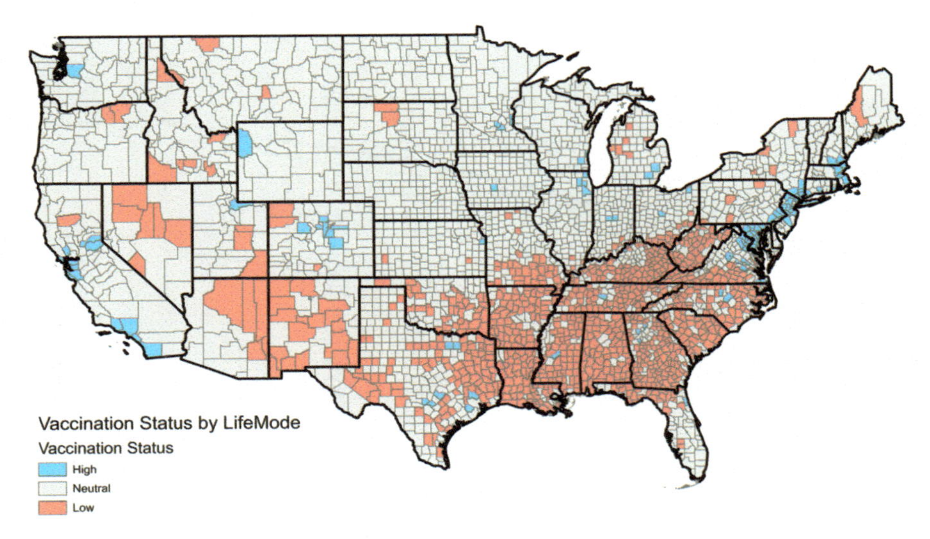

FIGURE 12.3 Vaccination status by county across the United States, estimated using the dominant LifeMode assigned to each county.

TABLE 12.3
Model Coefficients and Significance Levels for the Predictor Variables (Principal Components 1 through 7)

	Estimate	Z-value	P-value (* indicates significance)
Intercept	−0.05318	−258	< 0.001*
PC1	0.1263	2150	< 0.001*
PC2	0.07037	591	< 0.001*
PC3	−0.0154	−198	< 0.001*
PC4	−0.06104	−672	< 0.001*
PC5	0.06659	627	< 0.001*
PC6	0.01086	94	< 0.001*
PC7	−0.00005	−0.378	0.705

the population there, limiting the opportunities for viral spread, residents may feel that COVID-19 mitigation is not a priority in their lives. Several clusters of negative residuals (real vaccination rates higher than modeled) are also apparent. One, in the Southwest, consists of rural counties that are inhabited mostly by Native Americans (for the AZ/NM counties) or Hispanics (for the counties along the TX border). What we are seeing here appears to be racial/ethnic minorities adopting the vaccine at higher rates than would be expected based on their lifestyle classifications. Additionally, looking closely at other parts of the West such as the Dakotas and Montana, counties that contain Native American reservations have higher-than-expected vaccination

TABLE 12.4
Logistic Regression Model Measures of Fit

RMSE	8.84%
R-Squared	0.401
MAE	0.0652

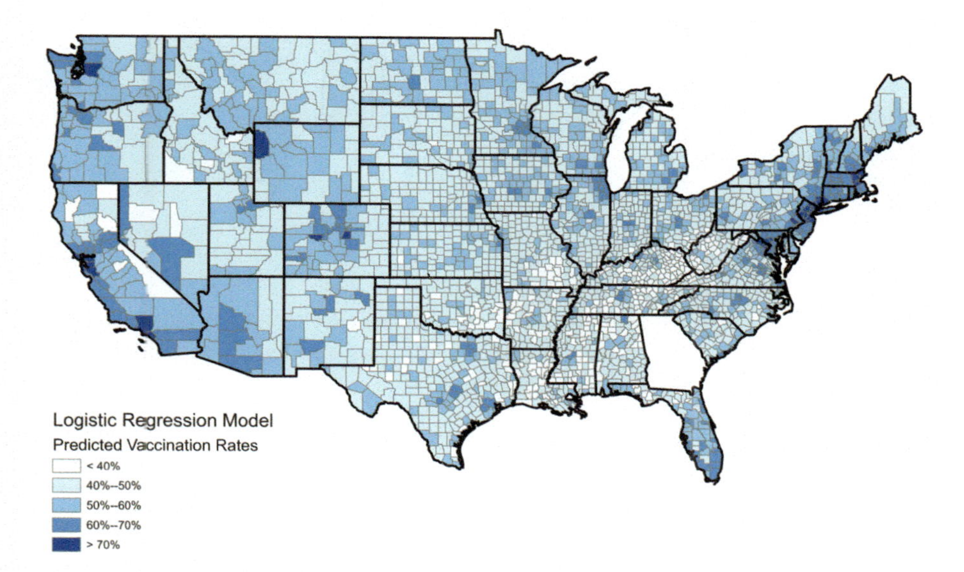

FIGURE 12.4 Predicted vaccination rates by county according to logistic regression modeling.

rates. Altogether, it appears that the model is not accounting for race/ethnicity to the extent that it should, which is surprising because the model did contain minority population as a variable. Perhaps in the calculation of the principal components, the aspect of minority population is being overshadowed by the variations of lifestyle types. The random forest model, which is not based on the principal components, may provide more accurate results.

12.4.3 Random Forest Analysis

The random forest model slightly outperformed the logistic regression model, yielding 59.7% of variance explained (logistic: 40.1%). The residual standard error was 7.23% (logistic: 8.84%). Figure 12.6 (below) provides insight into what variables were most important in the model. The most important variable by far was Republican voting habits, with Republican counties less likely to take the vaccine. The other variables in order of importance were: College Educated, Rustic Outposts, Minority Race/

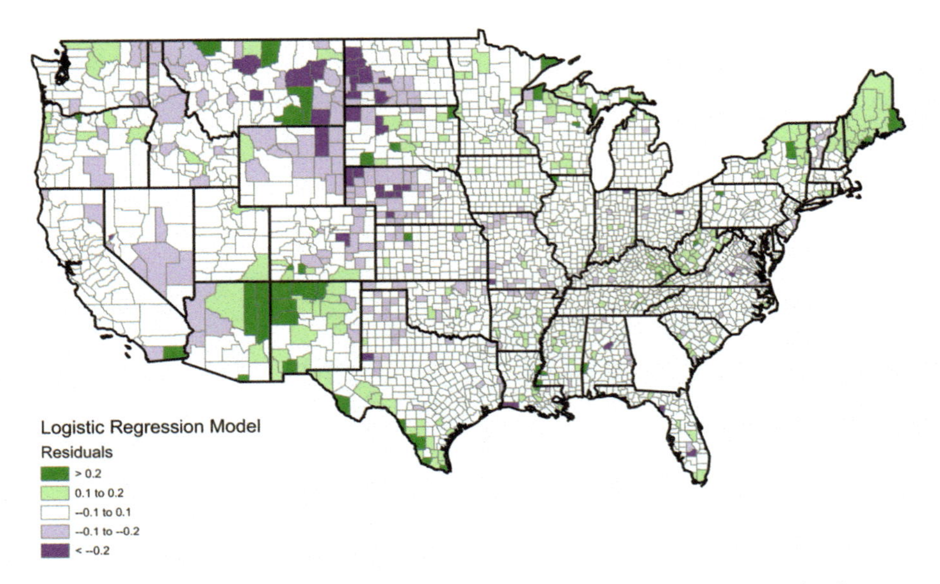

FIGURE 12.5 Residuals from logistic regression model. Green indicates vaccination rates were higher than expected; purple indicates vaccination rates were lower than expected.

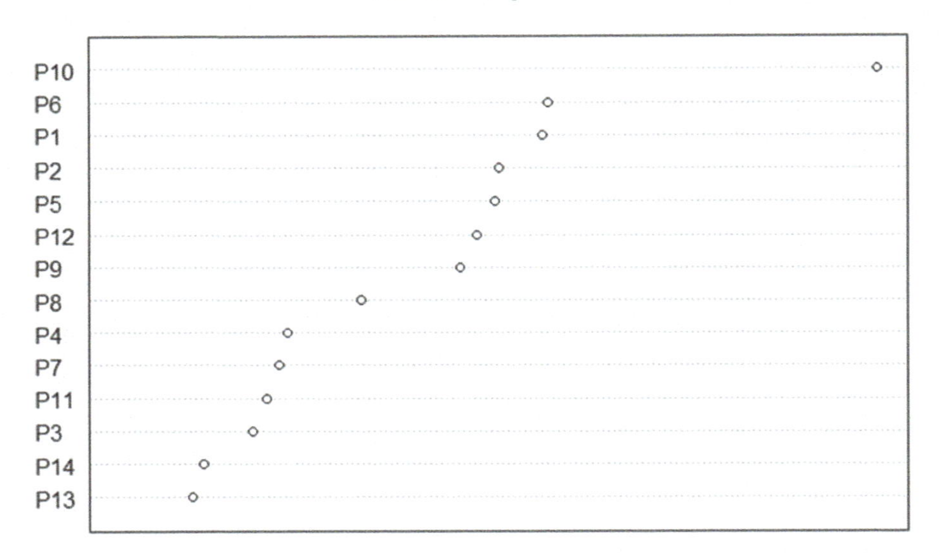

FIGURE 12.6 Variable Importance in the random forest model.

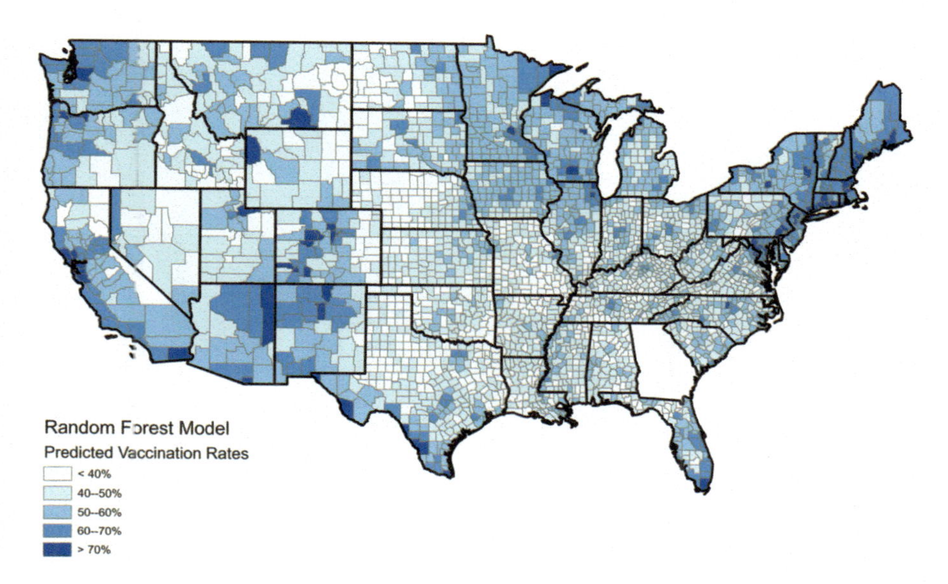

FIGURE 12.7 Predicted vaccination rates by county according to the random forest model.

Ethnicity, Median Income, Median Age, Upscale Avenues, Cozy Country Living, Affluent Estates, Senior Styles, GenXUrban, Hometown, Middle Ground, Family Landscapes, Uptown Individuals, Midtown Singles, Sprouting Explorers, Next Wave, and Scholars and Patriots.

Figures 12.7 and 12.8 show the random forest resultant maps and convey the effectiveness of the random forest model compared to the logistic regression model. Figure 12.8 (random forest residuals) corresponds exactly with Figure 12.5 (logistic regression residuals) using identical color schemes. While Figure 12.4 shows that the logistic regression model had numerous counties, which suffered an error of more than 20%, the random forest model (Figure 12.7) had no such counties. The random forest model produced an error of less than 10% for the overwhelming majority of counties. Counties with errors of greater than 10% are few and far between and are all rural and sparsely populated. The model still underestimates vaccination rates in several western counties with Native American reservations, but these are the only places where such errors are observed. Based on these results, it is clear that the random forest model should be preferred for the modeling of vaccination rates.

12.5 DISCUSSION

Ensuring access to COVID-19 vaccines and improving vaccination rates for at-risk populations can help address the disparate health effects of the virus (CDC, 2022). The common approach by public health officials is to educate people about why vaccination is a good idea (WHO, 2021). However, there is more nuance to vaccine hesitancy than we may realize (Salmon et al., 2015; McClure, 2018; Dubé, 2013; Hamel et al., 2020; King et al., 2021a; King et al., 2021b). The formation of attitudes about

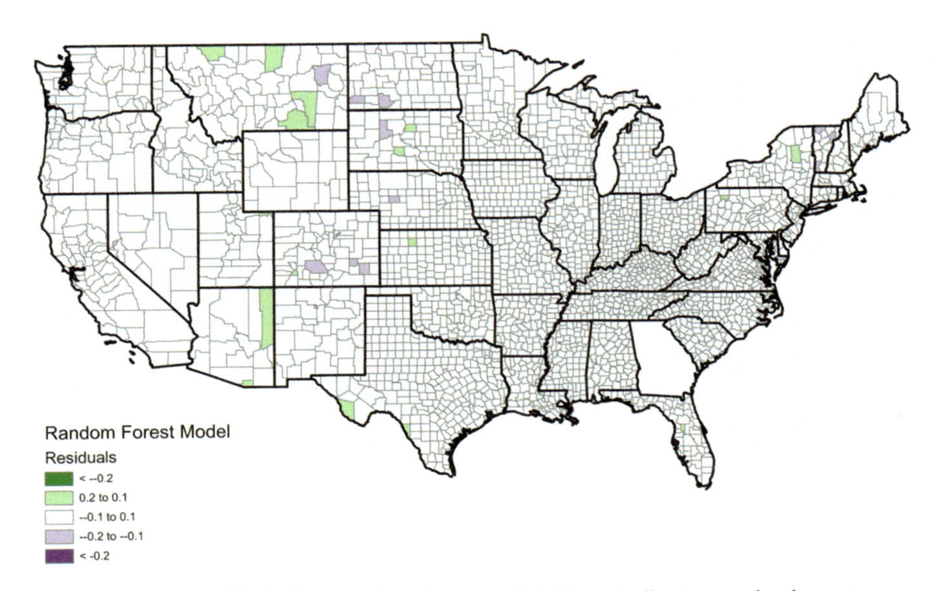

FIGURE 12.8 Residuals from random forest model. Green indicates vaccination rates were higher than expected; purple indicates vaccination rates were lower than expected.

Note that the predictions are much more accurate in general than those of the logistic regression model.

vaccination is complex. In this chapter, we describe how aspects of the human behavioral context, such as lifestyles, led to geographically targeted at-risk populations based on their lifestyle traits. By mapping the LifeModes according to their vaccination status, we can reach the most hesitant subgroup of Americans and predict the households where vaccine hesitancy is high, and interventions can be considered.

Mapping the LifeModes according to their vaccination status is the cornerstone of public health policy and interventions to promote vaccination. To reach vaccination goals for everyone in the community, health-care providers, public health officials, and immunization partners should target their communications toward these households and consider their lifestyle preferences when tailoring messages and strategies to increase vaccination acceptance and uptake. Understanding the current and past vaccination trends over the last several months is important for projecting what may happen in the next couple months and helping policy makers identify vaccine-resistant pockets. The wide disparity in the acceptance of the vaccines decreased over time. Getting enough people vaccinated can reduce the speed of new variants spreading as well.

Our findings demonstrate that vaccine uptake appears to be highest in the urban corridors of the Northeast, the West Coast, and in the retirement communities of Arizona and Florida, and lowest in the rural areas of the Great Plains and the Southeast. Looking closely at other parts of the West such as the Dakotas and Montana, counties that contain Native American reservations have higher vaccination rates. Racial/ethnic minorities also adopt the vaccine at higher rates. The most effective predictor of vaccination hesitancy was Republican voting habits, with Republican counties less likely to take the vaccine. This finding really highlights the politicization of public health recommendations. The other predictors in order of importance were college education, minority race/ethnicity, median income, and median age. The following LifeModes were associated with high vaccination rates consistently above national average: Upscale Avenues, Next Wave, Affluent Estates, and Uptown Individuals. Residents of these LifeModes are affluent and well educated. Affluent Estates and Upscale Avenues represent health-conscious and early adopter households of new products and technology, so this explains their high vaccination rates. GenXUrban, Middle Ground, Sprouting Explorers, Senior Styles, and Scholars and Patriots closely follow nationwide vaccination rates, but are lower than the national average. There is a consistent increase in the lower-rate LifeModes, such as Hometown, Family Landscapes, and Midtown Singles. Cozy Country living exhibits the smallest increase. Rustic Outposts has the lowest vaccination rate.

Having an effective marketing strategy based on at-risk households' lifestyle preferences could lead to higher vaccination rates. For example, Family Landscape are successful young families, residing in suburban and semirural areas and are more likely to vaccinate than individuals without children. It could be that parents, with their regularly scheduled doctor's visits, children's immunization schedules, and concern about their children's welfare, are also more engaged in their own immunization health.

Meanwhile, Rustic Outposts represents the LifeMode with the lowest vaccination rate and it represents mostly poor, rural, southern households. This lifestyle is also heavily Republican in political preference, so to increase vaccine uptake in these communities, the most important step is to reduce the politicization of the vaccine. Specific strategies for reducing the politicization of the vaccine are outside the bounds of our expertise and thus beyond the scope of this chapter but would certainly be a worthwhile pursuit for political scientists.

The below-average vaccination adherence of the Midtown Singles and Scholars and Patriots LifeModes is also surprising from a political perspective, as these LifeModes tend to consist of younger people with more progressive political views. However, the young median age of the population is probably the key to understanding this phenomenon. Age was a significant factor in the random forest model, as older populations were more likely to be vaccinated than younger populations. The young median age associated with these LifeModes likely explains the lower-than-expected vaccination rates. Hesitancy among these groups could likely be reduced by education about the importance of vaccination in reducing community transmission, urging young people to obtain the vaccination for this purpose even if the disease itself poses a relatively small threat to these population groups. This lifestyle segmentation not only helps to geographically target at-risk populations, but also to conduct more effective intervention, prevention, and treatment by increasing vaccine confidence in at-risk households based on their lifestyle traits. For example, Rural Outposts exhibit a lack of trust in government and science and consequently in vaccines. Public health messages with better assistance regarding potential side effects and safety need to reach out to Rural Outposts households to increase vaccine confidence.

Improvement activities can possibly be implemented to increase adhesion in these at-risk communities where people live, work, learn, pray, play, and gather. Our ability to map high-risk lifestyle segments for geographical areas smaller than the county level allows for more targeted interventions and activities. Our findings could help public health policy makers narrow their focus on the most promising groups or lifestyle segments by households, zip codes, and block groups (Ozdenerol & Seboly, 2021). This finer-scale information could be used to improve vaccine delivery and operations that can effectively achieve the target populations. Staff and volunteers can be recruited to talk with people about vaccinations for themselves and their loved ones, answer their vaccine questions, and schedule their vaccination appointments. Vaccination events can be promoted, and vaccination information can be distributed to teachers, parents, and students through various communication means such as flyers, emails, and newsletters (CDC, 2022). Employers can be encouraged to help employees get vaccinated. Education and outreach efforts and pop-up vaccination clinics can be coordinated with faith-based organizations, community centers, recreation centers, and local parks. Geofencing and hyper-target digital marketing strategies through social media and virtual events can be implemented to increase vaccination uptake in these high-risk lifestyle segments (CDC, 2022).

The locations of high-risk LifeModes are clearly the areas in the United States where clinical trials of new vaccines for COVID-19 variants can recruit patients and advertise with the right message based on their lifestyle preferences and determine their overall motivation to engage in clinical research. We also found that the effect of proximity to highly infected areas plays a role in vaccination status and vaccine awareness. Educational tools, clinical trials campaigns, and public health messages could be issued to these at-risk lifestyle households in those areas where COVID-19 infections are high. For example, rural counties in the Southwest that are inhabited mostly by Native Americans (for the AZ/NM counties) or Hispanics (for the counties along the TX border) adopt the vaccine at higher rates than would be expected based on their lifestyle classifications. Additionally, looking closely at other parts of the West, such as the Dakotas and Montana, counties that contain Native American reservations have higher-than-expected vaccination rates. In contrast, because of the sparse, isolated nature of the population in areas of the northern Great Plains, which limits the opportunities for viral spread, residents may feel that COVID-19 mitigation is not a priority in their lives. In other words, low-trust individuals who live close to an affected area harbor more favorable views about vaccination for COVID-19. This implies that citizens who are skeptical of the CDC and similar institutions base their vaccination decision-making to some degree on whether a given disease occurs in close vicinity to their community.

The main limitations with our findings come from inherent difficulties with the vaccination adherence data. These data are reported on a state-by-state basis, so uniformity in counting methods across states is not guaranteed. Moreover, states do not provide complete details on their counting methodologies, so the exact reasons for these discrepancies cannot be uncovered. For example, as discussed earlier, the entire state of Georgia experienced unusually low vaccination rates compared to the nationwide average and even the immediately neighboring states. This was likely due to some aspect of Georgia's data collection process, introducing a negative bias. We remedied this by excluding Georgia's data from the model. While there may be discrepancies among other states, no other states demonstrated an obvious bias compared to the nation, and therefore no other states were removed.

12.6 CONCLUSIONS

We conclude that we can identify patterns of vaccination uptake and hesitancy by correlating COVID-19 vaccination data with lifestyle segments. We can predict the neighborhoods where vaccine hesitancy is high, and interventions can be considered. Linking segments to geographically identified patients (e.g., vaccinated, vaccine hesitant) could support the ability to estimate vaccination acceptance and uptake, help policy makers identify vaccine-resistant pockets, and predict demand for health-care delivery. These associations can lead to more efficient, more targeted, and more cost-effective vaccinations (Ozdenerol & Seboly, 2021). Our methodology for COVID-19 vaccination in this chapter, and in our previous papers (Ozdenerol & Seboly, 2021; Ozdenerol et al., 2021), is a model for transforming lifestyles into the incidence of diseases, along with other health metrics such as vaccination.

Market intelligence tools (i.e., segmentation tools) (Ozdenerol & Seboly, 2021) can assist researchers and key health-care stakeholders in exploring the causal pathways between drivers (e.g., lifestyle traits) and outcomes (e.g., vaccination rates). Our ability to profile a target population of LifeModes and lifestyle segments associated with vaccination hesitancy and their lifestyle traits can help inform continued efforts to promote vaccination uptake, improve vaccination rates, mitigate the harms from COVID-19, and ultimately lower hospital readmission rates related to COVID-19 and its underlying medical conditions.

Our findings illustrate the power of leveraging market intelligence tools to improve health outcomes and health-care delivery. Health-care delivery organizations of all kinds, from independent individual provider units to large integrated health systems and public health policy makers, need to market and distribute vaccines to these households and communities. By designating at-risk market segments down to the census block-group level by psychographics, demographics, and socioeconomics, they can better reach and engage their patients and tailor their policies and strategies in which patients are communicated with differently depending on their personalities and preferences.

There are a number of gaps in our knowledge around health consumer technology in research that follows from our findings and would benefit from further research, including patient technology advancements that were made during the pandemic and resulted in tremendous technological growth which meets the standards of care that patients want. Part of that involves a care plan which encompasses patient communication between themselves and their provider, remote monitoring, and check-ins on additional aspects of their care plan, such as clinical trials and vaccination status. Future research could explore the leadership role of market intelligence tools intertwined with patient communication. More methodological work is needed on how to robustly capture the morbidity data, maybe through claims data and health metrics (e.g., vaccination data), or a robust analysis and exploration of innovative uses of GIS in visualizing and integrating health data and electronic health information to better target populations or geographical areas in most need of public health interventions, ultimately improving population health outcomes. It would be helpful to further explore applying our concept with a real-time segmentation system that does not rely on survey data, which captures the intent with the live aspects of social media, web behavior, and real-world visitation.

REFERENCES

Acorn–The Smarter Consumer Classification. CACI. http://acorn.caci.co.uk/

American Community Survey (ACS). (2018_ www.census.gov/programs-surveys/acs

Bolcato, M., Rodriguez, D., Feola, A., Di Mizio, G., Bonsignore, A., Ciliberti, R., Tettamanti, C., Trabucco Aurilio, M., & Aprile, A. (2021). COVID-19 pandemic and equal access to vaccines. *Vaccines, 9*(6), 538. https://doi.org/10.3390/vaccines9060538

Callaghan, T., Moghtaderi, A., Lueck, J. A., Hotez, P. J., Strych, U., Dor, A., Franklin Fowler, E., & Motta, M. (2021). Correlates and disparities of intention to vaccinate against COVID-19. *Soc Sci Med., 272*, 113638. DOI: 10.1016/j.socscimed.2020.113638. Epub 2021 Jan 4. PMID: 33414032; PMCID: PMC7834845.

Centers for Disease Control and Prevention (CDC). (2020). Ways health departments can help increase COVID-19 vaccinations. www.cdc.gov/vaccines/covid-19/health-departments/generate-vaccinations.html

Centers for Disease Control and Prevention (CDC). (2021). US COVID-19 Vaccination Rates. https://data.cdc.gov/Vaccinations/COVID-19-Vaccinations-in-the-United-States-County/8xkx-amqh

Centers for Disease Control and Prevention (CDC). (2022). *Science Brief: COVID-19 State of Vaccine Confidence Insights Report. 2021*. www.cdc.gov/vaccines/covid-19/downloads/SoVC-report-8-508.pdf

Claritas MyBestSegments. (2017). https://segmentationsolutions.nielsen.com/mybestsegments/Default.jsp?ID= 7020&menuOption=learnmore&pageName=PRIZM%2BSocial%2BGroups&segSystem=CLA.PNE

Cliff, A., & Ord, J. (1973). *Spatial autocorrelation*. Pion: London. [The original edition of *Spatial Processes* 1981].

Dubé, E., Laberge, C., Guay, M., Bramadat, P., Roy, R., & Bettinger, J. A. (2013). Vaccine hesitancy: An overview. *Human Vaccines & Immunotherapeutics*, *9*, 1763–1773.

Eccleston-Turner, M., & Upton, H. (2021). International collaboration to ensure equitable access to vaccines for COVID-19: The ACT-Accelerator and the COVAX facility. *Milbank Q, 99*(2), 426–449.

Esri. ACS–American Community Survey Demographic Variables. (2015–2019). www.arcgis.com/home/ item.html?id=30e5fe3149c34df1ba922e6f5bbf808f

Esri Data–Current Year Demographic & Business Data–Estimates & Projections. www.esri.com

Esri–Tapestry. www.esri.com/landing-pages/tapestry

Experian. "Segmentation." www.segmentationportal.com

Food and Drug Administration (FDA). (2022). Fact Sheet for Health Care Providers Administering Vaccine. www.fda.gov/media/153713/download

Hamel, L., Kirzinger, A., Muñana, C., & Brodie, M. (2020, December 15). KKF COVID-19 Vaccine Monitor: December 2020. The Kaiser Family Foundation. www.kff.org/coronavirus-covid-19/report/kff-covid19-vaccine-monitor-december- 2020/

Harris, R., Sleight, P., & Webber, R. (2005). *Geodemographics, GIS and Neighborhood Targeting*. London: Wiley.

King, W. C., Rubinstein, M., Reinhart, A., & Mejia, R. J. (2021a). Time trends and factors related to COVID-19 vaccine hesitancy from January–May 2021 among US adults: Findings from a large-scale national survey. *PLoS One, 16*(12), e0260731. https://doi.org/10.1371/journal.pone.0260731. *medRxiv* 2021.

King, W. C., Rubinstein, M., Reinhart, A., & Mejia, R. (2021b). COVID-19 vaccine hesitancy January–March 2021 among 18–64 year old US adults by employment and occupation. *Preventive Medicine Reports, 24*, https://doi.org/10.1016/j.pmedr.2021.101569

Liaw, A., & Wiener, M. (2002). Classification and regression by random forest. *R News, 2*, 18–22.

McClure, C. C., Cataldi, J. R., & O'Leary, S. T. (2018). Vaccine hesitancy: Where we are and where we are going. *Clinical Therapy, 39*, 1550–1562.

McGormick, J., Leganza, G., & Hennig, C. (2020). The Forrester Wave: Location Intelligence Platforms, Q2 2020 Tools and Technology: The Digital Intelligence Playbook. https://reprints.forrester.com/#/assets/2/595/RES157271/reports

Mosaic USA Consumer Lifestyle Segmentation by Experian. www.experian.com/marketing-services/ consumer-segmentation.html

New York Times. US County Level Election Results. https://github.com/tonmcg/US_County_Level_Election_ Results_08-20

Ozdenerol, E., Bingham-Byrne, R. M., & Seboly, J. D. (2021). The effects of lifestyle on the risk of Lyme disease in the United States: Evaluation of market segmentation systems in prevention and control strategies. *International Journal of Environmental Research and Public Health, 18*(24). https://doi.org/10.3390/ijerph182412883.

Ozdenerol, E., & Seboly, J. (2021). Lifestyle effects on the risk of transmission of COVID-19 in the United States: Evaluation of market segmentation systems. *International Journal of Environmental Research and Public Health, 18*(9). https://doi.org/10.3390/ijerph18094826

P2 People & Places. Geodemographic Classification P2 People & Places 2017. www.p2peopleandplaces.co.uk/

Salmon, A., Dudley, M. Z., Glanz, J. M., & Omer, S. B. (2015). Vaccine hesitancy: Causes, consequences, and a call to action. *Vaccine, 33* (Suppl. 4), D66–D71.

Siegler, A. J., Luisi, N., Hall, E. W., Bradley, H., Sanchez, T., Lopman, B. A., & Sullivan, P. S. (2021). Trajectory of COVID-19 vaccine hesitancy over time and association of initial vaccine hesitancy with subsequent vaccination. *JAMA Network Open, 4*, e2126882.

Stein, F. (2021). Risky business: COVAX and the financialization of global vaccine equity. *Global Health, 17*(112).

Survey of the American Consumer®. www.mrisimmons.com/solutions/national-studies/survey- american-consumer/

Tolles, J., & Meurer, W. (2016). Logistic regression relating patient characteristics to outcomes. *JAMA, 316*, 533–534.

World Health Organization (WHO). (2021). Global COVID-19 Vaccination-Strategic Vision for 2022. https://cdn.who.int/media/docs/default- source/immunization/sage/covid/global-covid-19-vaccination-strategic-vision-for-2022_sage-yellow-book.pdf

13 A Gendered Approach to Examining Pandemic-Induced Livelihood Crisis in the Informal Sector

The Case of Female Domestic Workers in Titwala

Sujayita Bhattacharjee and Madhuri Sharma

CONTENTS

13.1 INTRODUCTION

Informality is an integral part of the present global economy. Informal workers are mostly engaged in activities such as rag picking, construction work, street vending, domestic work, and the like (Mitra, 2006; Sharma, 2017). Such forms of livelihood are termed informal because people working in these sectors are not covered under any formal laws and regulations. In other words, such forms of employment are not secured under formal contracts, and hence "those involved in such work practices

DOI: 10.1201/9781003227106-13

don't enjoy any form of workers' benefits or workers' representation and they are also not covered under social protection schemes" (FAO, 2020). The informal sector employs about 60% of the world's workforce, or roughly two billion people, and accounts for 90% of employment in developing nations (ILO, 2018a). The informal economy is often viewed as peripheral or marginal and is given less significance in comparison to the formal sector. It is, however, a fundamental component of the present economic system and is associated with modern capitalistic developments, economic growth, and global integration given that the globalization-induced neoliberal economy has made greater shares of economy and the labor therein informal (Chen, 2007; Coe et al., 2012). Although at a superficial level, the informal sector may appear not to be related to the formal sector, both the sectors are intertwined across different levels of economic exchanges since several formal sectors have increasingly depended on contractual and informal labor, given the ease of hire-n-fire policies in the neoliberal economy. The interaction between both the sectors results from the existence of back-and-forth linkages created by labor and capital mobility, as well as the movement of commodities and services across them (Mughal & Schneider, 2020). Mitra (2006) suggests that the growth of informal sector activities is a possible outcome of the growth of activities in the formal sector as well as the increasing pressure of the population. It has been noted that developing economies are generally dominated by the presence of a large informal sector (Blades et al., 2011; Porta & Shleifer, 2014; Shonchoy & Junankar, 2014). The informal sector holds a significant place in India's economy. While it absorbs a large chunk of the population, which otherwise would have been left unemployed due to insufficiency of opportunities in the formal sector, the informal economy is also responsible for exacerbating income inequality and economic divide in the country (Punia, 2020).

The informal sector comprises the main sector of employment for women in developing countries (Chen, 2007; Khan & Khan, 2009; UN Women, n.d.). Due to various reasons, women in the informal sector are represented disproportionately and it accounts for more than 60% of the women employed in the developing nations, and its intensity has been increasing ever since the increased participation of women in the low-waged service industries such as the ready-made garment manufacturing in an increased globally integrated economy (Beneria & Floro, 2005; Coe et al. 2012). Globally, the level of education has been found to exert a profound impact on the level of informality as with the increase in the former, the latter is seen to decrease (ILO, 2018a). According to a report published by the International Labour Organization (ILO; 2018b, p. 21), "women in the informal economy are more often found in the most vulnerable situations, for instance as domestic workers, home-based workers or contributing family workers, than their male counterparts." Low levels of education and poor socioeconomic conditions often compel women, especially in the urban areas, to earn their livelihood by working in the informal sector, particularly as domestic workers. According to the ILO, domestic workers comprise a significant part of the global workforce in informal employment and are among the most vulnerable groups of workers, especially when we consider the developing countries in context.

Domestic work comprises activities such as cooking, dishwashing, laundry, childcare, eldercare, and the like at the household level, which may be performed as unpaid or paid work, both depending on the circumstances. Domestic workers are those who perform paid domestic work in one or more than one household. The significance of domestic workers in the metropolitan lifestyle has increased with time as more and more people are getting involved in outside work irrespective of gender. As per the official statistics, there are 4.75 million domestic workers in India, of whom three million are females (ILO, n.d.). This, however, is believed to be a severe underestimation and the actual numbers are estimated to range between 20 million to 80 million workers (ILO, n.d.). The female domestic workers of India are employed as part-time employees or stay-at-home employees and they meet their household expenses from the incomes earned by them (Sumalatha et al., 2021).

The informal sector forms an important key area where livelihood challenges have intensified with the initiation of the COVID-19 pandemic (Workie et al., 2020). According to several scholars, the socioeconomic impact of the COVID-19 pandemic in India is well evident across its informal sector, and its disproportionate impacts are felt much more through loss of livelihoods for those engaged in its informal economic sectors (Srivastava & Shukla, 2021). This can be attributed to the fact that workers engaged in the informal and unregulated sectors comprise largely manual laborers, and the lockdown's disproportionate effects on unregulated workers remain largely unquantified (ibid.). Domestic workers comprise a vast majority of informal workers. However, due to the absence of a national policy guiding informal sector workers, they are exploited as this sector is largely unorganized. Many of these workers are at the mercy of their employers and middlemen/agents, and hence suffer from abject poverty, lack of education, the competing demand for jobs, and the oversupply of domestic workers, which results in depressed wages (EPW Engage, 2018). The vulnerability of this population has been further aggravated by the COVID-19 pandemic and the lockdown. While the plight of the migrant workers heading back home to the villages from the cities of India has gained a lot of attention from media as well as academia in recent times, not much attention has been given to the plight of the domestic workers who have lived and worked in the urban areas of the country for ages, and hence have no place to go back to in rural India. However, the loss of employment faced by such domestic workers during the lockdown is well known to almost all urbanites in India. Taking this aspect into consideration, this study takes up the case of female domestic workers in Titwala and explores the crisis of livelihood faced by them under the pandemic-induced lockdown of 2020.

This study, thus, seeks to answer the following research questions: What is the emerging nexus of the pandemic, gender, and livelihood? How has the pandemic-induced lockdown affected the informal sector of India, especially domestic workers? What has been the predicament of the female domestic workers of Titwala because of the pandemic-induced lockdown? By delving into the case of the domestic workers of Titwala, this study attempts to contribute to the urban studies that go beyond the metropolitan cores of India, which have been going through significant urban growth and transformation in recent times. These areas have recently provided employment of various informal types to numerous migrant workers flocking into the urban villages of India.

13.2 LITERATURE REVIEW

Informality today is viewed as a defining character of developing economies. Avis (2016) associates it with the rapid urbanization in the developing world. Benería and Floro (2005) note that studies which were conducted during the 1970s and 1980s on the informal sector in most cases assumed the modern economy would most likely absorb it. Since then, the enormous development of informality has flipped these assumptions, pointing to the global economy's strong tendency to generate precarious work with uncertain wages (Benería & Floro, 2005). Informal employment is especially growing across the developing world even under conditions of accelerated growth with the emergence of various nonstandard work practices (Carr & Chen, 2001; Rada, 2009). Not governed by the laws of formal employment, informal workers are prone to various vulnerabilities, which leads them to suffer under abject poverty. RoyChowdhury (2004) stresses that the problems of the informal sector are not limited to merely low wages but also employment insecurity, long and unregulated working hours, absence of medical or life insurance coverage, lack of retirement benefits, and the like. The informal sector, however, also has certain benefits, such as the ease of earning opportunities for the lesser qualified and lower capital needs, which support employment even during a recession, and while working in the informal sector, if desired, one can acquire the necessary skill and knowledge to transition into the formal sector at a later date (Pardo, 2019; Sharma, 2017). However, the mobility of workers from the informal to the formal sector takes place in rare instances. Several studies suggest that the formal sector's poor growth and sluggish job opportunities make the transition from the informal to the formal sector extremely difficult (Banerjee,1986; Mitra, 2001, 2003, 2008; Papola, 1981). In discussing the migrant workers serving in the urban informal sector, Mitra (2006) argues that even when not finding any formal employment, these migrants continue to reside in urban areas because they experience upward income mobility in the informal sector itself.

Carr and Chen (2001) note that women working in the informal economy are more prone to poverty than men. There also exists a gender gap in income and wages within the informal economy as women are underrepresented in higher-income jobs within the informal sectors and are overrepresented in lower-income jobs (Carr & Chen, 2001; Islam & Sharma, 2021). India's labor market exhibits a complex character embedded with a high degree of informality and precarity, where women form a significant share of the informal workforce (Samantroy & Sarkar, 2020), with a vast majority of them working as domestic workers, especially in the urban areas of the country. Ghosh (2013) notes that two-thirds of India's domestic workers are in urban areas. However, legally the domestic workers are neither included in the category of "workmen" nor are their workplaces considered as "establishments," as a result of which they are excluded from the industry regulations (EPW Engage, 2018). Their workplace is unlike any office, factory, or construction site; it is someone else's home, and by referring to these settings as "households," we avoid conjuring up images of workplaces (Mohan et al., 2022). This camouflages their existence in the economic landscape and makes their voices unheard in an already stratified informal work environment (Mohan et al., 2022). No wonder the women employed in the informal sector as domestic workers are denied social protection (Islam & Sharma,

2021) as well as labor law coverage (Samantroy & Sarkar, 2020). Nirmala (2009) notes that domestic workers usually work under exploitative conditions where they are paid less, provided no employment security, and are exposed to all kinds of economic turmoil. Although many women have been able to enter the labor market and gain economic independence due to their employment as a domestic worker, this has not helped in shrinking the prevalent gender inequalities (Sharma, 2021a, b), except for most likely providing them with some food on their tables (Islam & Sharma, 2021; Sharma, 2017).

Along with the existing vulnerability of women workers, the ongoing COVID-19 pandemic has added a new dimension to comprehending informality (Samantroy & Sarkar, 2020). Domestic workers across the world have emerged as the worst victims of the COVID-19 pandemic, losing employment and being forced to work more hours than any other sector (UN News, 2021). In India, the pandemic has added to the socioeconomic woes of domestic workers as well as has contributed to an increase in inequalities and the violation of rights mostly in the urban areas (Sumalatha et al., 2021). Since women make up the vast majority of domestic workers, the economic effects of the pandemic have led to serious gender-based consequences on this group (Mohan et al., 2022). During the lockdown, most households did not allow domestic workers to continue with their work and in several cases they were even denied their monthly wages (Samantroy & Sarkar, 2020). As a result, "these women underwent a period of extreme economic shock which led to their increased poverty and hunger" (Mohan et al., 2022).

Women in the informal economy of India were already in a marginalized state due to a variety of forms of discrimination deeply prevalent in the labor market (Samantroy & Sarkar, 2020). This discrimination is rooted in the patriarchal notions prevalent in the society at large, and much more so in South Asian economies like Bangladesh (Islam & Sharma, 2021). Singh and Kaur (2021) note that patriarchal conventions prevalent in Indian society often culminate in women having a lack of skills and training, limited access to start-up finance, and an inability to work in the formal sector. As a result, many women from poor households end up working as domestic workers in the urban areas of India. In other words, paid domestic work becomes their only source of decent employment. Also, due to the prevalence of patriarchal notions, domestic work is often considered the sole responsibility of women (Islam & Sharma, 2021; Sharma, 2021b). The domestic work sector mostly comprises those belonging to the poorest segments of society with limited education and little or no access to other forms of employment (ILO, 2012). Domestic workers are not given enough value, and they are underpaid as domestic work is considered unskilled labor (Chadha, 2020). Thus, with the pandemic, the already discriminated against domestic workers started facing even more discrimination and marginalization, which has made their lives even more excruciating.

13.3 METHODOLOGY

The study focuses on the predicament of female domestic workers of India under the pandemic and the subsequent lockdown. This study specifically explores the case of

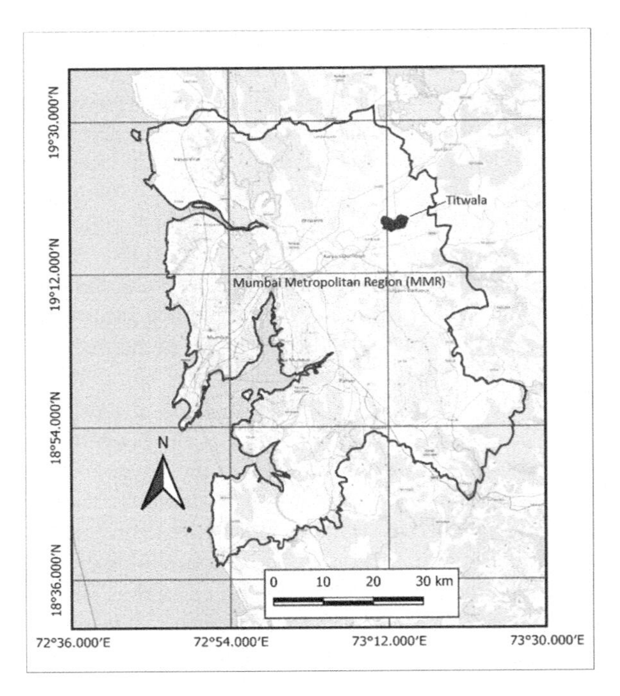

FIGURE 13.1 Location of Titwala in Mumbai Metropolitan Region. (Source: Prepared using maps provided by "Spatial Data of Municipalities (Maps) Project by Data {Meet}"—made available under the Creative Commons Attribution-ShareAlike 2.5 India.)

the female domestic workers of Titwala, an extended suburb of Mumbai, located in the state of Maharashtra in India. It forms part of the district of Thane and is administered under the A-Ward of the Kalyan-Dombivili Municipal Corporation (KDMC). Titwala is considered an outer suburb of Mumbai, which has emerged as one of the preferred residential areas of the Mumbai Metropolitan Region (MMR) (Figure 13.1 and Figure 13.2). With the growth of real estate and the increasing number of households in Titwala, many domestic workers have found employment there. The real-estate growth in Titwala has led to the moving-in of a mostly service sector genre who are employed primarily in Mumbai, Thane, and Navi Mumbai. These new residents are among the millions of middle-class yuppie residents of the MMR, who commute daily from various suburban localities to their places of work in the cities, thus creating high demand for housemaids and domestic help. This emerging character of Titwala, along with its location in Maharashtra, comprises the highest number of domestic workers among all the states of India (ILO, 2016), and hence makes it an ideal location for a microlevel study. However, since our focus is on a small portion of one of the municipal wards of KDMC, the insufficiency and inaccessibility of data from sources such as the Census of India form a major problem that we had to confront while carrying out our research. In addition, the pandemic has delayed the 2021

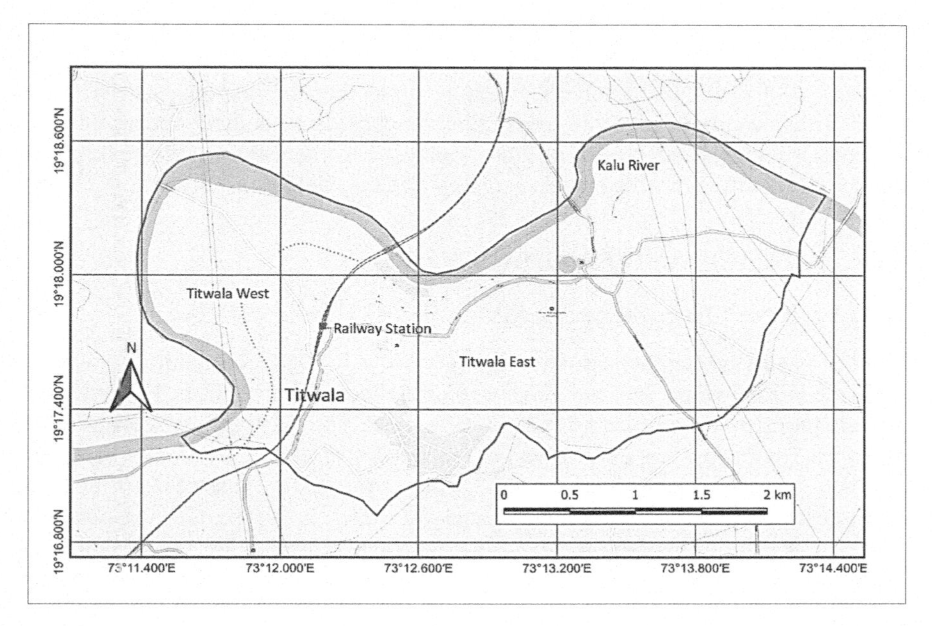

FIGURE 13.2 Titwala—the study area. (Source: Prepared using map given at KDMC's Fire Response Hazard and Mitigation Plan (2010, p. 10) and Mishra (2009, p.14).)

census count, and as a result, we are left with the 2011 census data for secondary references, which in the face of the unprecedented urban growth of MMR would not add much meaning to our research. This is more so in the case of Titwala because much of its urban growth has taken place within the last ten years.

Thus, we had to rely on employing a qualitative/survey method as the best-suited method for collecting our data for this study. Accordingly, a sample survey was conducted in the low-income residential areas of Titwala, where most domestic workers lived. The samples for interviews were drawn randomly and we were able to interview a total of 38 respondents. With no specific recent population data to rely on for selecting the sample size, the criteria of "saturation" was employed for determining the sample size. Data collection is said to reach saturation when it is enough to cover the themes of that particular research and when continuing the collection of data any further is not likely to bring out any new information (Fofana et al., 2020; Morse, 1995). During the process of conducting interviews, we figured that all these respondents who were domestic workers were females, belonging to the economically poorer sections of society. They were mostly married women, lived in the low-income neighborhoods of the area in households comprising of two to eight members, which often included their spouse, young children, and in-laws. Some of these women were the sole earners of their families. The number of earning members in these households ranged from one to four, whereas the number of non-earning members in these households also ranged from one to four. The secondary data for this study were collected from journals, reports, books, and news reports. Both types of data were analyzed using mixed methods—with quantitative analysis including descriptive statistics, and qualitative analysis including dissecting the respondents' narratives along with meaningful debates from the secondary data content. Accordingly, we present our findings and the status of these domestic workers along with the following subthemes that illustrate the current scenario of their livelihoods' demise.

13.4 ANALYSIS AND FINDINGS

13.4.1 Basic Overview of the Respondents

In all, this survey cum semistructured interview was conducted with a total of 38 respondents, of whom 24 respondents (63.2%) were from Titwala East, whereas the remaining 14 respondents (36.8%) were from Titwala West. Based on simple descriptive statistics, it is evident that the mean size of the family from where these domestic workers came was about 4.63, and almost an average of 2.16 members from each household were earning members (Table 13.1). The mean dependency ratio—a measure of burden on every single worker to provide for others within the household—was very high, with a mean value of 1.46; in many cases, it was as high as 4.00. The descriptive statistics also suggest that there were zero workers with employment during the lockdown, and they had no income during the lockdown. During these times, almost every household had to take loans for their survival as most families got no salary from their employing households. There was a decline in the number of domestic workers employed all throughout.

Considering the earnings before and after the lockdown, the mean earnings declined from INR 3,874 before the lockdown period to INR 1,816 after the lockdown, a significant decline at an average of INR 2,058. In some cases, the decline was as steep as INR 7,698, a serious financial decline posing a threat to the survival of domestic workers. We also found that every single worker had no income during the lockdown, and every single one of them suffered through an enormous decline in income even when they were able to gain back some employment after the end of the lockdown. Sadly, almost every worker ended up borrowing money for survival during the lockdown, and the average borrowing per household was about INR 3,211, maxing at almost INR 10,000 in a few cases.

13.4.2 Pandemic, Gender, and Livelihood

The economic effect of the COVID-19 pandemic has led to severe repercussions on the labor market. The pandemic has intensified the preexisting inequalities, particularly gender inequalities, along multiple dimensions (Profeta, 2021). Studies have suggested that the pandemic has aggravated the loss of livelihoods among women compared to men (Fisher & Ryan, 2021). Even though the SARS-CoV-2 virus (responsible for causing COVID-19) does not discriminate among people, its impacts have been extremely gendered in nature. The pandemic has placed the livelihoods and economic security of women at far greater risk compared to men as women largely engage in low-waged service sectors of economy across the world (Islam & Sharma,

TABLE 13.1
Basic Overview of Respondents

Variables	Mean	Range	Min.	Max.	35th	50th	65th
Total Family Members	4.63	5	2	7	5.00	5.00	5.00
Earning Members	2.16	3	1	4	2.00	2.00	2.00
Non-earning Members	2.53	3	1	4	2.00	3.00	3.00
Dependency Ratio—Nonworking vs. Working HH-Member	1.46	3.50	0.50	4.00	0.75	1.00	1.50
No. of Households Employed Before Lockdown	3.37	6	1	7	2.00	3.00	4.00
Income Prior to Lockdown (INR Per Month)	3874	6800	1200	8000	2800	3200	4000
No. of Households Employed During Lockdown	0.00	0	0	0	0.00	0.00	0.00
No. of Households Employed After Lockdown	1.84	5	0	5	1.00	2.00	2.00
Pct-Decline, No. of-HH-Employed-before-vs.-After Lockdown	-0.35	1.67	-1.00	0.67	-0.50	-0.50	0.00
Income During Lockdown (INR Per Month)	0.00	0	0	0	0.00	0.00	0.00
Income After Lockdown (INR Per Month)	1816	4000	0	4000	1500	1600	1800
Decline in Income (INR Per Month)	-2058	7698	-6998	700	-3000	-1400	-900
No. of Employers Paying Salary during Lockdown	0.26	2	0	2	0.00	0.00	0.00
If Yes, Amount Borrowed (INR)	3211	10000	0	10000	0	3000	5000

Source: Sample survey conducted for the study

2021; Patel, 2021; Sharma, 2021b). Azeez et al. (2020) and others have noted that migrant female workers are particularly susceptible to various facets of deprivation due to their poverty as well as for their occupation in the informal sector (Islam & Sharma, 2021; Sharma, 2017). A report published by the United Nations Women and the Gender and COVID-19 Working Group (2020) describes how the pandemic has contributed to the precariousness of economic in security among women, especially among those employed in the informal sector. The report points out that 740 million women employed in the informal sectors of the economy had their income drop by 60% during the first month of the pandemic (UN Women and the Gender and COVID-19 Working Group, 2020). According to UN Women (2020), for domestic workers, a vast majority of whom are women, the situation has been particularly devastating, with a 72% job loss in this category across the world. Thus, the gender disparities correlate with informal conditions and the informal nature of workspaces, resulting in distinct forms of marginalization for the women associated with informal livelihoods (Singh & Kaur, 2021).

13.4.3 Loss of Informal Livelihoods in India—the Plight of Domestic Workers

Loss of livelihood in India, especially among informal workers, has been a major effect of the COVID-19 pandemic. The imposition of the lockdown brought severe misfortune to those engaged in the informal sector as many of these occupations involve in-person interaction. The urban economies in India are highly dependent on the services provided by informal workers. Despite this, the onset of the pandemic and subsequent imposition of the lockdown made them the worst victims of the crisis. The informal sector and its workers comprise the backbone of the economy in the cities of India, by providing much-needed services such as drivers of private vehicles, domestic workers, construction workers, waiters, and the like (Bhattacharjee & Sattar, 2021; Sharma, 2017). The disproportionate impact of the pandemic on the informal sector has shrouded these informal workers in a continuum of uncertainties and disrupted incomes (Bhargava & Bhargava, 2021).

The exodus of migrant workers from the cities back to their native places in rural India has been extensively covered by media, revealing the crisis emergent on the informal workers (*India Today*, 2020; *Indian Express*, 2020; *News18*, 2020; *Economic Times*, 2020). Many domestic workers who were also migrant workers became part of the great exodus where thousands of migrant workers moved from the cities to their native villages due to the existential crisis induced upon them by the sudden imposition of the lockdown and the consequent loss of employment. However, not all domestic workers employed in the urban areas of India had a place in rural India to go back to as several of them were those already settled in the informal settlements and slums of the urban areas. Also, for various reasons, some migrants opted to stay back rather than embarking on an uncertain path. The lives of such domestic workers in the crisis have been no less difficult.

In most cases, the domestic workers lost their employment suddenly with the imposition of the lockdown. Goel et al. (2020) note that during the period of lockdown, not only were most domestic workers cut off from the households where they were employed, but most of them were also not given any salary for the period. During the lockdown in Mumbai, several domestic workers who resided in the slums and worked in upper-middle-class homes were sent back on unpaid leave (Parth, 2020). In Delhi, many domestic workers faced a loss of employment, harassment from employers, and struggled under an economic crisis (Chandra, 2020). Domestic workers in Bengaluru had a similar fate, where a majority of them were asked not to come to work during the lockdown (Ram, 2020). As such, the unexpected unemployment and lack of income potential impacted the livelihood prospects of many domestic laborers, adversely affecting the capability of their families for meeting even basic needs (Goel et al., 2020; Sumalatha, 2021). These were aggravated since not all employers/households where these domestic workers worked had their own livelihoods safeguarded in these economically vulnerable times. A majority of the households in India work in private jobs, and the pandemic had impacted their livelihoods as well, making it very difficult for them to continue paying their domestic help when they were themselves not earning their regular income and were suffering through economic hardship. Too many low- to low-middle-class families and small business owners in India had become un/under-employed and/or were on partial salaries due to huge losses suffered in the private and public sectors alike. These had their consequential impacts on domestic workers as well since the cycles of income earning potentials were disrupted at various levels.

13.4.4 Empirical Analysis of Household Surveys

13.4.4.1 Livelihood Crisis Faced by the Domestic Workers of Titwala

The livelihood crises faced by the domestic workers in this study area following the advent of the pandemic and the imposed lockdown were felt along multiple dimensions. The analysis of the sample data revealed that before the imposition of the lockdown, about 15.8% of the domestic workers in Titwala were employed in one household, 18.4% were employed in two households, 21.1% in three households, 15.8% in four households, 10.5% in five households, 7.9% in six households, and 10.5% in seven households. The imposition of the lockdown led them to lose their jobs. During the period of lockdown, all of them were in a jobless state. The sample data revealed that after the lockdown was lifted, 10.5% of them could not find employment, whereas in terms of others with employment, the overall effect was felt through severe underemployment. Of the rest 89.5% of domestic workers, about 36.8% found employment in one household, 26.3% in two households, 13.2% in three households, 7.9% in four households, and 5.3% in five households (Figure 13.3).

This transition from fuller employment to partial and/or underemployment status impacted their income levels. Before the lockdown, their monthly earnings ranged between INR 1,000 and 8,000, while under lockdown they had no earnings. After the lifting of the lockdown, the earnings of the sample group dropped down to the range of INR 0 to 4,000 only.

Until the imposition of the lockdown, a majority of domestic workers, that is, about 47.3%, earned between INR 2,000 and 4,000, followed by 21.1% with

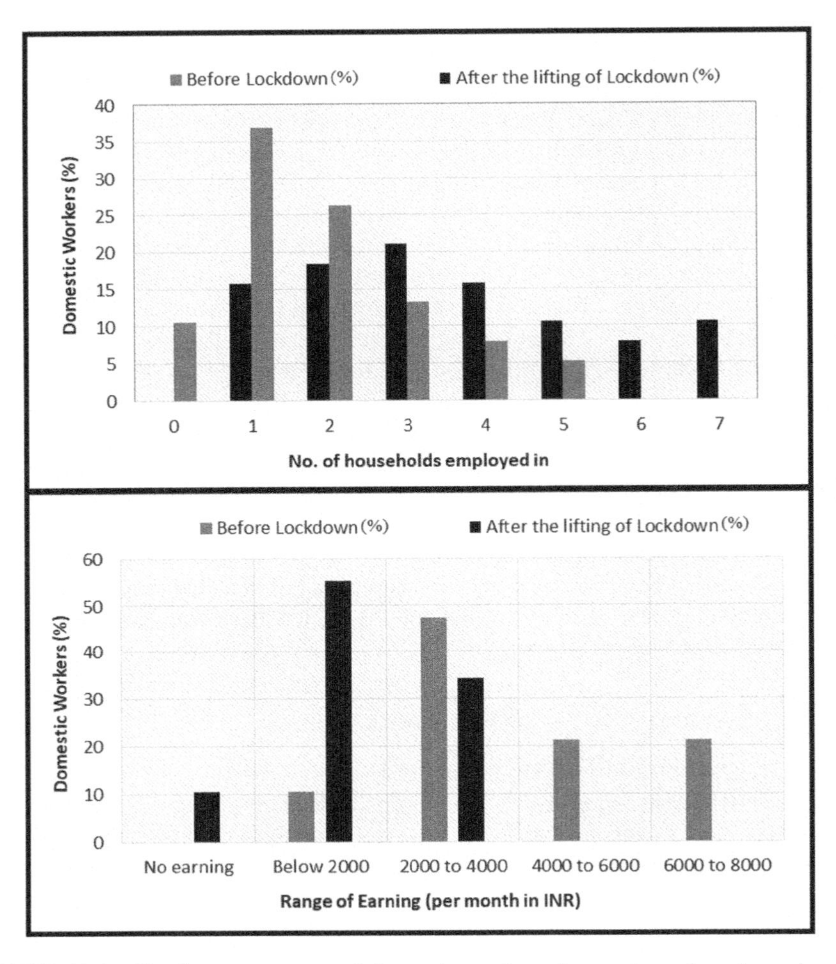

FIGURE 13.3 Employment status of domestic workers (upper image) and earnings of domestic workers (lower image) before and after the lockdown. (Source: Survey sample data from 38 respondents.)

earnings between INR 4,000 and 6,000, 21.1% with earnings between INR 6,000 and 8,000, and 10.5% who earned below INR 2,000. However, after the lockdown, many of these domestic workers struggled with finding employment again. Some of them could not find any employment, while others had to make compromises with their earnings to get employed. Thus, after the lockdown, 10.5% of these domestic workers were not earning at all, and the monthly income earned by a majority of the domestic workers, that is, 55.3%, stood below INR 2,000. Moreover, 34.2% of the domestic workers earned between INR 2,000 and 4000. None of the domestic workers in the study area earned more than INR 4,000 in the post-lockdown scenario (Figure 13.3).

One of the respondents from Indira Nagar, located in Titwala East, narrated her earnings bargaining potential as follows:

> *Bargaining has gotten a lot easier for our employers. They've started hiring people who can work for less money. They don't realize we have a family to look after as well. Two of my former employers have hired others because they agreed to work for less money.*

Another respondent from Panchshil Nagar, located in Titwala West had the following to share:

> *Many people continue to be afraid of the virus. They believe that if they re-hire us, we will infect them. So far, I've worked in two households, and when I enter their homes, they spray disinfectants and sanitizers on me every day. It makes me feel as though I'm the virus!*

We also conducted simple correlations analysis using the data gathered in our surveys. We found a strong positive correlation (r = 0.887) between the number of households where the domestic workers were employed before the lockdown and their monthly earnings before the lockdown. Similarly, a strong positive correlation (r = 0.832) existed between the number of households the domestic workers were employed in after the lockdown and their monthly earnings after the lockdown. As noted from the strength of the r-coefficients, the values declined from 0.887 (pre-lockdown) to 0.832 (post-lockdown). This implies that the decrease in the number of households in which they were employed is strongly related to the decrease in their amount of income per month. This is essentially because of their inability to find alternative sources of employment due to their lack of necessary skills.

The declaration of a lockdown came as a sudden hit to the domestic workers as they were cut off from their employers' households where they worked. Various news articles have reported that across various areas of India, domestic workers were instructed by their employers not to come to work following the declaration of a lockdown. Similar conditions emerged in Titwala, too, where the domestic workers were instructed by their employers to refrain from coming to work until further instructed. Since working as a domestic worker is not a secured form of employment in India and there are not sufficient laws protecting the rights of those working in the informal sector, the meaning of not being able to go to work during the lockdown not only meant lack of payment but also permanent loss of employment in several cases.

Sindwani (2020) notes that in India it is not only a common practice among employers to cut pay for the days of work missed by the domestic workers, but also there are hardly any provisions of granting weekly holidays or other benefits to them. Hence, during the lockdown, the government of India appealed to its citizens not to cut the pay of domestic workers, car cleaners, and other such informal workers employed by the households. Yet in the study area, only 21.1% of the domestic workers received a partial salary, and that too from select households only. As a result, 84.2% of the domestic workers had to borrow money for their survival during the lockdown. Many of them also purchased household essentials on credit from the nearby shops. However, some of them were denied further credit after a while as they were unable to repay their debts. Under such circumstances, they had to borrow money from local moneylenders at very high rates of interest.

According to one of the domestic workers from Deshekar Nagar, located in Titwala East,

> *My husband and I were both laid off as a result of the lockdown. He used to work as a wage worker on a building site, but the construction work was halted due to the lockdown. Initially, we expected things to return to normal in two weeks. The lockdown, however, lasted for months. For a while, we were able to borrow money and buy food on credit, but our inability to repay the debts made it tough. We were unable to feed ourselves and our children due to financial constraints.*

It is noteworthy that with little or no education, the other members of their households also usually find employment in the unorganized and informal sectors. As a result, most other members of the families of the domestic workers also became unemployed during the COVID-19 lockdown.

Some nonprofits and other associations also distributed food items to the poorer sections of the society. The sample survey revealed that 52.6% received support in some form or other. However, these supports were neither consistent nor did all the poor households receive these supports.

Also, not all organizations or groups came with the best intention of doing good for the poor people. Most of these organization were engaged in brief advertising stunts and did not truly care for the betterment of the poor.

One of the respondents, who also resides in Indira Nagar, in the eastern half of Titwala, pointed out,

> *Some groups visited once or twice, handed out rations, and took pictures of the donations being made. After that, they were never seen in the area.*

So, more than the charity, these were mostly publicity stunts. Although, the genuine help provided by some groups cannot be denied, in the absence of proper dissemination of information and less social contact under the lockdown, many people who were in much need of such help did not come to know about these opportunities and hence could not avail themselves of them.

The predicament of their children during the lockdown added to their woes. It was not only the lack of food or other essentials but also the halt caused to their education under the pandemic that was a grave concern. Now, one may argue that educational institutes were functioning online; in fact, even now online classes are going on, so how could these children have suffered? The reality is that availing oneself of online education requires access to certain devices such as computers, the Internet, or smartphones. Most of the domestic workers and their children do not have access to such devices. Most domestic workers have a basic phone, which they keep for staying in contact with their employers. From the sample survey, it was found that 95% of the domestic workers of Titwala did not own smartphones. Only 22% of the domestic workers had smartphones and none of the households had any computers.

A narrative provided in this regard by one of the respondents from Panchshil Nagar in the western half of Titwala exposes the bitter truth. She told us:

> *I don't have any phone but my husband has a smart phone, which could be used for the online classes by our children. But we have two children. One is in 5th standard and the other is in 2nd standard. Their classes are held at the same time, but by having only one smart phone we couldn't make both of them attend their classes side by side. Also, left with no work, it gradually became difficult for us to purchase data packages every month that was needed for online access and connectivity. Basic survival needs became our foremost necessity, and education had to be neglected. We are not educated but we thought at least our children will be educated someday, but now even that seems like a distant dream.*

In the face of this harsh reality, initiatives such as "Digital India" and the provisions of "equal rights to all" have become simple slogans. In addition to not having access to Internet-based education, many of these children also suffered from hunger amid the lockdown since the mid-day school meals were the only source of food for many kids during normal school days. Due to the lockdown, these children, especially in the poorer school districts, were deprived of basic mid-day meals as well.

In short, while the government has numerous programs that tout a digital India and equal access to all, the ground reality is very different wherein none of these existential concerns could be addressed in a timely manner during this crisis. It was not only a net loss of livelihoods for the domestic workers, but also ripple effects were felt for the young school-age children who were deprived of both basic nutrition and education. Even now, the ongoing waves of COVID-19 have restricted access to in-person teaching, thus allowing limited access to children in the impoverished pockets of India.

13.5 CONCLUSION

The pandemic and the lockdown have exposed the vulnerabilities of those employed in the unorganized sectors, such as the domestic worker. The unorganized/informal sector has always been less recognized in India, even though almost 75% of the urban population is engaged in some type of informal economy. The World Health Organization (WHO) suggests that informal economy workers are particularly vulnerable because many of these workers lack social protection and access to quality health care due to the informal nature of these workers, who are not protected by formal labor laws, and hence they have lost access to many productive assets that are generally available to others (WHO, 2020). Informal workers have been lacking basic worker protections such as paid leave, notice period, severance pay, and the like, even when the COVID-19 pandemic was not in the picture (UN Women, 2020). The pandemic has, however, exposed these insecurities that are an integral part of informal livelihoods.

In addition, we suggest that the patriarchal notions prevalent in Indian society also led to the largely prevalent assumption that domestic work (both paid and unpaid) is the sole responsibility of women, and it is hardly even considered in the category of work. Hence, the hard work and toil of domestic workers are rarely given the value they deserve. In a country where there have been no laws regulating the vast segment of the informal economy, domestic work probably lies far behind in the realm of recognition and addressal, particularly since these are largely conducted by women.

The hardships faced by the domestic workers of Titwala during the pandemic, and the subsequent lockdown exposed the plight of millions of domestic workers across India. It is evident from the emerging scenario that it is of utmost necessity to give due recognition to domestic workers as well as all those who earn their bread and butter from the informal sector. The need exists in the country to protect their rights by imposing strong rules and laws. It is not merely them but also the younger generations, that is, their children, who have suffered educational, nutritional, and socioeconomic deprivation during the pandemic and lockdown. Granting them the necessary rights and ensuring their well-being would also mean that their children can have a more secure future. It is high time that their work and their lives both be given the value they deserve. In this regard, the first step would be to "build a comprehensive statistical base on various dimensions of the informal economy as an integral part of the national statistical system," as was planned many years ago (Government of India | Ministry of Labour and Employment, 2015).

We hope our interviews-based insights into the realities of domestic workers' plight create some motivations in the Ministries of Labor and of Human Rights to take appropriate actions such that the lives of the deprived segments could be improved somehow. The fact that almost 75% of India's labor is engaged in informal economy, is important enough to draw the attention of policy makers to comprehensively improve this sector. Given the increasing trends of informal work and informal labor even in the formal sectors of economy in an increasingly neoliberal India, it is critical that adequate and timely action be taken to provide the necessary laws and support system to improve the lives of informal workers all over India.

REFERENCES

Avis, W. (2016). *Urban Governance* (p. 28). Birmingham: Topic Guide. https://gsdrc.org/wp-content/uploads/2016/11/UrbanGov_GSDRC.pdf

Azeez, A., Negi, D. P., Rani, A., & Senthil Kumar, A. P. (2020). The impact of COVID-19 on migrant women workers in India. *Eurasian Geography and Economics, 62*(1), 93–112. https://doi.org/10.1080/15387216.2020.1843513

Banerjee, B. (1986). *Rural to urban migration and the urban labour market: A case study of Delhi*. Bombay: Himalaya Publishing House.

Benería, L., & Floro, M. (2005). Distribution, gender, and labor market informalization: A conceptual framework with a focus on homeworkers. In N. Kudva & L. Benería (Eds.), *Rethinking informalization—poverty, precarious jobs and social protection* (pp. 9–27). Cornell University.

Bhargava, R., & Bhargava, M. (2021). *COVID-19 is creating a hunger catastrophe in India—here's an opportunity to break the cycle*. World Economic Forum. www.weforum.org/agenda/2021/06/covid-19-pandemic-hunger-catastrophe-india-poverty-food-insecurity-relief/

Bhattacharjee, S., & Sattar, S. (2021). COVID-19 pandemic and the poor in the urban spaces of India with special reference to Mumbai. *Equality, Diversity and Inclusion: An International Journal* (ahead-of-print). https://doi.org/10.1108/edi-07-2020-0196

Blades, D., Ferreira, F., & Lugo, M. (2011). The informal economy in developing countries: An introduction. *Review of Income and Wealth, 57*(Special Issue), S1–S7. https://doi.org/10.1111/j.1475-4991.2011.00457.x

Carr, M., & Chen, M. (2001). *Globalization and the informal economy: How global trade and investment impact on the working poor* (pp. 1–142). WIEGO. www.wiego.org/sites/default/files/publications/files/wcms_122053.pdf

Chadha, N. (2020). *Domestic workers in India: An invisible workforce* [Ebook]. https://sprf.in/wp-content/uploads/2021/02/Domestic-Workers-in-India-An-Invisible-Workforce.pdf

Chandra, J. (2020, October 18). Rights of domestic workers in focus post-lockdown. *The Hindu*. www.thehindu.com/news/national/rights-of-domestic-workers-in-focus-post-lockdown/article32887623.ece

Chen, M. A. (2007). Rethinking the informal economy: Linkages with the formal economy and the formal regulatory environment. DESA Working Paper No. 46. www.un.org/esa/desa/papers/2007/wp46_2007.pdf

Coe, N. M., Kelly, P. F., & Yeung, H. W. C. (2012). *Economic geography: A contemporary introduction*. 2nd ed.. Wiley.

Economic Times. (2020, May 23). *4 crore migrant workers in India; 75 lakh return home so far: MHA*. https://economictimes.indiatimes.com/news/politics-and-nation/4-crore-migrant-workers-in-india-75-lakh-return-home-so-far-mha/articleshow/75921049.cms

Endashaw, W., Mackolil, J., Nyika, J., & Sendhil, R. (2020). Deciphering the impact of COVID-19 pandemic on food security, agriculture, and livelihoods: A review of the evidence from developing countries. *Current Research in Environmental Sustainability, 2*, 100014, https://doi.org/10.1016/j.crsust.2020.100014

EPW Engage. (2018, November 2). Where are the laws to protect the rights of domestic workers in India? *Economic and Political Weekly*. www.epw.in/engage/article/domestic-workers-rights

Fisher, A. N., & Ryan, M. K. (2021). Gender inequalities during COVID-19. *Group Processes & Intergroup Relations, 24*(2), 237–245. https://doi.org/10.1177/1368430220984248

Fofana, F., Bazeley, P., & Regnault, A. (2020). Applying a mixed methods design to test saturation for qualitative data in health outcomes research. *PLoS One, 15*(6), e0234898. https://doi.org/10.1371/journal.pone.0234898

Food and Agriculture Organization of the United Nations (FAO). (2020). Impact of COVID-19 on informal workers. www.fao.org/3/ca8560en/CA8560EN.pdf

Government of India. Ministry of Labour and Employment. (2015). *Report on employment in informal sector and conditions of informal employment (2013-14)*. Chandigarh. https://labour.gov.in/sites/default/files/Report%20vol%204%20final.pdf

Government of Maharashtra. Law and Judiciary Department. (2019). Maharashtra Act No. I of 2009. The Maharashtra Domestic Workers Welfare Board Act, 2008. Mumbai: Government Printing, Stationery and Publications, Maharashtra State. https://bombayhighcourt.nic.in/libweb/acts/2009.01.pdf

Ghosh, J. (2013). The plight of domestic workers in India. *Frontline, 30*(2), 48–52.

Goel, S., Sen, P., Dev, P., & Vijayalakshmi, A. (2020, June 4). During coronavirus lockdown, women domestic workers have struggled to buy essentials, says survey. Scroll.in. https://scroll.in/article/963519/during-coronavirus-lockdown-women-domestic-workers-have-struggled-to-buy-essentials-shows-survey

Hamzah, F. B., Lau, C., Nazri, H., Ligot, D. V., Lee, G., Tan, C. L., ... & Chung, M. H. (2020). CoronaTracker: Worldwide COVID-19 outbreak data analysis and prediction. *Bulletin of the World Health Organization, 1*(32), 1–32.

International Labour Organization (ILO). (n.d.). About domestic work. www.ilo.org/newdelhi/areasofwork/WCMS_141187/lang--en/index.htm

International Labour Organization (ILO). (2012). *Effective protection for domestic workers: A guide to designing labour laws*. www.ilo.org/travail/areasofwork/domestic-workers/WCMS_173365/lang--en/index.htm

International Labour Organization (ILO). (2016). *Formalizing employment in domestic work: Converging systems for decent work for domestic workers*. Ilo.org. www.ilo.org/newdelhi/whatwedo/projects/WCMS_421110/lang--en/index.htm

International Labour Organization (ILO). (2018a) More than 60 per cent of the world's employed population are in the informal economy. Ilo.org. www.ilo.org/global/about-the-ilo/newsroom/news/WCMS_627189/lang--en/index.htm

International Labour Organization (ILO). (2018b). *Women and men in the informal economy: A statistical picture* (3rd ed.). www.ilo.org/global/publications/books/WCMS_626831/lang--en/index.htm

Indian Express. (2020, December 25). The long walk of India's migrant workers in Covid-hit 2020. https://indianexpress.com/article/india/the-long-walk-of-indias-migrant-workers-in-covid-hit-2020-7118809/

India Today. (2020, March 26). Hit by lockdown, stranded on roads: Migrant labourers walk for days to reach home. www.indiatoday.in/india/story/coronavirus-outbreak-lockdown-migrant-workers-condition-1659868-2020-03-26

Islam, F.B., & Sharma, M. (2021). Gendered dimensions of unpaid activities: An empirical insight into rural Bangladesh households. *Sustainability, 13*(12), 6670. https:/doi.org/10.3390/su13126670

Kalyan Dombivli Municipal Corporation. (2010). *Fire response hazard and mitigation plan* [Ebook]. https://mahafireservice.gov.in/directorate/mitigation/cities/Kalyan.pdf

Khan, T., & Khan, R. E. A. (2009). Urban informal sector: How much women are struggling for family survival. *Pakistan Development Review, 48*(1), 67–95. www.jstor.org/stable/41260908

Mandi, J., Chakrabarty, M., & Mukherjee, S. (2020). How to ease Covid-19 lockdown? Forward guidance using a multi-dimensional vulnerability index. *Ideas for India*. www.ideasforindia.in/topics/macroeconomics/how-to-ease-covid-19-lockdown-forward-guidance-using-a-multidimensional-vulnerability-index.html

Mishra, S. (2009). *An exploration of environmental capital in context of multiple deprivation: A case of Kalyan Dombivili, India* [Master's thesis, International Institute for Geo-information Science and Earth Observation, Enschede, Netherlands]. https://webapps.itc.utwente.nl/librarywww/papers_2009/msc/upm/mishra.pdf

Mitra, A. (2001). Employment in the informal sector. In A. Kundu & A. N. Sharma (Eds.), *Informal sector in India: Perspectives and policies* (pp. 85–92). New Delhi: Institute for Human Development and Institute of Applied Manpower Research.

Mitra, A. (2003). *Occupational choices, networks and transfers: An exegesis based on micro data from Delhi slums*. New Delhi: Manohar.

Mitra, A. (2006). Labour market mobility of low income households. *Economic and Political Weekly, 41*(21), 2123–2130.

Mitra, A. (2008). Social capital, livelihood and upward mobility. *Habitat International, 32*(2), 261–269. https://doi.org/10.1016/j.habitatint.2007.08.006

Mohan, D., Mistry, J., Singh, A., Mittal, V., Sekhani, R., & Agarwal, S. (2021, October 9). How Covid-19 pandemic has pushed India's female domestic workers further to the margins.

Scroll.in. https://scroll.in/article/1006469/how-covid-19-pandemic-has-further-pushed-indias-female-domestic-workers-to-the-margins
Morse, J. (1995). The significance of saturation. *Qualitative Health Research, 5*(2), 147–149. https://doi.org/10.1177/104973239500500201
Mughal, K., & Schneider, F. (2020). How informal sector affects the formal economy in Pakistan? A lesson for developing countries. *South Asian Journal of Macroeconomics and Public Finance, 9*(1), 7–21. https://doi.org/10.1177/2277978719898975
News18. (2020, May 20). Mapping accidents that killed over 100 migrant workers on their way to home during lockdown. www.news18.com/news/india/mapping-accidents-that-killed-over-100-migrant-workers-on-their-way-to-home-during-nationwide-lockdown-2627947.html
Nirmala, V. (2009). Indian informal sector labour market: The formalizing problems. Conference Paper, The Special IARIW-SAIM Conference on "Measuring the Informal Economy in Developing Countries," Kathmandu, Nepal, International Association for Research in Income and Wealth. www.iariw.org/papers/2009/6 Nirmala.pdf
Papola, T. S. (1981). *Urban informal sector in a developing economy*. New Delhi: Vikas Publishing House.
Pardo, D. (2019, October 30). The informal economy and developing countries: An engine of growth and a source of abuse. Tomorrow.city. https://tomorrow.city/a/informal-economy-in-developing-countries
Parth, M. N. (2020, April 22). Dumped by Mumbai's upper middle and middle class, domestic workers survive on charity to stay afloat during COVID-19 lockdown. *First Post*. www.firstpost.com/health/coronavirus-outbreak-dumped-by-mumbai-upper-middle-and-middle-class-domestic-workers-survive-on-charity-to-stay-afloat-during-covid-19-lockdown-8287141.html
Patel, V. (2021, March 5) Gendered experiences of COVID-19: Women, labour, and informal sector. *EPW Engage, 56*(11). www.epw.in/engage/article/gendered-experiences-covid-19-women-labour-and
Porta, R., & Shleifer, A. (2014). Informality and development. *Journal of Economic Perspectives, 28*(3), 109–126. https://citeseerx.ist.psu.edu/viewdoc/download?doi=10.1.1.655.2183&rep=rep1&type=pdf
Profeta, P. (2021) Gender equality and the COVID-19 pandemic: Labour market, family relationships and public policy. *Intereconomics, 56*, 270–273. https://doi.org/10.1007/s10272-021-0997-2
Punia, K. (2020). Future of unemployment and the informal sector of India. Observer Research Foundation. www.orfonline.org/expert-speak/future-of-unemployment-and-the-informal-sector-of-india-63190/
Rada, C. (2009). *Formal and informal sectors in China and India: An accounting-based approach*. Salt Lake City: University of Utah Press.
Ram, T. (2020, May 27). "They think we carry the virus": Scores of domestic workers in Bengaluru lose jobs. *The News Minute*. www.thenewsminute.com/article/they-think-we-carry-virus-scores-domestic-workers-bengaluru-lose-jobs-125390
RoyChowdhury, S. (2004, January 3). Globalisation and labour. *Economic and Political Weekly, 39*(1), 105–108.
Samantroy, E., & Sarkar, K. (2020, September 7). Violence in times of COVID-19: Lack of legal protection for women informal workers. *Economic and Political Weekly*. www.epw.in/engage/article/violence-times-covid-19-lack-legal-protection.
Sharma, M. (2017). Quality of life of labor engaged in informal economy in the National Capital Territory of Delhi, India. *Khoj: The International Peer Reviewed Journal of Geography, 4*, 14–25. https://doi.org/10.5958/2455-6963.2017.00002.9

Sharma, M. (2021a). Gender disparity and economy in U.S. Counties: Change and continuity, 2000–2017. In E. Ozdenerol (Ed.), *Gender inequalities—GIS approaches to gender analysis* (pp. 73–96). Abingdon, UK: CPC Press.

Sharma, M. (2021b). Multiple dimensions of gender (dis)parity: A county-scale analysis of occupational attainment in the USA, 2019. *Sustainability, 13*(16), 8915. https://doi.org/10.3390/su13168915

Shonchoy, A., & Junankar, P. (2014). The informal labour market in India: transitory or permanent employment for migrants? *IZA Journal of Labor & Development, 3*(1). https://doi.org/10.1186/2193-9020-3-9

Sindwani, P. (2020, March 20). Modi appeals people to not cut the pay of domestic help and others who can't come to work. *Business Insider India.* www.businessinsider.in/india/news/modi-appeals-people-to-not-cut-the-pay-of-domestic-help-and-others-who-cant-come-to-work/articleshow/74716721.cms

Singh, N., & Kaur, A. (2021, October 27). The COVID-19 pandemic: Narratives of informal women workers in Indian Punjab. *Gender, Work & Organization,* 1–20. https://doi.org/10.1111/gwao.12766

Srivastava, P., & Shukla, P. (2021, May 29). Domestic workers' struggles during the pandemic and beyond: Crisis behind closed doors. *Economic and Political Weekly, 56*(22). www.epw.in/journal/2021/22/commentary/crisis-behind-closed-doors.html

Sumalatha, B. S., Bhat, L. D., & Chitra, K. P. (2021). Impact of Covid-19 on informal sector: A study of women domestic workers in India. *The Indian Economic Journal, 69*(3), 441–461. https://doi.org/10.1177/00194662211023845

UN News. (2021, June 15). Domestic workers among hardest hit by COVID crisis, says UN labour agency. https://news.un.org/en/story/2021/06/1094022

UN Women. (n.d.). *Women in the changing world of work—Facts you should know*. Interactive. unwomen.org. https://interactive.unwomen.org/multimedia/infographic/changingworldofwork/en/index.html

UN Women. (2020). COVID-19 and its economic toll on women: The story behind the numbers. www.unwomen.org/en/news/stories/2020/9/feature-covid-19-economic-impacts-on-women

UN Women and the Gender and COVID-19 Working Group. (2020). *Will the pandemic derail hard-won progress on gender equality?* www.unwomen.org/sites/default/files/Headquarters/Attachments/Sections/Library/Publications/2020/Spotlight-on-gender-COVID-19-and-the-SDGs-en.pdf

Workie, E., Mackolil, J., Nyika, J., & Ramadas, S. (2020). Deciphering the impact of COVID-19 pandemic on food security, agriculture, and livelihoods: A review of the evidence from developing countries. *Current Research in Environmental Sustainability, 2*, 100014. https://doi.org/10.1016/j.crsust.2020.100014

World Health Organization (WHO). (2020). Impact of COVID-19 on people's livelihoods, their health and our food systems. www.who.int/news/item/13-10-2020-impact-of-covid-19-on-people's-livelihoods-their-health-and-our-food-systems

Index

A

B

C

J

K

L

M

N

O

P

Q

R

S

T

U